D1745001

Vicia faba:

AGRONOMY, PHYSIOLOGY AND BREEDING

WORLD CROPS:

PRODUCTION, UTILIZATION, DESCRIPTION

Volume 10

Related titles previously published:

1. Stanton WR, Flach M, eds: SAGO. The equatorial swamp as a natural resource. 1980. ISBN 90-247-2470-8
2. Pollmer WG, Phipps RH, eds: Improvement of quality traits of maize for grain and silage use. 1980. ISBN 90-247-2289-6
3. Bond DA, ed: *Vicia faba:* Feeding value, processing and viruses. 1980. ISBN 90-247-2362-0
4. Thompson R, ed: *Vicia faba:* Physiology and breeding. 1981. ISBN 90-247-2496-1
5. Bunting ES, ed: Production and utilization of protein and oilseed crops. 1981. ISBN 90-247-2532-1
6. Hawtin G, Webb C, eds: Faba bean improvement. 1982. ISBN 90-247-2593-3
7. Margaris N, Koedam A, Vokou D, eds: Aromatic plants: Basic and applied aspects. 1982. ISBN 90-247-2720-0
8. Thompson R, Casey R, eds: Perspectives for peas and lupins as protein crops. 1983. ISBN 90-247-2792-8
9. Saxena MC, Singh KB, eds: Ascochyta blight and winter sowing of chickpeas. 1984. ISBN 90-247-2875-4

Vicia faba:

Agronomy, Physiology and Breeding

Proceedings of a Seminar in the CEC Programme of Coordination of Research on Plant Protein Improvement, held at the University of Nottingham, United Kingdom, 14–16 September 1983

Sponsored by the Commission of the European Communities, Directorate-General for Agriculture, Coordination of Agricultural Research

edited by

P.D. HEBBLETHWAITE, T.C.K. DAWKINS, M.C. HEATH
School of Agriculture
University of Nottingham
United Kingdom

G. LOCKWOOD
Plant Breeding Institute
Cambridge
United Kingdom

1984 **MARTINUS NIJHOFF/DR W. JUNK PUBLISHERS**
a member of the KLUWER ACADEMIC PUBLISHERS GROUP
THE HAGUE / BOSTON / LANCASTER

for the Commission of the European Communities

Distributors

for the United States and Canada: Kluwer Boston, Inc., 190 Old Derby Street, Hingham, MA 02043, USA
for all other countries: Kluwer Academic Publishers Group, Distribution Center, P.O.Box 322, 3300 AH Dordrecht, The Netherlands

Library of Congress Cataloging in Publication Data

Main entry under title:

Vicia faba.

(World crops ; v. 10)
"Proceedings of a seminar in the CEC Programme of
Coordination of Research on Plant Protein Improvement,
held at the University of Nottingham, United Kingdom,
14-16 September 1983; sponsored by the Commision of the
European Communities, Directorate-General for Agriculture,
Coordination of Agricultural Research."
    1. Faba bean--Congresses. I. Hebblethwaite, P. D.
II. CEC Programme of Coordination of Research on Plant
Protein Improvement. III. Commission of the European
Communities. Coordination of Agricultural Research.
IV. Series.
SB351.F3V53 1984   633.3   84-4181
ISBN 90-247-29 64-5

ISBN 90-247-2964-5 (this volume)
ISBN 90-247-2263-2 (series)
EUR 9002 EN

Book information

Publication arranged by: Commission of the European Communities, Directorate-General Information Market and Innovation, Luxembourg

Copyright/legal notice

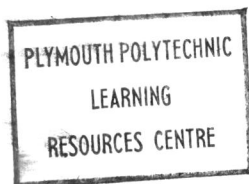

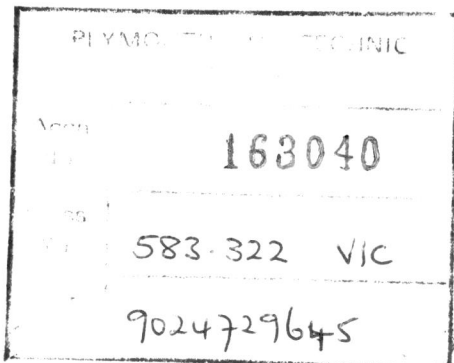

CONTENTS

PREFACE

This is the fourth major publication on Vicia faba reporting
proceedings of seminars organised through the Commission of the European
Communities in the context of the E.E.C. Common Research Programme on
Plant Protein Improvement.

The previous three volumes report proceedings from the seminar in
Bari in 1978 (Some current research on Vicia faba in Western Europe), and
from Cambridge in 1979 (Vicia faba : Feeding value, processing and
viruses) and in Wageningen in 1980 (Vicia faba : Physiology and
Breeding).

The theme of this seminar, held at the University of Nottingham from
14th to 16th September 1983 was selected to examine current research on
agronomy, physiology, plant breeding and nutrition.

84 delegates from 15 European countries attended. Throughout the
seminar there was a spirit of friendliness and co-operation. Everyone
seemed dedicated to doing real justice to the faba bean crop.

The organisation of this seminar would not have been possible
without the help of my secretary, Mrs. Jeanne Rodwell who undertook most
of the administrative and secretarial work.

P.D. Hebblethwaite

COMPONENTS OF THE YIELD AND YIELD OF VICIA FABA

Ph. Plancquaert, J.L. Raphalen

Institut Technique des Cereales et des Fourrages
8 Avenue du President Wilson,
75116 Paris, France.

ABSTRACT

Preliminary experiments have indicated the main factors affecting the yield of winter (6 trials) and spring faba beans (5 trials) : information is presented on the development of pods and flowers, yield and yield components and grain protein content at different locations in France.

INTRODUCTION

The yield of Vicia faba is often variable. Climatic conditions have a considerable effect on growth and development and influence the development of diseases and pests.

The objectives of these experiments are :
- to evaluate development of the components of yield
- to identify the climatic parameters which influence growth and development at different growth stages.

This paper presents provisional results from experiments carried out by I.T.C.F. in conjunction with I.N.R.A., school of agriculture, breeders and funded by U.N.I.P. and the Ministry of Agriculture.

MATERIALS AND METHODS

- Cultural conditions : in each experiment, Vicia faba was grown under optimum conditions : pests and diseases were controlled regularly. The varieties used were Ascott (spring) and Survoy (winter).
- Observations :
    - number of plants after emergence and post winter from 4 x 1 m^2 areas.
    - development of flowers and ovules by numbering, every 5 to 6 days, flowers and pods on 3 x 20 consecutive plants.

P.D. Hebblethwaite, T.C.K. Dawkins, M.C. Heath and G. Lockwood (eds.)
Vicia faba: Agronomy, Physiology and Breeding. ISBN 90-247-2964-5.
© 1984, Martinus Nijhoff/Dr W. Junk Publishers. Printed in The Netherlands.

Further observations were made on ripening and enabled the effects
on the yield at different stages to be quantified (number of pods, of
grains and 1000 seed weight) :
- accumulation of dry matter : Five weeks after the beginning of
flowering until ripening, weekly samples were taken from 2 m^2 x 3 to 4
replications to measure the fresh and dried weights of pods, grains and
plants without pods and to measure the 1000 seed weight, and the nitrogen
content.
- Characteristics of experiments
     5 experiments were carried out in 1981 and 6 in 1982. Some details
of the experiments are shown in Table 1

TABLE 1   Some characteristics of experiments

Trial	35	21	30	28	60	31
Department	Ille et Vilaine	Cote d'Or	Gard	Eure et Loir	Oise	Haute Garonne
Commune	Rennes	Dijon	Nîmes	Chartres	Rouvroy les M.	Villefranche de L.
Responsible	ENSA	INRA	CNA BRL	Lycée	ITCF	ITCF
Type of soil	sandy loan	clay	clayed loan	loaned clay	clayed loan	loaned clay
Depth	deep	50 cm		deep	50 cm	

RESULTS
Development of flowers and ovules (Table 2).
     Number of flowers seems independent of number of plants. Number of
young pods varied in the same way as number of flowers except in 1982 at
Villefranche for winter faba bean (WF) and at Nimes for spring faba bean
(SF). Number of pods at harvest seems almost independent of number of
young pods and flowers.
     Pod drop begins as ovule development begins and continues until
ripening. On average 11% of flowers produce pods with WF, 15 to 18%
with SF. Ovule development on the basis nodes was very good for SF in
1982, the year during which total radiation was 26% higher than 1981. On
the other hand, the considerable inter-variability must be noted.

TABLE 2   Development of plants

Trial	Plant /m²	Flowers /m²	Young pods % flowers	Young pods /m²	Final pods % flowers	Final pods /m²
Spring Faba bean 1981						
35	44	1825	47	861	19	347
21	53	1869	36	673	17	323
30	42	2401	41	984	14	334
60	63	1831	51	928	24	436
Spring Faba bean 1982						
35	51	1595	27	430	13	208
21	52	1724	27	463	16	277
30	48	2336	37	857	8	175
28	55	2592	50	1294	20	521
60	63	2278	45	1020	20	445
Winter Faba bean 1982						
35	20	3412	31	1046	11	358
21	29	2553	27	685	13	336
30	28	3096	29	902	8	257
28	23	3564	53	1880	14	485
60	27	2074	34	709	17	345
31	23	4088	16	667	5	183

The upper nodes had the most flowers. Ovule development and persistence of pods was best on the middle nodes but their contribution yield was less marked for SF than for WF.

Winter Faba bean (WF)   Spring Faba bean (SF)

▲ Flowers
● Young pods
■ Pods

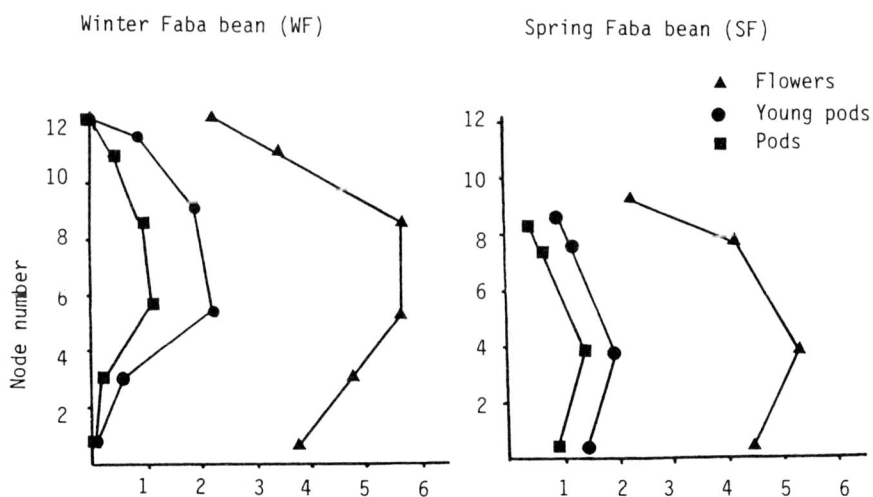

Fig. 1   Diagram of development of flowers and ovules (Dijon 1982)

4

Reference stages

Development stages may be indicated by dates, by day after sowing or accumulated temperature. The development stages are sowing, flowering (beginning, 50%, end), dry matter content in the grain (50%...).

TABLE 3   Accumulated temperature (°C) and stages of growth

	Winter Faba 1982	Spring Faba 1981	1982
Sowing - Beginning blooming	1126	828	751
Sowing - 50% blooming	1360	961	910
Beginning Bl - 50% dried matter	1275	1114	1018

These average sums mask the difference between locations and years. Total radiation in 1982 was 26% higher than that of 1981 and partially explains the differences between years for SF. For WF, the sum of temperature from sowing does not completely explain the date of flowering: it seems that flower induction of these plants also depends on photo period.

1000 - seed weight (1000 S.W.)

In order to facilitate comparison (Fig. 2), the different 1000 seed weights are given in % of maximum obtained 1000 S.W.

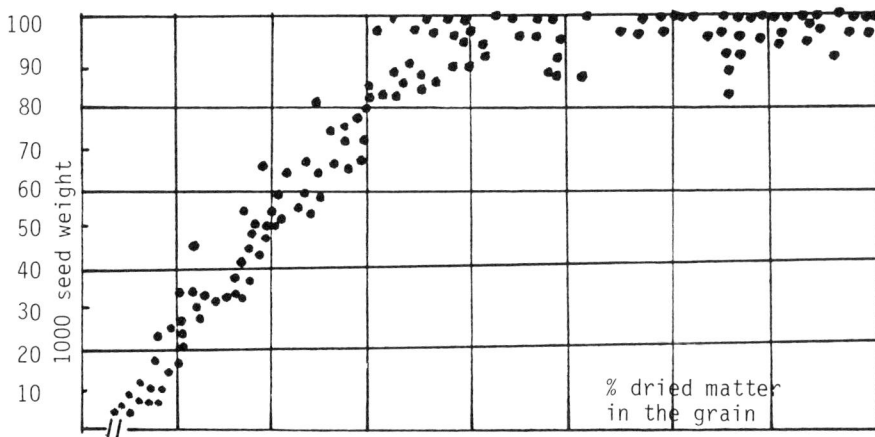

Fig. 2   Evolution of 1000 seed weight with dry matter content

It seems that there is a constant relationship between dry matter content of grain and 1000-seed weight: this is maximum when dried matter content is about 45-50% and independent of location, type, year....

Development of the whole plant

The study of development began 4 to 6 weeks after the commencement of flowering.

The weight of whole plant (vegetative-part + grain) increases until the dried matter content of grain is about 50%. The weight of the vegetative part (stems + leaves + pods) first decreases quickly from 15-20% DM content (milk stage) to 50% DM content of grain, then stabilizes, except there is a degradation of plant tissue.

This decrease is in relation to the transfer of reserves to the grain but is also probably partly due to the degradation of tissues, respiration losses, damage by diseases and pests.

Yield of vegetative part and whole plant (stems + leaves + pods + grains) may be followed between 2 dates:

D 1 : 15 - 20% of DM of grain = Milk stage of grain

D 2 : 50% of DM of grain = maximum 1000 S.W.

assuming there is no degradation of plant tissue.

It is possible to indicate the proportion:

P1 = acquired reserves in grain at D1 date

P2 = acquired reserves in plant at D1, but not yet transferred into the grain.

P3 = reserves from photosynthesis between D1 and D2.

Table 4 shows that at 50% DM (D2), 50 to 60% of potential yield of grain (P1 + P2) is obtained from which less than 25% (P1) is aready in the grain.

TABLE 4   Transfer of reserves in %

	% DM at D1	at D2	P1	P2	P3
W. Faba	16	51	12	45	43
S. Faba	19	50	22	28	50

Yield and yield components

Yield was estimated by harvesting 4 plots of 10 m^2.

TABLE 5   Yield in each trial (q/ha, at 15% DM., 1000 seed weight
           in parentheses

Trial	1981 Spring Faba	1982 Spring F	Winter F
35	45.4 (409)	24.8 (413)	40.7 (430)
21	53.4 (473)	40.5 (428)	47.9 (549)
30	39.2 (325)	6.6 (184)	17.4 (290)
28	38.0 (389)	42.9 (401)	46.9 (522)
60	45.6 (409)	45.7 (386)	58.3 (556)
31	-	-	42.1 (530)

To explain the differences in yield, one must know how this yield is obtained. The component in number are as follows:

1. plants/m^2
2. pods/plant
3. pods/m^2

4. grains/pod
5. grains/m^2
6. 1000 seed weight

and there are two yields :

7. biological yield

8. machine yield

The components 1 x 2 x 4 x 6 give the "biological yield" which is correlated to "machine yield" but is about 15% higher.

For each component, we calculated the average obtained with 5 or 6 experiments, and gave the value observed in each, in % of this average (Table 6).

In 1981 in compensation, components 1 - 2 and 4 gave almost the same number of grains/m^2 in Rennes (35), Nîmes (30), Rouvroy (60) and Dijon (21) but final yield depended on the 1000 seed weight. In Nîmes (30) several factors are unfavourable for 1000 seed weight :

- early and heavy lodging;
- high ETP during formation of grain : 8.5 mm/day from 10th to 20th June, instead of 3.5 to 4.5 in other experiments;

- maximum temperature more than 28°C for 8 days;
- important vegetation : 13 t/ha DM of plant without pods the 16th June instead of 6 to 8 in other experiments.

In 1982, for spring faba, the classification of grain yield was the same as number of pods per m^2. In Chartres (28) and Rouvroy (60) there was abundant flowering, good ovule development and persistence of pods. In Rennes (35) and Dijon (21) flowering was average and % ovule development poor (27%) but persistence was better in Dijon than in Rennes. In Nimes (30) abundant flowering was followed by a poor ovule development (37%) and very little persistence perhaps due to damage caused by poorly controlled Botrytis.

TABLE 6   Components of the yield in trials

Trial / Component	35	21	30	28	60	31	Value of 100
Spring Faba bean 1981							
1	94	113	89	70	134	-	47
2	105	89	114	116	77	-	6.94
3	101	104	105	84	106	-	316
4	105	99	101	98	96	-	2.94
5	107	103	105	82	102	-	929
6	102	118	81	97	102	-	401
7	109	122	85	80	104	-	37.30
8	102	120	88	86	103	-	37.70
Spring Faba bean 1982							
1	95	97	89	102	117	-	54
2	81	120	43	132	125	-	5.10
3	75	113	37	132	143	-	281
4	98	102	99	102	99	-	2.89
5	73	115	37	135	140	-	814
6	115	117	55	107	105	-	368
7	79	127	19	137	139	-	37.40
8	77	126	21	134	142	-	32.10
Winter Faba bean 1982							
1	80	116	112	92	108	92	25
2	125	95	40	126	108	106	11.75
3	102	113	46	119	120	100	286
4	95	95	120	97	99	94	2.86
5	99	109	56	118	122	96	802
6	91	112	71	105	112	110	496
7	89	116	39	122	131	103	48.60
8	96	113	41	111	138	100	42.20

For winter faba bean, the classification of yield is also connected with the number of pods per m^2, differences of level depending on the 1000 seed weight.

The low yield in Nimes came from a very low number of pods/m^2 together with a very low 1000 seed weight. The fall of pods is certainly increased by damage from poorly controlled Botrytis.

Protein content

Samples taken at 4 different dates (beginning of flowering to ripening), enable us to follow the nitrogen content of different parts of the plant: Table 7 shows the average development of protein content in dried matter. This content varying according to the experiment, is high until milk stage, decreases quickly to until the 50% dried matter content stage (end of the growing of grain) and then is stable or slightly decreases.

TABLE 7   Protein content

Date	Stem + leaf	Pod	Grain
1	12.1	25.0	37.2
2	10.1	17.7	32.2
3	8.2	13.0	31.2
4	7.1	9.8	31.2

This is very clear for winter faba bean for which there were few variations in 1982 between trials; for spring faba bean this variability is greater.

CONCLUSIONS

These provisional results are obtained over 1 year for the winter faba bean, and 2 for the spring bean. We must analyse results from 1983 with its different climatic conditions, before finding the climatic parameters affecting the components and the yield of this species.

PROPOSAL OF A GROWTH STAGES KEY FOR VICIA FABA

R. Stülpnagel

Gesamthochschule Kassel
Fachbereich Landwirtschaft
D-3430 Witzenhausen

ABSTRACT

The basic concept of standardising the description of
the stages of plant development used in the Federal Republic
of Germany is to divide the development of plants into 10
macro stages and micro stages having a binary code from 0 to
99 for computerization. To describe other plant species used
in agriculture a proposal for the stages of Vicia faba development is presented including a description of lateral branching. This proposal was tested during 1983 with 86 inbred lines
from different regions (germplasm collection) and 5 local
varieties. Results with average, minimum and maximum number of
days from micro stage to micro stage and from macro stage to
macro stage are presented, showing greater variabilities between inbred lines than between local varieties. Results, difficulties and possible errors in handling this proposal are
discussed.

INTRODUCTION

For registration of pesticides, field investigations,
plant breeding, extension and practical agriculture it is
necessary to have a description of the stages of Vicia faba
development avaible similar to those in use for cereals, rape
and potatoes. Until now, two proposals for the stages of V.
faba development had been published: The first is from Drennan
(1979) dividing the development of V.faba into vegetative(v),
flowering(f),and pod-bearing(p) stages, complete with a
numerical system to describe the number of vegetative, flowering and pod-bearing nodes exactly. The second proposal (Jahn
and Baeseler,1981) describes the development of the variety
'Fribo' (German Democratic Republic) which is divided into 6
growth periods and defines 18 stages.

Both proposals are very different from each other and are
not in accordance with the basic concept aspired in the "Bio-
logische Bundesanstalt"(BBA; Braunschweig, Federal Republic
of Germany) to standardise the description of the stages of
development for plant species used in agriculture (including
weeds). The basic concept is to divide development into 10

P.D. Hebblethwaite, T.C.K. Dawkins, M.C. Heath and G. Lockwood (eds.)
Vicia faba: Agronomy, Physiology and Breeding. ISBN 90-247-2964-5.
© 1984, Martinus Nijhoff/Dr W. Junk Publishers. Printed in The Netherlands.

Table 1: Proposal of a growth stages key for Vicia faba

Code	Definition
00	Germination
01	Dry seed
03	Start of imbibition(embryo distinctly visible under the seed coat)
05	Radicle emerged from seed
07	Shoot length about 1/2 the length of seed
09	Shoot length about twice the length of seed
10	Emergence
11	Seedling emerged at the soil surface
15	First leaf unfolded (= pinnate leaf with 2 leaflets)
20	Leaf and stem development
23	Second leaf unfolded; shoot begins to elongate
25	First transition leaf unfolded (= pinnate leaf with 3-4 leaflets)
27	First sequence leaf unfolded (= pinnate leaf with more than 4 leaflets)
30	-
40	-
50	Bud formation
53	First flower racemes visible at the tip of the shoot (buds still green)
57	First petals appear on the first flower racemes (petals grown out of sepals)
60	Flowering
61	First flower racemes in bloom (buds opened, i. e. flag of the flower steeply erected)
63	More than 3 flower racemes/plant in bloom
65	More than 5 flower racemes/plant in bloom (full-blossom)
67	End of full-blossom (distinctly reduced flowering intensity at the tip of the stem)
69	Flowering complete
70	Pod development
72	First pods visible in the lower inforescences (pods visible = pods longer than 2cm)
74	Pods visible in the middle inflorescences
76	Pods visible in the upper inflorescences
78	Pods in the lowest inflorescences fully developed in size - seed fully developed in size and distinctly contrasted in the pod (green ripeness)
80	Ripeness
81	First pods loose green colour (start of the darkening stage)
83	First pods darkly coloured
85	1/3 of all pods darkly coloured
87	2/3 of all pods darkly coloured
89	All pods darkly coloured; seed in the upper pods can still be cut with fingernail
90	Dying off
92	Complete straw ripeness; seed in the upper pods completely hard; straw dry

For assessments in the field:

Date of emergence: When 75 % of the beans have emerged
Start of flowering: When 10 % of the beans are flowering
End of flowering: When 10 % of the beans are still flowering

If two growth stages occur simultaneously, record the later stage.

TABLE 2 Average, minimum and maximum number of days from
micro stage to micro stage and from macro stage to macro
stage in inbred lines and the varieties Herz Freya, Herra,
Minica, Narbor and Kristall.

code	average	inbred lines		varieties	
		minimum	maximum	minimum	maximum
11 - 15	5.2	2	8	4	8
15 - 23	3.3	2	7	2	4
23 - 25	27.8	20	35	21	33
25 - 53	6.5	2	20	5	12
53 - 57	4.3	2	7	3	5
57 - 61	3.5	1	5	2	4
61 - 63	4.1	2	7	4	8
63 - 65	4.2	2	8	5	8
(65 - 69)	(25.1)	(13)	(37)	(17)	(27)
65 - 72	9.9	3	18	4	9
72 - 74	8.3	4	16	9	12
74 - 76	6.5	3	13	5	10
76 - 78	6.5	2	12	2	8
78 - 81	15.1	8	24	13	19
81 - 83	3.2	1	12	2	5
83 - 85	2.7	2	5	2	5
85 - 87	3.9	1	11	2	5
87 - 89	7.9	2	14	3	10
89 - 92	7.7	2	17	7	10
01 - 11	33.2	30	38	30	36
11 - 15	5.2	2	8	2	4
15 - 53	37.6	29	44	32	40
53 - 61	7.8	6	12	6	8
(61 - 69)	(33.4)	(21)	(46)	(32)	(37)
61 - 72	18.2	13	25	16	20
72 - 81	36.4	29	44	36	43
81 - 92	25.5	15	38	24	27
01 - 92	163.9	144	178	159	168

macro stages with further micro stages having a binary code
from 0 to 99 for computerization. All stages of development
are described. The basic concept requires that if a macro
stage in a plant species does not occur it is omitted.

PROPOSAL FOR VICIA FABA

The proposal for the stages of V.faba development is
presented in Table 1. The proposal contains 8 macro stages and
is further divided in 27 micro stages. The macro stages 30 and
40 are omitted because they are reserved for stem elongation
(cereals or rape; 30) and for crop cover (potatoes; 40). This
proposal was tested in the summer of 1983 with 86 inbred lines
(I_4 and I_5) of V.faba from different regions (obtained from
the germplasm collection at Braunschweig) belonging to a
selection programme for virus resistance. The varieties Herz
Freya, Herra, Minica, Narbor and Kristall were used as standards. From each inbred line and variety 16 plants were sown
on 17th of March. Before plants started flowering they were
covered with a mosquito net to prevent cross-fertilization.
Plants were observed every second day beginning at plant
emergence (it was not possible to record the stages from germination to emergence in this investigation). The results of
these observations are summarized in Table 2 with the average
number of days from micro stage to micro stage and from macro
stage to macro stage for the inbred lines and the 5 varieties.
The variability expressed as the range between minimum and
maximum number of days (Table 2) is greater between inbred
lines from different regions than between investigated
varieties. This variability is of interest to plant breeding,
for example the duration of the flowering (stage 61 - 69), pod
development (61 - 72 and 72 - 81), ripeness (81 - 92) and
dying off (89 - 92) periods. However, there are two types of
variability which have to be considered when interpreting
results. First, each result from a line or variety is an
average from plants differing by one or more days from one
another and differences increased slightly during plant
development. The second type of variability arose in the difficulty to say exactly whether some described stages of plant

TABLE 3 Description of branching of <u>Vicia faba</u>.

Code	Definition
10	Unbranched (lateral branches max. 2 cm long)
20	Lateral branches visible (branches up to a maximum length of 5 cm)
	<u>Lateral branches longer than 5 cm</u>
21	Equivalent branching with slight lateral branch formation (1-2 branches)
22	Equivalent branching with moderate lateral branch formation (2-4 branches)
23	Equivalent branching with strong lateral branch formation (more than 4 branches)
26	Unequivalent branching with slight lateral branch formation (1-2 branches)
27	Unequivalent branching with moderate lateral branch formation (2-4 branches)
28	Unequivalent branching with strong lateral branch formation (more than 4 branches)

<u>Equivalent branching</u>: Development of the main branch does not exceed the development of lateral branch(es).

<u>Unequivalent branching</u>: Development of the main branch distinctly exceeds the development of lateral branches, i. e. the main branch is at least 1/3 longer than the lateral branch(es).

TABLE 4 Macro stages where lateral branches were longer than 5 cm in 86 inbred lines and 5 varieties.

macro stage		10	20	50	60	70	80	90
number	inbred lines	--	20	5	35	6	--	--
of	varieties	--	1	1	3	--	--	--

development occured at a definite day and not sooner or later. For these reasons stages 27 and 67 in Table 2 were left out. Moreover, we experienced difficulties with this proposal when for example top-flowering mutants were obtained during the reproductive period. Beginning with stage 61 we had to change this proposal slightly or make modifications. This will,therefore, not be the final version of this system.

DESCRIPTION OF BRANCHING OF VICIA FABA

Where V.faba forms one or more branches in addition to the main one it may be necessary to describe this lateral branching. To standardise this description of branching a proposal for a two-figure code is presented in Table 3. In this code lateral branching is classified as unbranched (lateral branches max. 2cm long), lateral branches visible (branches up to a maximum length of 5cm) and lateral branches longer than 5 cm. If lateral branches are longer than 5 cm branching is divided into equivalent and unequivalent branching wherein slight, moderate and strong lateral branch formation is recognised (Table 3). When this two-figure code is added to the stages of development, a precise description of Vicia faba is possible for special investigations.

A survey of lateral branching in the investigated material is presented in Table 4. The number of lines or varieties were listed only when the lateral branches exceeded 5 cm. Most branching occured at the beginning of the vegetative (stage 23) and reproductive periods (61 - 63). At all macro stages slight, moderate and strong lateral branch formation occured.

REFERENCES

Drennan, D.S.H. 1979. A suggestion for classification of Vicia faba growth stages. Fabis Newsletters,1, 15.

Jahn, K. and Baeseler, G. 1981. Vorschlag zur Einschätzung der Entwicklungsstadien der Ackerbohne (Vicia faba L.). Nachrichtenblatt Pflanzenschutzdienst DDR, 35, 188 - 190.

VASCULAR DEVELOPMENT IN THE REPRODUCTIVE TISSUES OF VICIA FABA L.

Gretel White, P. Gates and D. Boulter

Botany Department, University of Durham, DH1 3LE.

ABSTRACT

A detailed study of dry weight changes in the reproductive tissues of Vicia faba L., combined with anatomical studies, has shown that there is a rapid and massive increase in vascular development in peduncles and pedicles after fertilisation, preceding pod development and seed growth. Deposition of strengthening tissue in the vascular system can be detected soon after fertilisation, and the vascular and supportive tissues of the raceme are likely to be a major primary sink for assimilates at this stage. Genotypic differences exist in the rate of reproductive vascular development. Implications of these data for the understanding of inter-pod competition, and for the development of various plant ideotypes, are discussed.

INTRODUCTION

Yield loss in Vicia faba L., due to reproductive failure, can occur as a result of bud abortion, flower shedding, pod or ovule abortion (Gates, Smith, White and Boulter 1983). When high levels of pod set are attained at individual nodes on the plant, it is commonly observed that those formed last on the raceme fail to develop to maturity.

In defining improved ideotypes for the faba bean it is important to understand the causes of inter-pod competition, to avoid projecting plant models which are unattainable due to anatomical or physiological constraints within the plants. By analysing in detail the anatomical and physiological changes occurring during the transition from the flower to the active, assimilate-importing sink, a better understanding of the processes controlling the partitioning of assimilates, and therefore yield, may be obtained.

MATERIALS AND METHODS

Dry weights

100 plants each of genotypes Maris Bead (MB), independent vascular supply inbred line G (IVSG) and a rapid maturing inbred line (STW were grown in Levington Universal Compost in 15 cm pots in the glasshouse during April 1983. Flowers were tripped and date tagged at anthesis. When flowers at flowering node six reached anthesis 10 plants were harvested, with racemes of the same age being pooled. Racemes were

P.D. Hebblethwaite, T.C.K. Dawkins, M.C. Heath and G. Lockwood (eds.)
Vicia faba: Agronomy, Physiology and Breeding. ISBN 90-247-2964-5.

dissected into peduncle, pedicel, pod and seed components. Petals were
discarded, and seeds were dissected from ovaries and counted. The plant
parts were then placed on pre-dried, pre-weighed papers and dried in an
oven at 105°C until constant weight was attained. Harvests were repeated at
regular intervals to establish a growth pattern from anthesis to maturity.

Dry weights per unit of tissue were obtained, and for peduncles
weight of peduncle/pod was calculated as this parameter is dependent on
number of pods/node.

Fluorescence microscopy

Fresh tissues were hand-sectioned and stained in 0.1% w/v Calcofluor
White-M2R (Polysciences Inc.), mounted in distilled water and examined
using a Nikon microscope with TMD-EF epifluorescence attachment fitted
with a violet (405mm) filter casette. Specimens were photographed using
Ilford XP1 film.

RESULTS

Two clear trends are evident from figure 1. Firstly, there is a
developmental sequence in the four tissues examined. Development began
immediately after tripping in pedicels and peduncles, whereas rapid pod dry
weight increase only begun between days 8 and 12, and similar increases in
seed weight were delayed until days 20-24. Secondly, dry weight gain in
line STW, particularly in pedicels, is more rapid than in genotype MB and
IVSG. This latter trend is confirmed in Table 1, where rate of dry weight
increase in line STW is shown to be not only faster, but also the final dry
weight increase is relatively higher than in the other genotypes compared.
Similar trends are evident in pods and seeds in this line, although in these
tissues final dry weight increase is not significantly different from other
genotypes. Reference to Table 2 shows that the ratio of dry weights of seed
tissue to non-seed tissue differs little between genotypes.

Table 1 records the massive dry weight increases which take place in
the reproductive tissues, particularly in pedicels and seeds. The major
contributory factor to dry weight increase in reproductive tissues is
illustrated in figure 2, where fluorescence micrographs of peduncle tissue
at anthesis and 40 days after tripping reveal that approximately 40% of the
cross-sectional area is composed of densely packed, highly lignified cells,
surrounded by a narrow zone of phloem, with thin-walled parenchyma cells

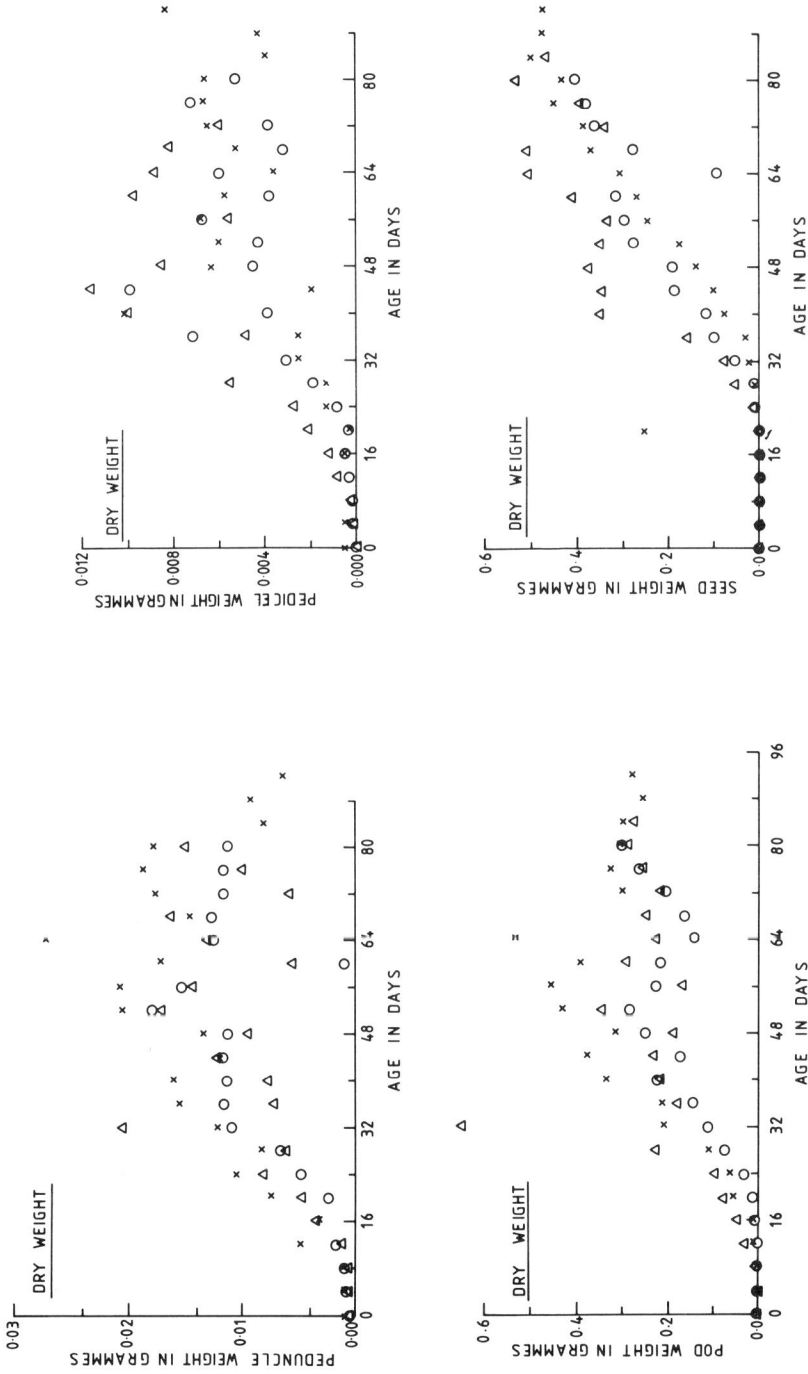

FIG.1.   Dry wt. increase in reproductive tissues of genotypes STW (△), IVSG (○) and MB (X).

DAYS AFTER TRIPPING

Tissue	Genotype	8	16	24	32	40	48	56	64	72	80
Peduncle	IVSG	0	1.10	2.44	2.78×10^1	4.38×10^1	5.06×10^1	3.16×10^1	1.65×10^1	2.03×10^1	1.87×10^1
	STW	1.54	1.00×10^2	7.65×10^2	8.08×10^3	2.25×10^3	1.09×10^3	2.87×10^3	2.16×10^3	3.54×10^2	3.21×10^3
	MB	0	2.35	1.3×10^1	1.87×10^1	4.16×10^1	2.25×10^1	2.05×10^1	1.13×10^2	6.05×10^1	4.38×10^1
Pedicel	IVSG	1.99	3.32×10^1	4.83×10^2	2.48×10^3	2.16×10^3	5.79×10^3	4.50×10^3	4.21×10^3	2.06×10^3	1.67×10^3
	STW	1.55×10^3	1.50×10^5	1.02×10^6	2.22×10^6	2.98×10^6	3.22×10^6	5.41×10^6	1.56×10^7	6.47×10^6	1.57×10^7
	MB	1.03×10^1	2.07×10^1	2.57×10^2	1.01×10^3	3.56×10^2	1.04×10^4	1.23×10^4	2.70×10^3	1.69×10^4	1.30×10^4
Pod	IVSG	0	1.07	4.91×10^1	2.10×10^2	3.28×10^2	1.43×10^3	5.25×10^2	1.80×10^2	5.05×10^2	4.01×10^2
	STW	9.68	2.05×10^3	9.77×10^3	5.54×10^4	5.45×10^4	1.66×10^5	3.79×10^2	7.58×10^2	6.70×10^2	4.70×10^2
	MB	0	1.22	4.23×10^1	5.19×10^2	1.41×10^3	2.25×10^3	3.99×10^3	2.72×10^3	1.27×10^3	1.81×10^3
Seed	IVSG	0	7.72×10^1	2.69×10^5	7.07×10^6	2.29×10^7	8.08×10^7	1.28×10^8	8.76×10^6	2.08×10^8	2.69×10^8
	STW	2.85	2.72×10^3	2.03×10^5	5.80×10^6	1.42×10^8	4.80×10^8	3.52×10^8	9.2×10^8	5.26×10^8	1.05×10^9
	MB	2.88	1.31×10^2	5.91×10^3	9.44×10^5	2.50×10^6	8.80×10^6	1.82×10^8	3.02×10^8	5.68×10^8	6.57×10^8

Table 1. Increase in dry wt. relative to initial dry weight in the reproductive tissues of STW, MB and IVSG.

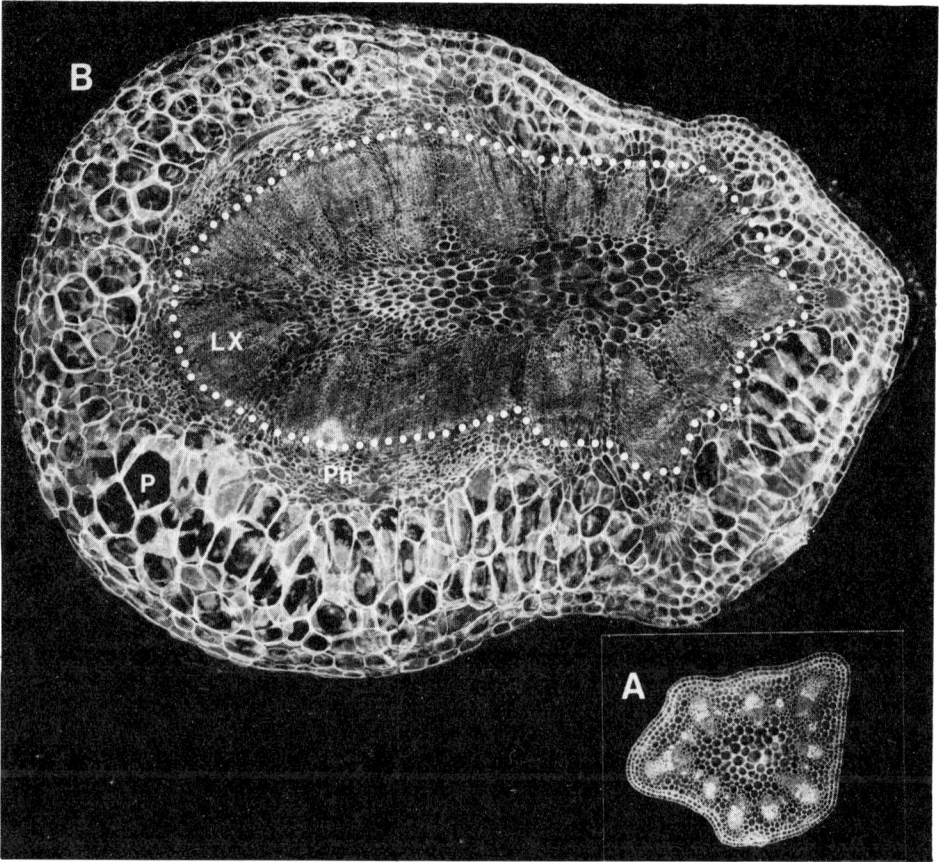

FIG. 2.   Cross sections of peduncles of line STW at anthesis (A) and 40
           days after tripping (B) stained with Calcofluor M2R. LX =
           LIGNIFIED XYLEM (outlined); Ph = PHLOEM; P = PARENCHYMA X20.

TABLE 2.   Dry wt. of pedicel, peduncle and pod per gramme of seed tissue.

Genotype	
Maris Bead	0.314
STW	0.320
IVSG	0.336

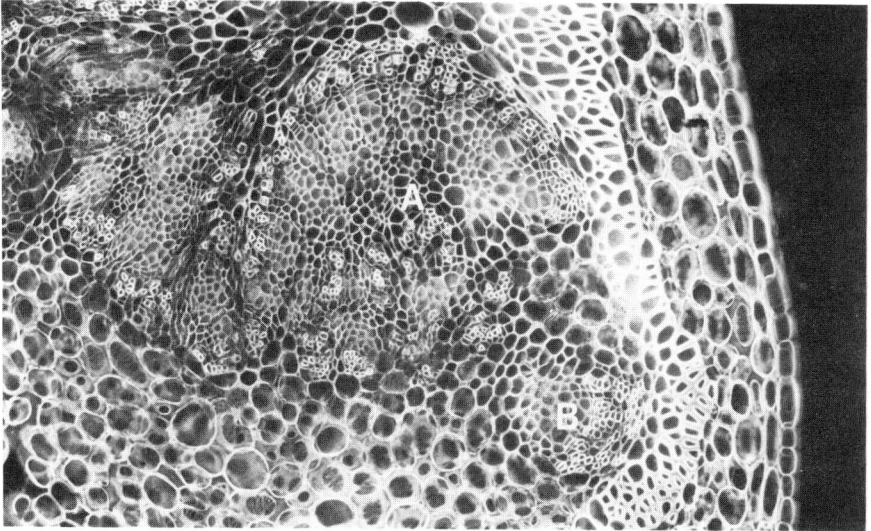

FIG. 3.   Cross section of dorsal region of pod of line STW, stained with
Calcofluor M2R. Vascular bundle A supplying developing seed,
vascular bundle B supplying aborted seed.   X20.

lying outside.

DISCUSSION

The massive accumulation of lignified xylem, which is characteristic
of peduncle (Fig. 2) and pedicel tissues accounts for the high rates of dry
weight increase recorded in Fig. 1 and Table 1.   At anthesis the cross-
sectional area of pedicel vascular bundles is small, lignification is
negligible, and the pedicel/peduncle connection is fragile (Gates et al.
1981). Post-fertilisation development in these tissues is characterised by
inter-fascicular meristematic activity generating a thick annulus of
lignified xylem, surrounded by a thinner annulus of phloem, whilst development of the outer parenchyma is restricted to cell expansion (Fig. 2).

The sequence of dry weight changes recorded in Fig. 1 indicate that
seed dry weight increase is dependent on sequential vascular development in
the supporting vascular tissues, and it has been observed that aborted seeds
are generally supplied by vascular bundles which exhibit arrested development (Fig. 3). It is proposed therefore that seed abortion is a result of
vascular developmental failure, which may be in turn the result of
differences in time of fertilisation of individual ovules (Gates et al. 1983).

The growth rates of reproductive tissues supporting the ovules (Table 1) indicate that they are important assimilate sinks. After fertilisation there is a lag phase in pod development coinciding with severe competition between young pods and vegetative sinks, and pods do not receive more than 10% of ^{14}C fixed until flowering has ended (Jacquiéry and Keller 1980). During this period rapid vascular tissue development takes place, coinciding with the disappearance of large quantities of starch deposited in the reproductive tissues before anthesis (Gates et al. 1981). It is probable that this non-structural carbohydrate (NSC) is remobilised and deposited in lignified cell walls during the lag phase early in pod development. Intense Golgi body activity associated with cell wall formation has been observed in pedicels after pollination (Gates and Harris, unpublished). It has been reported by Jacquiéry and Keller (1978) that the stem is a major accumulator of dry matter during flowering, whilst Müntz, Schalldach and Manteuffel (1976) have demonstrated the role of pod walls in temporary starch storage. The strategic storage of NSC at sites of rapid growth may be characteristic of the indeterminate growth form of V. faba. The mobilisation of NSC for early reproductive development implies i) that reproductive success may depend on pre-anthesis storage of starch and ii) that depletion of stored NSC by environmental conditions favouring vegetative growth should be considered as a cause of reproductive failure.

Table 1 and Fig. 1 indicate that genotypic differences exist in the rate of growth of reproductive tissues. STW is known to exhibit extremely early maturity compared with other genotypes, and the rapidity of reproductive tissue development in this line may be an important contributory factor to this agronomically important trait. Table 2 shows that the enhanced rate of development is not accompanied by an increase in the relative amount of non-seed reproductive tissue, indicating a basic physiological difference in this line. This is currently under investigation in this laboratory.

BIBLIOGRAPHY

Müntz, K., Schalldach, I. and Manteuffel, R. 1976 . The function of the pod for protein storage in seeds of Vicia faba L. IV. Seed dependent reactivation of stored starch from the pods. Biochem. Physiol. Pflanzen 170, 465-477.
Jacquiéry, R. & Keller, E.R. 1980 . Beeinflussung des fruchtansatzes bei der Ackerbohne (Vicia faba L.) durch die verteilung der assimilate (Teil II). Angew. Botanik 54. 29-39.

Jacquiéry, R. & Keller, E.R. 1978 . Beeinflussung des fruchtansatzes bei der Ackerbohne (Vicia faba L.) durch die verteilung der assimilate (Teil I) Angew Botanik 52, 261-276.

Gates, P.J., Smith, M.L., White, G. and Boulter, D. 1983 . Reproductive physiology and yield stability in Vicia faba L. In the Physiology, Genetics and Nodulation of Temperate Legumes. Ed. D.R. Davies and D.G. Jones. Pitman Books, London. In press.

Gates, P.J., Yarwood, J.N., Harris, N., Smith, M.L. and Boulter, M.E. 1981 . Cellular changes in the pedicel and peduncle during flower abscission in Vicia faba. In Vicia faba, Physiology and Breeding, World Crops Vol. 4, pp. 299-316. Ed. R. Thompson. Martinus Nijhoff, The Hague.

GROWTH HABIT IN RELATION TO ASSIMILATE PARTITIONING AND SOME
CONSEQUENCES FOR FIELD BEAN BREEDING

D.A. Baker, G.P. Chapman, M.J. Standish and M.P. Bailey

Wye College, University of London, Wye,
Ashford, Kent. TN25 5AH.

ABSTRACT

Analyses indicate that although there are not major differences in the
two types of field bean, determinate and indeterminate in terms of photosynthetic potential the leaves of determinate plants have a delayed senescence and remain photosynthetically active for a longer period of time. The
partitioning of $^{14}CO_2$ fed to individual leaves of determinate and indeterminate plants indicates that the genetic removal of indeterminate stem
growth alters the number of competing sinks but does not increase the percentage of assimilates moving to developing pods. The implications of
these results for further genetic manipulation of the source:sink relationships of the field bean are discussed.

INTRODUCTION

Although *Vicia faba* has been the subject of a number of assimilate

partitioning studies (see Crompton, Lloyd-Jones and Hill-Cottingham, 1981)

to date only a limited number of studies have been conducted on determinate

material (Austin, Morgan and Ford, 1981; Baker, Chapman, Standish and

Bailey, 1983). These investigations showed that the inherent advantage of

the terminal inflorescence may be lost when tillers and axillary branches,

which develop asynchronously with the main shoot, remain infertile. Under

such conditions these branches compete for assimilates as an alternative

sink, in much the same manner as does the continued growth of the apex in

the indeterminate habit. Furthermore, the determinate material is no more

efficient at fixing CO_2 than the indeterminate, but the individual leaves

on the determinates are photosynthetically active for a longer period due

to delayed senescence (Baker *et al.*, 1983).

In this paper the source:sink relationships of the two growth habits

have been further characterised to provide physiological guidelines for

future breeding strategies, in particular to remedy the effective source

limitation of the determinate habit.

MATERIALS AND METHODS

Indeterminate (Maris Bead) and determinate (Wye 19280/1 and Wye

19260/19) plants of *Vicia faba* L. were grown under glass in individual

pots containing potting compost.

P.D. Hebblethwaite, T.C.K. Dawkins, M.C. Heath and G. Lockwood (eds.)
Vicia faba: Agronomy, Physiology and Breeding. ISBN 90-247-2964-5.
© 1984, Martinus Nijhoff/Dr W. Junk Publishers. Printed in The Netherlands.

During the development of the plants the chlorophyll content and photosynthetic potential were determined for individual leaves. Chlorophyll content was calculated after extraction with acetone, partitioning the extract with water and petroleum ether and determining the absorbance at 645 nm and 663 nm on a spectrophotometer. Photosynthetic potential was determined for individual leaves by enclosing four 13.5 mm leaf discs in a plastic petri dish equilibrating for one hour and then liberating known volumes of $^{14}CO_2$ [25 µl 1.01 M Na$_2$ $^{14}CO_3$, 1.48 GBq (^{14}C)/1 to give 0.8% CO_2 (0.52 MBq/1) in the chamber]. After 30 mins exposure to $^{14}CO_2$ the leaves were solubilised, mixed with scintillant and activity determined by liquid scintillation counting.

Photosynthetic rates of attached leaves were also measured. This was achieved by enclosing individual leaves in a sealed acrylic chamber and passing air at a known flow rate over them. The change in CO_2 concentration of the air passing through the chamber was measured with an IRGA. Flow rates were adjusted to maintain a concentration change in the range 20 to 25 µl/1 where possible while keeping the flow rate greater than 3 ml /s. This was performed at 7 levels of light irradiance from 0 lux to 9500 lux, at a temperature of 19°C. The dark respiration rate, compensation point and photosynthetic rate at 8000 lux were calculated by fitting a quadratic regression line to a graph of photosynthetic rate. Measurements were conducted on three or four leaves (apical, sub-apical, first podding node and basal) on the first stems of both determinate and indeterminate plants, for both intact plants and those on which all flower buds were removed, once a week from 35 days after planting to plant death.

Individual leaves on determinate and indeterminate plants were allowed to assimilate $^{14}CO_2$ for one hour (0.925 M Bq/plant). These plants were then harvested 24 hours or 6 weeks later and the percentage of ^{14}C determined in various parts after combustion of the samples in a tissue oxidiser. The leaves of indeterminate plants were fed $^{14}CO_2$ at the first, seventh and last flowering node, and leaves of determinate plants at the first, second and third flowering nodes, in each case at initial flowering, young pod development and rapid pod fill.

Whole plants were allowed to assimilate $^{14}CO_2$ for 30 mins [100 µl 1.01 M Na$_2$ $^{14}CO_3$, 1.48 G Bq(^{14}C)/1 to give 0.08% CO_2 (0.52 M Bq/1) in the chamber, which contained four plants]. The plants were then harvested either immediately or 24 hours later and the percentage of ^{14}C determined in various parts after combustion of the samples in a tissue oxidiser. Ethanol soluble sugars were extracted from parallel samples with 80% ethanol and acid-extractable polysaccharides were recovered from the residue by boiling for one hour in 3% HCl. The extracts were then neutralized, and decolourized, as described by Munns and Pearson (1974), mixed with scintillant and the activity determined by liquid scintillation counting.

RESULTS

The efficiency of the two growth habits in terms of leaf area duration and photosynthetic capacity are presented in Table 1 where it can be seen that there is no inherent difference in the two types in terms of their photosynthetic capacity. The greater yield of the indeterminate on an individual plant basis (Table 2) merely reflects the greater plant size, the determinate being one quarter the size and producing one quarter the amount of seed of the indeterminate type. The actual leaf efficiency in producing seeds may be estimated as seeds per square metre of leaf per day which gives a value of 1.845 for the determinate and 1.655 for the indeterminate, while mg of seed per square metre per day is 705.6 and 785.8 respectively. Thus it can be seen that while determinate plants are slightly more efficient at producing seed number, the smaller size of these seeds makes them slightly less efficient in terms of total yield.

The general pattern obtained for whole plant feeding is presented in Table 3 where it can be seen that the initial fixation and distribution for both growth habits is similar. Twenty-four hours after feeding the exported ^{14}C-labelled materials are found within the stem in both types of plant. However more of the fixed carbon is exported in the indeterminate type and an increased proportion is present in the flowers and developing apex, indicating the greater number and hence potency of these sinks in the indeterminate habit.

The potential of terminal pods in the determinate habit for enhanced photosynthesis has been considered. Fixation of $^{14}CO_2$ by the pods as a percentage of that of the subtending leaf is presented in Table 4 for the two plant types. In neither type does the pod make an appreciable contribution to the photosynthesis of the plant and in these terms there does not appear to be an advantage with terminal podding although cessation of terminal growth and freedom from lodging remain practical advantages.

TABLE 1   Leaf attribute durations (estimated from modelled plants)

	Determinate	Indeterminate
Area (square metre.days)	7.6	28.4
Dry weight (mg.days)	207	946
Chlorophyll content (migrogram.days)	3440	14900
Photo-capacity (dps.days)	8590	231.00

TABLE 2.   Pod and seed yield per plant [Mean (SD)]

	Determinate	Indeterminate
Pods	10.364 (11.475)	20.903 (24.066)
Seeded Pods	9.205 (10.689)	19.323 (22.581)
Seeds	14.023 (16.611)	47.000 (51.686)
From (plants)	44	31
Seed dry weight	382.4 (83.8)	474.8 (93.8)
Seed wt. on plant	5362.4	22315.6

TABLE 3   Whole plant feeding of Maris Bead and 19260/19 and percentage distribution of ^{14}C at initial flowering

	Maris Bead	19260/19
a) Harvest 0hrs		
Leaves	81.2 (2.3)	84.8 (2.9)
Stems + stipules	14.6 (1.6)	12.7 (2.5) including tillers
Flowers	2.2 (0.3)	1.9 (0.5)
Apex	1.4 (0.5)	-----
Roots	0.6 (0.2)	0.6 (0.1)
b) Harvest 24hrs		
Leaves	22.3 (0.8)	33.9 (4.0)
Stems + stipules	47.7 (2.7)	50.2 (5.5) including tillers
Flowers	15.8 (0.8)	6.6 (1.7)
Apex	8.6 (0.6)	----
Roots	4.8 (0.5)	9.3 (1.2)

Mean of 3 replicates   SE in brackets

TABLE 4   Photosynthesis of pods as a percentage of subtending leaf
           during pod fill

	Totals		Per unit area of leaf	
	Maris Bead	19260/19	Maris Bead	19260/19
Per cent	15.3 (2.4)	4.2 (1.1)	46.9 (4.6)	9.3 (1.5)
Range	9.4-26.3	1.9-8.8	31.6-65.6	5.0-14.5

For each variety - Mean of six replicates
SE in brackets

DISCUSSION

The determinate field bean is in breeders terms a recent arrival and
has so far been seen in only a few well tested lines. The breeder is therefore perhaps unwise to go further without a careful recognition of the
physiological possibilities of this modified growth habit.

The results presented here provide some physiological quantification
for a number of the qualitative assumptions made in our current breeding
programmes. Neotenous plants described earlier (Chapman, 1977; Chapman
and Peat, 1978) utilised the ti mutant described by Sjodin (1971) and our
conclusions apply only to these lines. These are:

1. The supposed advantages of determinate types may be lost when secondary
    infertile branches develop asynchronously with the main stem.

2. The vegetative apex regarded as a competing sink is an oversimplifica-
    tion and other evidence has shown that its effectiveness as a source
    increases if growth factors are applied to the axillary inflorescences
    (Chapman and Sadjadi 1981).

3. Pods, in whatever position, do not fix large quantities of carbon
    although the pods on the indeterminate plant are slightly more efficient.
    This potential advantage is therefore best discarded.

4. The stem is a significant storage organ and a study of the turnover of
    carbon in this organ throughout the season should be undertaken.

5. Genetic alternatives such as 'determinate' show unsuspected physiolog-
    ical contrasts and as such might continue to provide insights into the
    working of 'normal' plants.

What direction should the breeder now follow? If 'determinate' and
'indeterminate' are regarded as 'telescoped' and 'successional' types

28

perhaps the essential difference for yield accumulation may be recognised. Since the indeterminate types are proving superior, the 'telescoped' types can be 'for comparison only' and effort should be concentrated on optimising the succession of physiological events underlying yield. Among these, the range of stem and stipule variation and the resulting physiological conseq-uences might be considered an important area especially if results presented here were subsequently shown to apply under field conditions.

It is recognised that modern indeterminate field bean varieties are closely similar due doubtless in part to the needs of the trade. However, it is worth asking if 'high yield' is also the outcome of a common syndrome of many characters few of which are recognised beside the conventional components. If this were so, then a combination of physiology and of the new genetic variation now increasingly available, might enable us to search for the 'components of efficiency'. Perhaps therefore it is now appropriate to develop predictive yield models (primarily for indeterminate types). If these became sufficiently versatile they might pin-point yield constraints within the plant's physiology, for which genetic alternatives could be available.

REFERENCES

Austin, R.B., Morgan, C.L. and Ford, M.A., 1981. A field study of the carbon economy of normal and 'topless' field beans (Vicia faba) in Vicia faba : Physiology and Breeding (ed. R. Thompson) pp. 60-77, Nijhoff.
Baker, D.A., Chapman, G.P., Standish, M.J. and Bailey, M.P., 1983. Assimilate partitioning in a determinate variety of field bean in The Physiology, Genetics and Nodulation of Temperate Legumes (eds. D.R. Davies and D.G. Jones), Pitman. In press.
Chapman, G.P., 1977. Restructuring the field bean (Vicia faba L.) Scot. Hort. Res. Inst. Bull. 15, 3-9.
Chapman, G.P., 1981. Determinate growth in Vicia faba : an opportunity for accelerated genetic turnover in Vicia faba : Physiology and Breeding (Ed. R. Thompson) pp. 236-243, Nijhoff.
Chapman, G.P. and Peat, W.P., 1978. Procurement of yield in field and broad beans. Outlook on Agriculture 9, 267-272.
Chapman, G.P. and Sadjadi, A.S., 1981. Exogenous growth substances and internal competition in Vicia faba L. Z. Pflanzenphysiol. 104, 265-273.
Crompton, H.J., Lloyd-Jones, C.P. and Hill-Cottingham, D.G., 1981. Trans-location of labelled assimilates following photosynthesis of $^{14}CO_2$ by the field bean, Vicia faba. Physiol. Plant. 51, 189-194.
Sjodin, J., 1971. Induced morphological variation in Vicia faba L. Hereditas 67, 155-180.
Munns, R. and Pearson, C.J., 1974. Effect of water deficit on transloca-tion of carbohydrate in Solanum tuberosum. Aust. J. Plant Physiol. 1. 529-537.

NITRATE REDUCTION IN *VICIA FABA* GROWN

AT DIFFERENT TEMPERATURES

M. Andrews, J. Sutherland, J.I. Sprent
Department of Biological Sciences
The University
Dundee, DD1 4HN

ABSTRACT

Nitrate reductase activities were determined for spring and winter cultivars of *Vicia faba* grown under a range of conditions. An *in vivo* assay was used with and without added substrate; activities measured being termed potential and actual respectively. Potential activity was greatest during the light period for cultivar Banner Winter grown in the field. Relative activities in leaves, stem and root of laboratory grown Banner Winter were greatly dependent on applied nitrate concentration. Root activity (per unit fresh weight) changed little on increased applied nitrate ($1-20$ mol m^{-3}) whilst activities in leaf and stem increased markedly. The root had the highest activity per organ at 1 mol m^{-3} NO_3^-, leaves and stem at 20 mol m^{-3} NO_3^-. Activities, especially in the shoot, were higher in Herz Freya (a spring cultivar) than Banner Winter, when grown under the same conditions. A comparison of activities in similar plant parts grown at different temperatures, showed that for both Banner Winter and Maris Bead (a spring cultivar), highest values occurred at the lowest temperature ($5^{\circ}C$): this finding may be related to growth temperature, plant maturity or water potential. Nitrate reductase activity versus temperature curves constructed for Maris Bead showed a discontinuity at approximately $10^{\circ}C$. In two experiments activities measured at $0^{\circ}C$ were 40 and 50% of those at $30^{\circ}C$ for plants grown at $5^{\circ}C$ and 20 and 37% for plants grown at $15/10^{\circ}C$.

INTRODUCTION

Vicia faba may be autumn or spring sown in the United Kingdom. The ability of cultivars to tolerate low temperatures varies greatly and varieties sown in spring usually have poor tolerance. If sown in the autumn these cultivars produce chlorotic seedlings which do not turn fully green until spring (usually late March in East Scotland) (Fyson, 1981). In cold years, even when sown at the recommended time spring cultivars may initially be chlorotic. Since the crop has a long growing season in the United Kingdom rapid growth in early spring is essential.

The results presented here come from preliminary studies in a project whose objectives are to determine why chlorosis occurs in spring cultivars of *V. faba* grown at low temperatures, and whether this chlorosis can be overcome by, for example, application of combined nitrogen.

P.D. Hebblethwaite, T.C.K. Dawkins, M.C. Heath and G. Lockwood (eds.)
Vicia faba: Agronomy, Physiology and Breeding. ISBN 90-247-2964-5.
© 1984, Martinus Nijhoff/Dr W. Junk Publishers. Printed in The Netherlands.

MATERIALS AND METHODS

Growth of Plants

(a) Plants of cv. Banner Winter used in the diurnal experiment were grown in the field at the Scottish Crops Research Institute, Invergowrie. Combined nitrogen was not applied to the plot.

(b) Plants of cv. Banner Winter and cv. Herz Freya (a spring cultivar) used in experiments where temperature was not a variable, were grown from seed in plastic trays containing sterilised coarse sand. The trays were placed in a Sherer Controlled Environment Chamber, (CEL8) and exposed to a 14 h light/10 h dark cycle. Photon flux density at tray surface was approx. 100 μE m^{-2} s^{-1}; temperature was $5^{\circ}C$. The nutrient medium contained $Ca(NO_3)_2.4H_2O$ (1 mol m^{-3}), $CaSO_4.2H_2O$ (4 mol m^{-3}), KH_2PO_4 (1.5 mol m^{-3}), $MgSO_4$ (3.1 mol m^{-3}), K_2SO_4 (0.75 mol m^{-3}), $MnSO_4.4H_2O$ (1.0 mmol m^{-3}), $CuSO_4.5H_2O$ (0.1 mmol m^{-3}), $ZnSO_4.7H_2O$ (0.1 mmol m^{-3}), H_3PO_3 (5.0 mmol m^{-3}), NaCl (10.0 mmol m^{-3}), $NaMoO_4.2H_2O$ (0.5 mmol m^{-3}), $CoSO_4.6H_2O$ (0.02 mmol m^{-3}) and $C_6H_5O_7$ Fe.$5H_2O$ (5 mmol m^{-3}). After six weeks the plants were transferred to pots (one plant per pot) and applied nitrate concentration was changed; half the plants of each cultivar were given $Ca(NO_3)_2.4H_2O$ (0.5 mol m^{-3}) and half $Ca(NO_3)_2.4H_2O$ (5 mol m^{-3}). Calcium concentration was maintained at 5 mol m^{-3} using $CaSO_4.4H_2O$.

(c) Plants of cv. Maris Bead (a spring cultivar) and cv. Banner Winter used in experiments where temperature was a variable were taken from the field (spring sown) potted as above and placed in growth cabinets at three different temperatures, $5^{\circ}C$, $15/10^{\circ}C$ and $25/15^{\circ}C$. Day-length, light intensity and applied nutrient medium were as above except that 10 mol m^{-3} $Ca(NO_3)_2.4H_2O$ was used. Plants were kept at these temperatures for at least one month before assay. Maris Bead was used as a representative spring variety in this case as Herz Freya is not now a recommended variety.

Nitrate Reductase Assay

An *in vivo* assay optimised for *V. faba* was used to measure nitrate reductase activity.

In most cases two estimates termed potential and actual were obtained. To estimate potential activity, 300-400 mg of plant strips (2-3 mm width) were vacuum infiltrated for 10 minutes with 10 ml of 100 mol m^{-3} phosphate buffer (pH 7.6) containing 3% n-propanol and 20 mol m^{-3} KNO_3. The mixture

was then incubated at 30°C for 20 minutes in a shaking water bath in the
dark. 0.5 ml samples taken at 0 and 20 min incubation time were analysed
for NO_2^- colorometrically as described in MacKereth, Heron and Talling
(1978). Actual activity was obtained by following the above procedure
with KNO_3 omitted from the assay buffer. In some assays in experiments,
where temperature was a variable, potential activity was estimated at
different incubation temperatures by sampling the assay medium at invervals
and changing the assay temperature during the time of sampling.

All data are presented together with standard error of the mean values
and number of replicates (n). Abbreviations used for plant parts in Tables
1-4 are as indicated in Fig. 1 by underlining.

RESULTS

Potential nitrate reductase activity in field grown cv. Banner Winter
(Fig. 1) showed a marked diurnal variation in stems and roots (maximum in
light period), but changed little in the leaves during the 24 h period.
In general, activity per unit fresh weight decreased in the order stem,
leaves, tap root, lateral root, during the day. At night activity was
highest in the leaves. In almost all cases actual activity was considerably less than 100 nmoles NO_2^- g fw^{-1} h^{-1} with values being highest in the
roots.

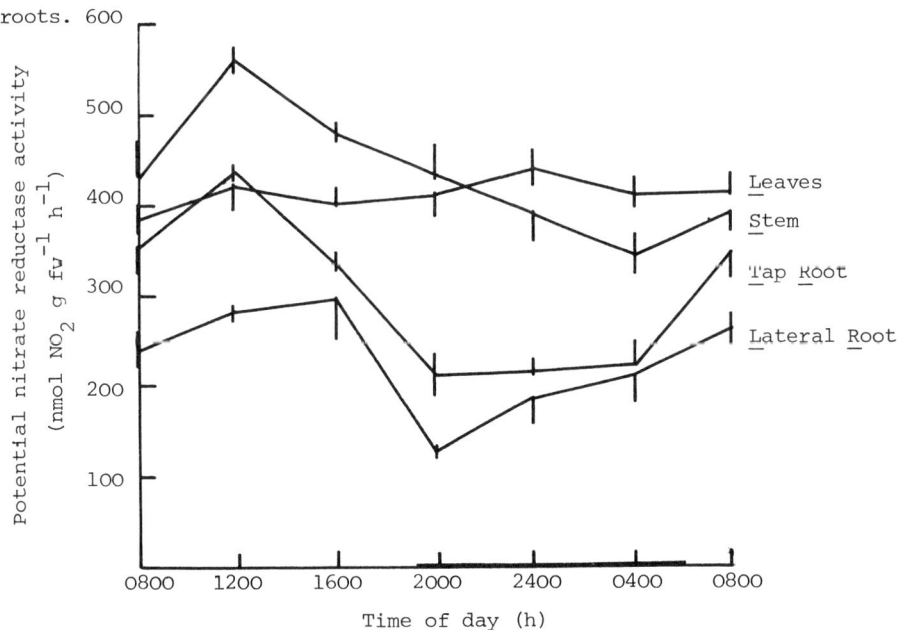

Fig. 1   Potential nitrate reductase activity in the component parts
of field grown cv. Banner Winter on 23-24 March 1983. n = 4.

The concentration of applied nitrate has been shown to greatly effect nitrate reductase activity in laboratory grown Banner Winter (Table 1).

TABLE 1   Nitrate reductase activity in the component parts of Banner Winter grown at three nitrate concentrations. n = 7.

		nmol NO_2^- g fw^{-1} h^{-1}			nmol NO_2^- organ^{-1} h^{-1}		
		1 mol m^{-3}	10 mol m^{-3}	20 mol m^{-3}	1 mol m^{-3}	10 mol m^{-3}	20 mol m^{-3}
L	P	115 (36)	298 (42)	1354 (207)	144 (39)	288 (51)	2626
	A	14 (7)	67 (34)	392 (34)	22 (13)	64 (33)	760
S	P	429 (54)	774 (57)	1558 (252)	322 (41)	597 (46)	2695
	A	28 (8)	209 (17)	562 (63)	22 (7)	158 (57)	972
T P	452 (66)	476 (65)			214 (44)	235 (26)	
R A	137 (29)	173 (39)	P 370 (11)		67 (16)	91 (20)	P 1093
L P	370 (63)	404 (70)	A 220 (36)		670 (141)	457 (50)	A 657
R A	134 (25)	141 (17)			202 (48)	139 (21)	

On increased nitrate concentration (1-20 mol m^{-3}) potential activity in the root remained constant whilst actual activity increased by approximately 60%. In contrast potential activity in the stem quadrupled and in the leaf showed an order of magnitude increase. In both stem and leaf actual activity increased more than 20-fold. The roots had the highest activity per organ at 1 mol m^{-3} nitrate; leaves and stem at 20 mol m^{-3} nitrate.

Nitrate reductase activity can vary considerably in different cultivars when grown under the same conditions (Table 2).

TABLE 2   Nitrate reductase activity (mid light period) in the component parts of Herz Freya and Banner Winter grown under the same conditions (10 mol m^{-3} NO_3^-). n = 6.

		nmol NO_2^- g fw^{-1} h^{-1}		nmol NO_2^- organ^{-1} h^{-1}	
		Herz Freya	Banner Winter	Herz Freya	Banner Winter
L	P	2129 (300)	298 (42)	1296 (91)	288 (51)
	A	397 (134)	67 (34)	334 (86)	64 (33)
S	P	868 (159)	774 (57)	546 (198)	597 (46)
	A	398 (127)	209 (17)	261 (165)	158 (51)
T	P	682 (122)	476 (65)	271 (31)	235 (26)
R	A	511 (129)	173 (39)	192 (24)	91 (20)
L	P	468 (53)	409 (70)	412 (78)	457 (50)
R	A	366 (50)	17 (17)	316 (56)	139 (28)

Potential activities in the component parts of Herz Freya are higher than in corresponding plant parts of Banner Winter. The increase is slight in root and stem but 7-fold in leaves. Actual activities are 2-7 times greater in the component parts of Herz Freya. Highest activity per organ occurred in the leaves of Herz Freya and the stem of Banner Winter.

The fresh weights of Banner Winter and Maris Bead used in temperature related studies are shown in Table 3.

TABLE 3   Fresh weight at the time of assay of the component parts of Banner Winter and Maris Bead grown at different temperatures. n = 6-7.

| | | | | Growth temperature | | | |
| | 25/15°C | | 15/10°C | | | 5°C | |
	B.W.	M.B.	B.W.	M.B.	B.W.	M.B.
L	3.44 (0.36)	-	3.62 (0.58)	4.09 (0.26)	1.94 (0.42)	2.47 (0.39)
S	3.36 (0.34)	-	5.59 (0.95)	7.84 (0.53)	1.73 (0.45)	3.74 (0.37)
R	3.54 (0.7)	-	4.72 (0.73)	3.45 (0.70)	3.16 (0.44)	3.50 (0.29)

Plant of both cultivars had considerably lower fresh weight when grown at 5°C than when grown at higher temperatures. In general nitrate reductase activity decreased in Banner Winter plants on increased growth temperature (Table 4). In particular leaves grown at 25/15°C had extremely low activity per unit fresh weight and per organ in comparison with leaves of plants from lower temperatures.

TABLE 4   Nitrate reductase activity in the component parts of Banner Winter grown at three different temperature regimes

| | | nmol NO_2^- g fw^{-1} h^{-1} | | | nmol NO_2^- organ^{-1} | | |
		25/15°C	15/10°C	5°C	25/15°C	15/10°C	5°C
L	P	129 (16)	868 (207)	1354 (207)	463	3162	2626
	A	44 (13)	427 (162)	392 (34)	151	1546	760
S	P	347 (57)	449 (87)	1558 (252)	1166	2509	2695
	A	249 (69)	253 (51)	562 (63)	836	1414	972
R	P	248 (31)	339 (22)	346 (11)	878	1600	1093
	A	238 (44)	220 (21)	208 (36)	842	1038	657

Nitrate reductase activity per unit fresh weight for leaves of Maris Bead grown at 5°C and 15/10°C was 1583 (134) and 1142 (152)

nmol NO_2 g fw h^{-1} respectively. Nitrate reductase activity versus temperature curves for these leaves constructed from the results of two experiments are shown in Fig. 2.

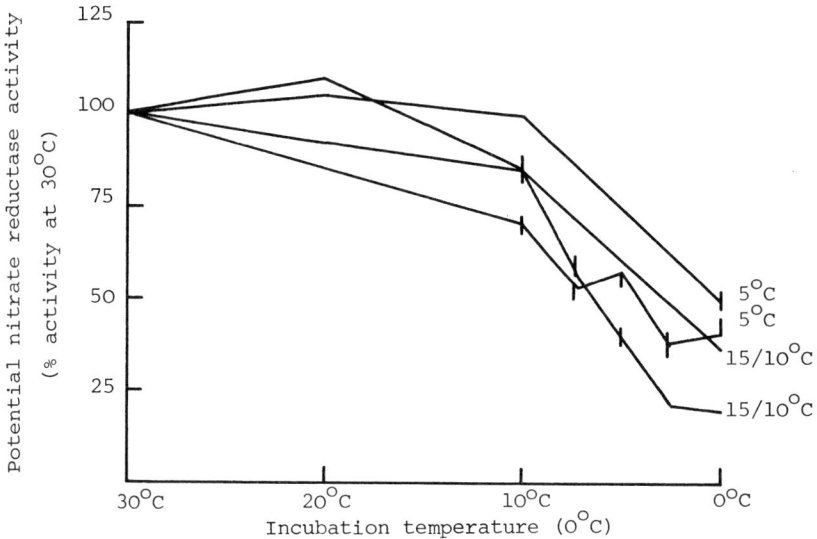

Fig. 2   The effect of assay temperature on potential nitrate reductase activity in leaves of Maris Bead grown at 5°C and 15/10°C. (n = 3-4).

In experiment 1 the temperature was taken from 30°C to 0°C in three stages (20°C - 10°C - 0°C) and experiment 2 five stages (30°C - 10°C - 7.5°C - 5°C - 2.5°C - 0°C). Activity in leaves from both growth temperature regimes increased slightly at 20°C but decreased from 10°C to 0°C: discontinuity occurs in the curves at approximately 10°C. Activities measured at 0°C were 40 and 50% of those at 30°C for plants grown at 5°C and 20 and 37% for plants grown at 15/10°C.

DISCUSSION

Nitrate reductase activity in *Vicia faba* shows a marked diurnal variation (Fig. 1) with highest activity occuring during the light period.

Although generally thought of as a "root reducer" (Pate, 1973) the results presented here (Fig. 1, Tables 1-4) indicate, that under certain conditions greatest nitrate reductase activity can occur in the shoot. It is accepted that the assay used here cannot show unequivocally that nitrate is reduced in the shoot, but if it is considered that (a) high actual activities can occur in stems and leaves thus indicating high levels of endogenous nitrate (b) shoot activity (actual and potential) drops to zero if nitrate application is stopped (Andrews, unpublished data), thus emphasising that the enzyme is induced by nitrate (c) increased applied nitrate concentration (1 mol m^{-3} - 20 mol m^{-3}) gives greatly increased growth rates (Andrews, unpublished data) but effects only slightly nitrate reductase activity in the root and (d) previous work on wheat (Brunetti and Hageman, 1976) and soybean (Crafts-Brandner and Harper, 1982) indicates that optimised *in vivo* nitrate reductase assays show a good correlation with reduction rates in root and shoot estimated using alternative methods; it seems highly probable that under certain conditions the shoot of *Vicia faba* reduces a large proportion of the nitrate taken up by the root.

Plants of the same age grown under different temperature regimes, after the same initial pretreatment, showed decreased nitrate reductase activity on increased temperature (Table 3): this was especially true of leaves grown at the highest temperature regime (25/15°C). Further work is necessary to resolve the interactions between growth temperature, maturity of plants and water potential effects (cf. Schrader and Thomas, 1981 and references therein) before conclusions can be reached as to the cause(s) of this decrease.

Few data are available concerning the responses to low temperatures, of the enzymes of N-metabolism in legumes. The data presented here indicates that nitrate reduction can take place in *Vicia faba* at low temperatures with activity at 0°C being some 20-50% of that at 30°C (cf. Bandana, Tripathi, Srivastava and Dixit, 1980). These values are very high in comparison with acetylene reduction rates measured for *V. faba* at low temperatures (Fyson, 1981). There is also evidence (Fig. 2) that plant growth temperature changes the response to nitrate reductase assay temperature (cf. Chopra, 1982). Activities measured at 0°C were 40 and 50% of those at 30°C for plants grown at 5°C and 20 and 37% for plants grown at 15/10°C. Further work in relation to this is being carried out.

Finally, a comparison of nitrate reductase activity in different parts of cv. Herz Freya and cv. Banner Winter grown under the same conditions (Table 2) show that like soybean (Crafts-Brandner and Harper, 1982), wheat (Brunetti and Hageman, 1976) and maize (Schrader and Thomas, 1980), marked intervarietal differences occur. Whether all spring varieties show higher activity than all winter varieties remains to be determined, but the values for Maris Bead (see results) fit that pattern. Herz Freya plants taken from the initial pretreatment were chlorotic, thus results indicate that such plants are able to take up and reduce nitrate at low temperatures. Whether these high activities in the leaf are as a result of the chlorosis, a phenomenon known to occur with herbicides which inhibit photosynthesis (Fedtke, 1981) and whether prolonged application of high nitrate concentrations would alleviate chlorosis has yet to be tested but it is encouraging that Maris Bead plants grown at $5^{\circ}C$ with 20 mM NO_3 were not chlorotic.

REFERENCES

Bandana, B., Tripathi, R.D., Srivastava, H.S. and Dixit, S.N. 1980. Temperature dependent chloramphenicol effect on dark incubated nitrate reductase activity in maize and black gram leaves. Nat. Acad. Sci. Letts. 51, 2-4.

Brunetti, N. and Hageman, R.H. 1976. Comparison of *in vivo* and *in vitro* assays of nitrate reductase in wheat (*Triticum aestivum*) seedlings. Plant Physiol. 58, 583-587.

Chopra, H. 1983. Effects of temperature on the *in vivo* assay of nitrate reductase in some C_3, C_4 species. Ann. Bot. 51, 617-620.

Crafts-Brandner, S.J. and Harper, J.E. 1982. Nitrate reduction by roots of soybean (*Glycine max* (L.) Merr) seedlings. Plant Physiol. 69, 1298-1303.

Fedtke, C. 1981. Nitrogen metabolism in photosynthetically inhibited plants. In "Biology of Inorganic Nitrogen and Sulphur" (Ed. H. Bothe and A. Trebst). (Springer Verlag, Berlin). pp. 260-265.

Fyson, A. 1981. Effects of low temperature on the development and functioning of root nodules of *Vicia faba* L. Ph.D. Thesis, University of Dundee.

MacKereth, F.J.H., Heron, J. and Talling, J.F. 1978. Water Analysis: some revised methods for limnologists. Scientific Publications of the Freshwater Biological Association 27.

Schrader, L.E. and Thomas, R.J. 1981. Nitrate uptake, production and transport in the whole plant. In "Developments in Plant and Soil Sciences", Vol. 3 Nitrogen and Carbon Metabolism (Ed. J.D. Bewley). (Martinus Nijhoff, The Hague). pp. 49-93.

EFFECT OF PLANT GROWTH REGULATOR COMBINATIONS ON FABA BEAN
DEVELOPMENT AND YIELD COMPONENTS

M. Kellerhals and E.R. Keller

Swiss Federal Institute of Technology
Department of Crop Science, Zürich, Switzerland

ABSTRACT

Faba bean development and yield can be influenced by plant growth regulator (PGR) combinations which might be more similar to the endogenous
hormonal balance of plants than single growth substances. The main objectives of our investigations with PGR combinations are to reduce the amount
of flower shedding and pod drop, and to increase lodging resistance. Due to
the plant architecture and the growth habit, the fruits of Vicia faba are
subject to different dominance and competitive phenomena.

Several field and greenhouse experiments have been conducted since
1981 using the traditional indeterminate variety Herz Freya. Striking
effects of growth substances on several parameters of vegetative and reproductive growth were detected. However, their influence on assimilate partitioning was too weak to induce grain yield increases. V. faba displayed
a strong compensation ability among different yield components.

A successful transformation of the plant type Herz Freya to a semi-
dwarf type was achieved by means of a growth regulator combination.

INTRODUCTION

With the utilisation of PGR combinations, we tried to study the yield
performance of the faba bean crop. Lawrence (1978) considers the balance
between various hormones at least as important as their absolute levels.
The recognition that two or more endogenous growth substances may affect a
specific aspect of growth and differentiation, has led to the concept of
hormonal balance, according to which normal growth and development depend
upon a quantitative relationship between the levels of interacting growth
substances (Wareing, 1978). Based on this concept, we continued our previous research on single growth substances (Keller et al., 1980) with the new
approach of applying PGR combinations to faba beans.

How could the application of combinations, considering time of application and concentration, lead to an improved yield in the faba bean crop?
Fig. 1 shows some possible modes. It becomes obvious that growth substance
application and plant development are interdependent. Intensive vegetative
and reproductive growth mostly occurs simultaneously, causing strong

P.D. Hebblethwaite, T.C.K. Dawkins, M.C. Heath and G. Lockwood (eds.)
Vicia faba: Agronomy, Physiology and Breeding. ISBN 90-247-2964-5.
© 1984, Martinus Nijhoff/Dr W. Junk Publishers. Printed in The Netherlands.

competition (Jaquiéry et al., 1978, 1980). The fruits of V. faba differ in their position on the raceme and on the plant, in their stage of development and in their relative distance to the source of assimilates. All these factors affect their competitive ability. Aufhammer et al. (1982) studied such dominance effects in wheat.

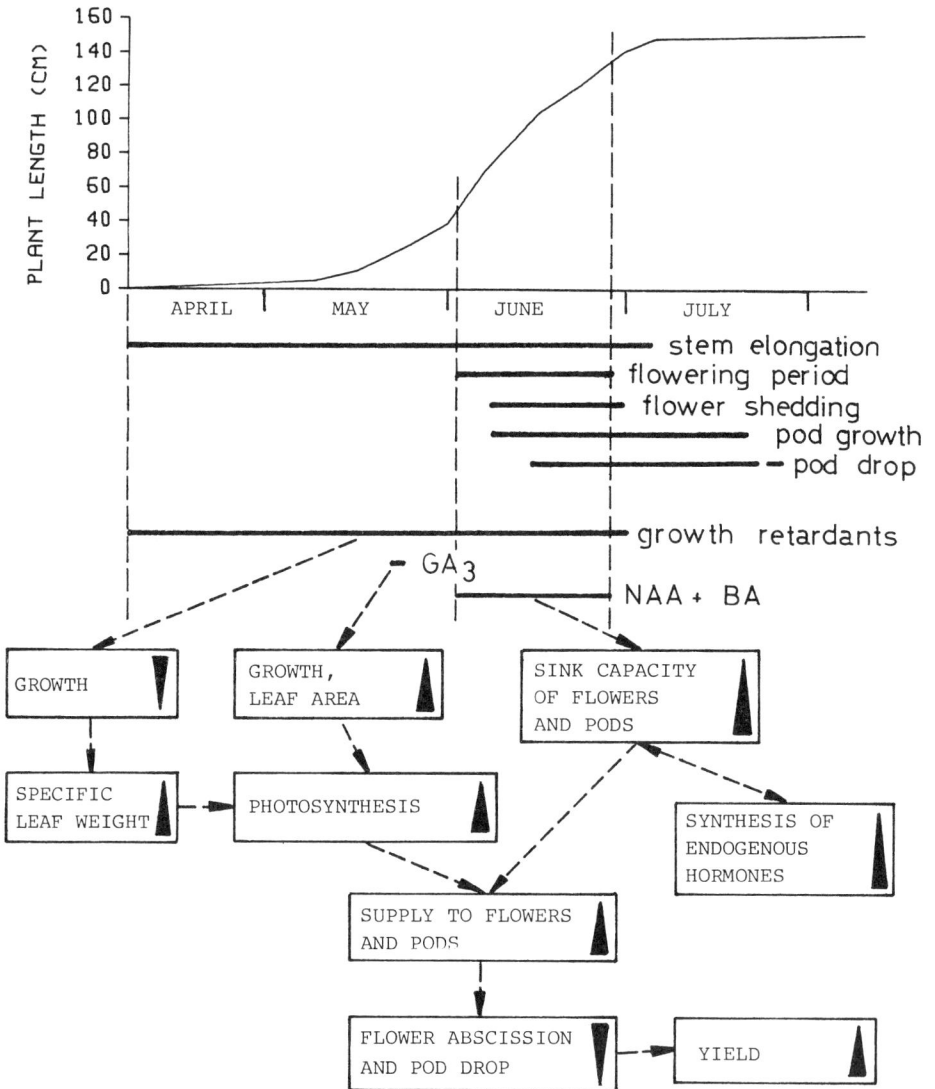

Fig. 1   Some possible effects achievable on Vicia faba by treatment with PGR combinations (derived from Morgan, 1981).

At present, our research focuses on two aspects of faba bean response to PGR combinations: 1) morphology, development and yield structure

2) endogenous hormonal balance

This paper considers the first aspect only.

MATERIALS AND METHODS

All our trials were conducted with the indeterminate one-stem variety Herz Freya.

Experiments 1981 and 1982

Our work with PGR combinations began with a field screening test in 1981 which included 70 different treatments of single or combined applied growth substances such as: GA_3, $GA_{4/7}$, BA, NAA, ABA, Atrinal and Alar. The experimental design was a randomized block, with 3 replications and a plot size of 5 m^2. Plant development and yield components were analysed. The most promising treatments were slightly modified and tested in more detail at two different locations in 1982.

Field trial 1983

Our research focused on three growth regulator combinations which also included two new growth retardants (Fig. 2). The aim was to induce an approach to the plant model postulated by Dantuma et al. (1983) i.e. a semi-dwarf small-seeded type. The experimental design is presented in Fig. 2. Run-off treatments of the growth substances were sprayed, by means of a back sprayer, as liquid solutions containing 0.2% Tween 20 as a wetting agent. In treatment 1 seeds were shaken for 40 min. in an aqueous solution containing Tetcyclacis.

In order to follow up the effects of the growth regulator combinations on the vegetative and reproductive development of the plants, 4 harvests were made during the growing season and a final harvest at maturity. For each harvest, 48 plants were chosen at random from each treatment i.e. 12 plants per plot.

An additional trial outdoors was designed in 1983 using plastic tubes to grow the faba bean plants. The aim was to study the root system and to confirm the results of the field trial.

Growth regulator combinations:

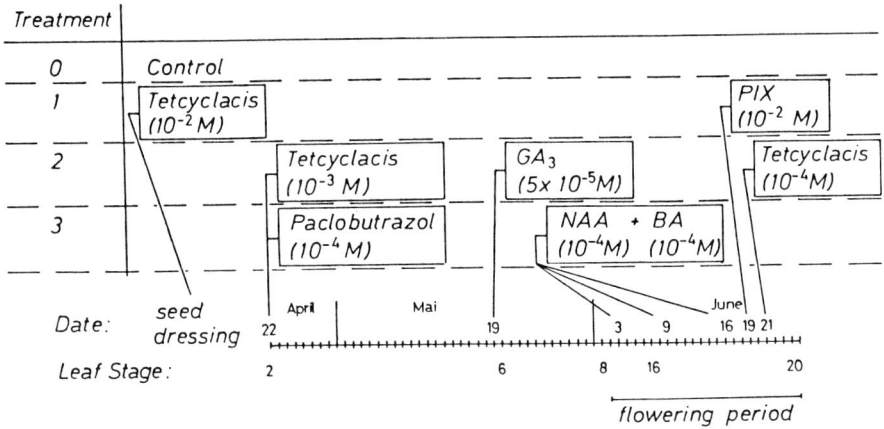

Tetcyclacis = BAS 106..W (BASF, Ludwigshafen, F.R.G.)
Paclobutrazol = PP 333 (ICI, Jealotts Hill Research Station, U.K.)
GA_3 = gibberellic acid   NAA = 1-Naphthylacetic acid (α)
BA = N6-Benzyladenine   PIX = mepiquat-chloride

Experimental design: randomized block with 4 replications, size of
plots: 18 m^2, sowing density: 33 plants/m^2
sowing date: 10.3.83

Fig 2   Field trial 1983

RESULTS AND DISCUSSION
    Results of the 1983 field trial are presented and discussed.

Effect on plant length development
    From Fig. 3 it is apparent that treatments 1 and 2 resulted in slightly
shorter plants as compared to the control. This effect can be mainly attributed to the application of Tetcyclacis as seedsoaking or spraying. The most
striking effects on plant length were induced by treatment 3. Paclobutrazol,
sprayed at the 2nd leaf stage, inhibited plant growth to the same extent as
did Tetcyclacis in treatments 1 and 2. The growth inhibition effect in
treatment 3 must be due mainly to the three NAA+BA sprayings during the
flowering period, since the growth of test plants treated with Paclobutrazol
only was reduced to a much lesser degree. A synergistic effect of Paclobutrazol and NAA+BA might be possible. After the NAA+BA sprayings, the
plants showed an epinastic growth habit, probably an indication that the
effect of NAA+BA is mediated by ethylene.

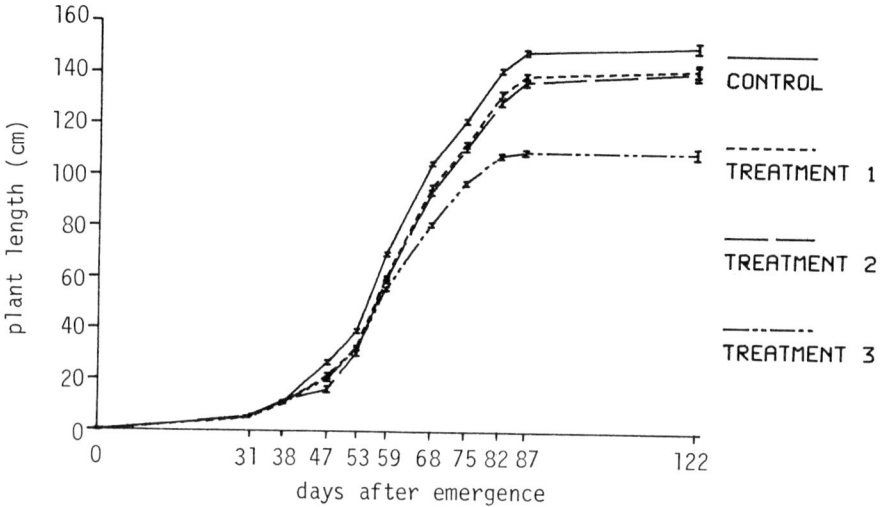

Fig. 3   Plant length development (bars represent confidence interval)

As both Tetcyclacis and Paclobutrazol interfere by blocking the pathway of gibberellin synthesis, a decrease in plant height of the treated plants was expected (Rademacher et al., 1981, Lever et al., 1982). This effect is reversible by the application of gibberellic acid (GA$_3$) tested in treatment 2. This treatment shows a short period of strong growth inhibition after the Tetcyclacis application up until the spraying of GA$_3$.

The observed PGR effects on plant length are due mainly to the elongation or shortening of the internodes.

Effect on leaf area and specific leaf weight

TABLE 1   Leaf area per plant in cm^2 (means of 48 plants)

treatment	harvest date			
	17.5.83	1.6.83	13.6.83	4.7.83
0	166.2	409.3	1275.9	2258.9
1	143.6	411.9	1231.8	2380.5
2	116.0	377.1	1142.5	2044.8
3	146.8	383.0	1219.3	2264.4
LSD 5%	16.2	80.3	144.4	382.3

Tab. 1 shows that for all harvest dates, either the control (17.5. and 13.6.) or treatment 1 (1.6. and 4.7.) has the highest leaf area. The growth inhibitors applied in treatments 1 to 3 induced a slight reduction in the leaf area. In treatment 1 this effect is of short term duration whereas in treatments 2 and 3 it was observed over a longer period of time. The decrease in vegetative growth in favour of the reproductive organs has already been studied by Gehriger et al., 1978. If we could achieve this aim without reducing the leaf area as the necessary source of assimilates, a yield increase might be possible. Treatment 3 was a partial solution to achieving this goal. The disadvantage of this treatment is delayed pod formation and therefore a later establishment of sinks. As we consider the early formation of sinks important for the inhibition of vegetative growth and for the activation of the source, we probably did not yet achieve the optimum result with treatment 3.

The influence of the growth regulator combinations on the specific leaf weight is shown in Fig. 4. Taking all treatments into consideration, a decrease in the specific leaf weight is detectable up to mid-June, when a slight increase is observed. The first harvest (17.5.83) showed an increase in the specific leaf weight of treatments 2 and 3 due to the growth

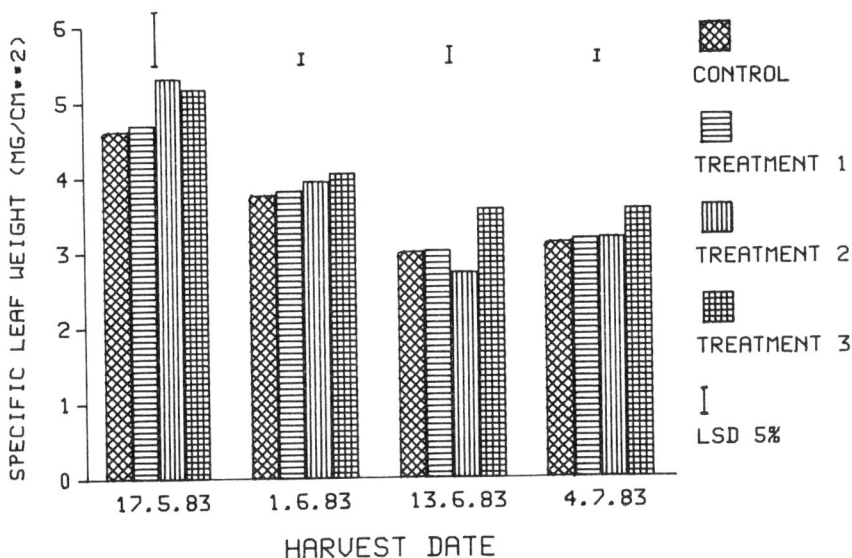

Fig. 4   Specific leaf weight (means of 48 plants)

inhibitors applied at the 2nd leaf stage. In treatment 3 this effect is re-inforced during the vegetation period by the spraying of NAA+BA.

We have not yet examined the influence of the specific leaf weight on photosynthesis in V. faba. It is suggested that an increase in specific leaf weight implies an increase in photosynthetic activity.

Effect on the abscission of reproductive organs

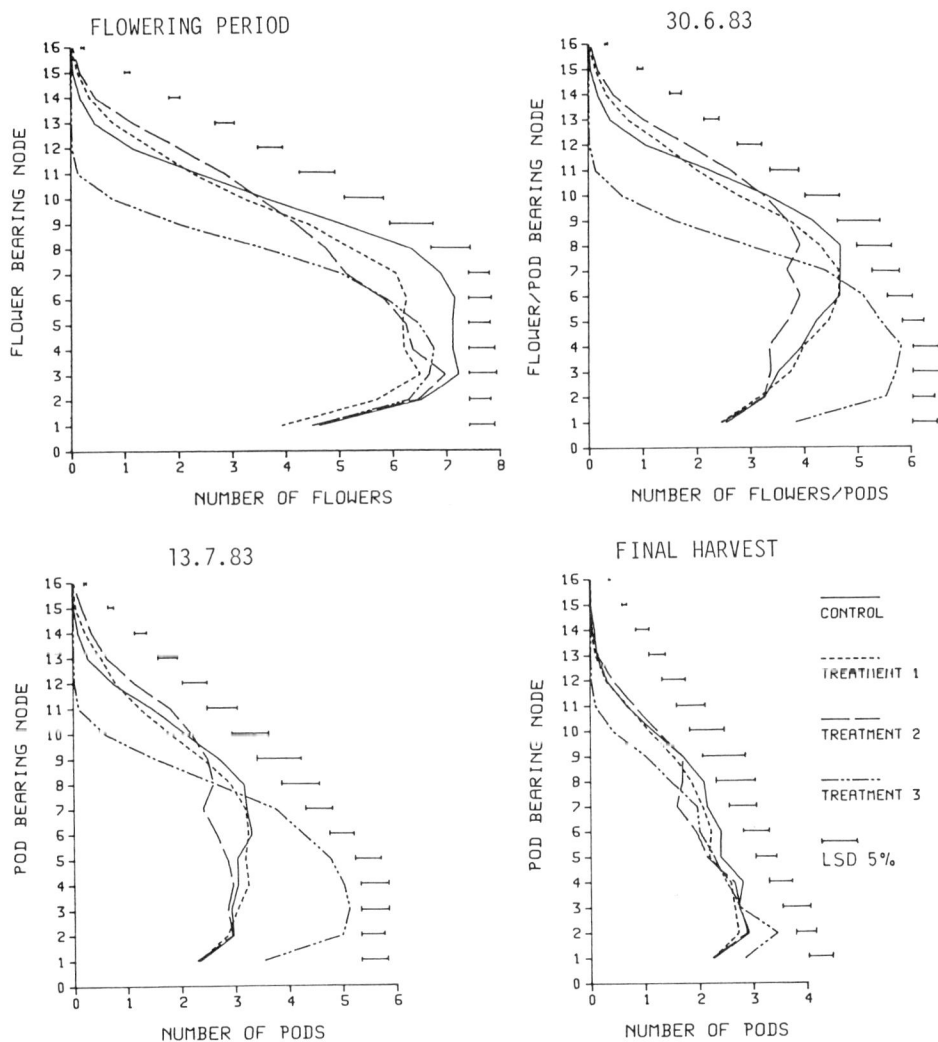

Fig. 5   Abscission of reproductive organs at four different stages of development (means of 48 plants).

Shedding of reproductive organs is considered to be an important aspect for the regulation of yield performance in V. faba, and hence our research is concerned with this phenomenon. Fig. 5 shows the effect of the PGR combinations on flower abscission and pod drop. The number of reproductive organs was determined at 4 different dates. It is obvious, that through treatment 3, a markedly shortened flower and pod bearing zone is compensated for by a better pod set on the remaining nodes. At maturity, this effect is confined to the two lowest podded nodes. Obviously, the abscission of reproductive organs was delayed but not reduced ultimately by treatment 3. However, our results show that growth substances must be involved in the abscission of reproductive organs. The shedding phenomena in V. faba might be a complex interplay of fertilisation, sequential development with competition for assimilates and growth substances.

Effect on grain yield

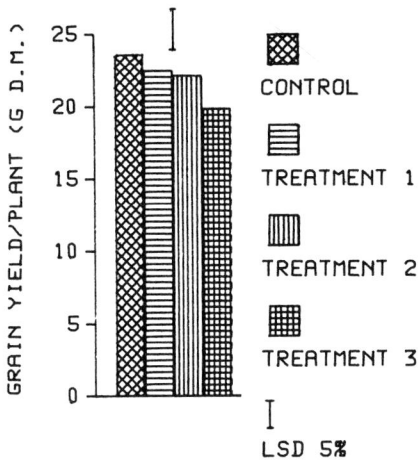

The PGR combinations had no positive effect on grain yield. The plants of treatments 1 and 2 revealed a similar grain yield to that of the control. However, treatment 3 caused a visible grain yield reduction. It seems that an increase in the sink activity of the reproductive organs due to the IAA+BA sprayings was delayed. On the other hand, an induction of ethylene may have caused abscission.

Fig. 6   Grain yield at maturity
(means of 48 plants)

FINAL CONSIDERATIONS

According to our results we conclude that growth regulator combinations may influence development, morphology and growth habit of the faba bean plant. However, increases in grain yield as compared with untreated plants are difficult to achieve.

A growth regulator combination which included Paclobutrazol, NAA and BA resulted in a reduction of plant height, improved lodging resistance, equal leaf area, delayed senescence and increased young pod set on a reduced number of pod bearing nodes as compared to the control. With this combination we successfully transformed the plant type Herz Freya to a semi-dwarf type, which may allow for greater population density.

REFERENCES

Aufhammer, W. and Zoschke, A. 1982. Beeinflussung der Beziehungen zwischen den einzelnen Kornanlagen von Weizenähren durch lokale Wirkstoff-applikationen. Z. Acker- und Pflanzenbau, 151, 338-353.

Dantuma, G., von Kittlitz,E., Frauen, M., and Bond, D.A. 1983. Yield, Yield Stability and Measurements of Morphological and Phenological Characters of Faba Bean (Vicia faba L.) Varieties Grown in a Wide Range of Environments in Western Europe. Z. Pflanzenzüchtung, 90, 85-105.

Gehriger, W., Bellucci, S., Keller, E.R. 1978. Influence of Decapitation and Growth Regulators on Yield Components and Yield of Vicia faba L. Commission of the European Communities. Eur 6244 EN, 421-435. Janssen Services, London.

Jaquiéry, R. and Keller, E.R. 1978. Beeinflussung des Fruchtansatzes bei der Ackerbohne (Vicia faba L.) durch die Verteilung der Assimilate. Teil I. Angew. Botanik 52, 261-276.

Jaquiéry, R. and Keller, E.R. 1980. Beeinflussung des Fruchtansatzes bei der Ackerbohne (Vicia faba L.) durch die Verteilung der Assimilate. Teil II. Angew. Botanik 54, 29-39.

Keller, E.R. and Bellucci, S. 1980. Anwendung von Phytohormonen und Wachstumsregulatoren bei Körnerleguminosen. Hohenheimer Arbeiten, 105, 94-111. (Eugen Ulmer, Stuttgart).

Lawrence, D.K. 1978. Hormone combinations as practical growth regulators. The effect of interactions between growth regulators on plant growth and yield. Monograph Nr. 2. Brit. Plant Growth Regulator Group.

Lever, B.G., Shearing, S.J., and Batch, J.J. 1982. PP 333 - A new broad spectrum growth retardant. Proceedings 1982 British Crop Protection Conference - Weeds.

Morgan, D.G. 1981. Factors affecting fruit and seed development in field beans and oil-seed rape. Monograph Nr. 6. Brit. Plant Growth Regulator Group.

Rademacher, W. and Jung, J. 1981. Comparative Potency of Various Synthetic Plant Growth Retardants on the Elongation of Rice Seedlings. Z. Acker-und Pflanzenbau, 150, 363-371.

Wareing, P.F. 1978. Some general aspects of growth substance interaction. Monograph Nr. 2, Brit. Plant Growth Regulator Group.

EFFECTS OF WATER STRESS ON SOME GROWTH PARAMETERS AND YIELDS OF FIELD BEAN CROPS

A.J. Karamanos

Laboratory of Crop Production
The Agricultural College
75 Iera Odos
Athens 301,Greece

ABSTRACT

The growth and yields in respect to the water relations of field bean crops (*Vicia faba* L. cv. Maris Bead) subjected to three irrigation regimes were examined. The well-watered crop showed significantly higher seed and biological yields than the intermediate and dry ones. Moreover, the wet plots exhibited significantly higher values of leaf area duration (LAD) than the intermediate and dry plots. Significant differences were found occasionaly between the wet and the other two treatments as regards the relative growth rates in dry matter and leaf area. The crop growth rate and net assimilation rate were mainly affected by plant development and other factors. LAD was highly correlated with the average plant water potential (Ψ), while the relative growth rates in dry matter and leaf area exhibited significant though less marked correlations. Seed yield was well correlated with the LAD only after flowering, but with Ψ already from the initial growth stages. Cell turgor appeared to have played an important role on both leaf area growth and final yields.

INTRODUCTION

In view of their significance as a plant protein source, field beans have being given special attention in the last 15 years. Considerable research work has been done concerning the growth of the crop and its responses to environmental factors (for a review see Keatinge and Shaykevich, 1977). The importance of water shortage as a factor limiting the growth and yield of field beans has been repeatedly emphasized (Brouwer, 1949;Reisch, 1952; Jones, 1963; Mériaux, 1972; Sprent et al., 1977) and thus considerable effort was made to understand the water relations of the plants (Kassam and Elston, 1974, 1976; Elston et al., 1976; Karamanos, 1978a). In addition, the effects of water stress on many aspects of plant growth and development have been examined (El Nadi, 1969, 1970; Sprent, 1972; Karamanos, 1978b; Farah, 1979,1981b; Karamanos et al., 1982).

Although the availability of water supply is known to affect decisively the yields of field beans (Farah, 1981a), a quantitative evaluation of the importance of water stress as a factor limiting crop yields is lacking. In this study, an effort is made to associate crop growth and yields with

P.D. Hebblethwaite, T.C.K. Dawkins, M.C. Heath and G. Lockwood (eds.)
Vicia faba: Agronomy, Physiology and Breeding. ISBN 90-247-2964-5.
© 1984, Martinus Nijhoff/Dr W. Junk Publishers. Printed in The Netherlands.

quantitative expressions of plant water status such as the water potential and its components. Such an approach makes the answer to the problem of yield dependence on water supply more definite, but it may also help to speculate the mechanisms which are likely to be involved in the determination of yields by plant water status.

MATERIALS AND METHODS

The experiment was carried out at the Reading University Farm (Sonning, Berks.) on a sandy loam soil. The crop (*Vicia faba* L. cv. Maris Bead) was sown on 6 May 1975 at a seed rate of 142 kg/ha in rows 25 cm apart. Fertilizer (0-20-20) was applied before sowing (7 April) at a rate of 152 kg/ha and incorporated three days later by means of a surface cultivation. Weeds were controlled chemically by spraying Simadex (Simazine 50%) on the soil surface at a rate of 1.7 kg/ha a.i. on 9 May. Insect pests (mainly weevils and aphids) were kept under control by spraying Lindane (2 l/ha a.i.) and Dimethoate (0.7 l/ha a.i.) on several occasions. The average crop density was 210000 plants/ha.

A randomized blocks design consisting of four replicates and three irrigation treatments, wet, medium, and dry, was chosen. In order to control plant water status, the crop was covered by polythene rainshelters and water was applied at different frequencies in the three irrigation treatments by means of surface irrigation. Irrigation was timed by the value of the plant water potential just before sunrise (Ψ_d). Thus, Ψ_d was allowed to fall to -0.3 MPa before water was applied in the wet, and to -0.5 and -0.8 MPa in the medium and dry treatments respectively. According to this scheme, the wet plots were irrigated 12 times starting from day 36, the medium ones only once (on day 64), while no irrigation was applied to the dry plots.

Starting from day 37 after sowing and at five day-intervals up to day 77, three plants per plot were cut on ground surface for growth analysis. Leaf area was determined in the laboratory with a leaf area photometer (EEL Unigalvo, Type 20). The plant material was ovendried at 65°C for 48 hours and weighed after cooling in a desiccator. Separate measurements were taken for leaves, stems, and pods.

Estimates of the final yields were taken when the pods became dark. The plants in the dry treatment matured first and were harvested on 18 August (105 days after sowing), while those in the wet and medium treatments were harvested on 12 September (130 days).

Plant water status was measured twice a week between days 34 and 77. Three plants per plot were sampled between 1500 and 1600 h by cutting the first fully expanded leaf (fifth to sixth unfolded leaf from the top of the plants). The leaves were then brought to the laboratory closed within air-tight humid vials for the determination of their water potential (Ψ). Ψ was measured with the length change technique which also makes possible the calculation of the solute (ψ_s) and pressure potentials (ψ_p) (Kassam, 1972). The plant water potential before sunrise (Ψ_d) which was used for irrigation scheduling was measured using the same sampling scheme with a pressure bomb designed according to Waring and Cleary (1967).

RESULTS AND DISCUSSION

Plant water status

The time courses of Ψ, ψ_s, and ψ_p in the three treatments are shown in fig. 1.

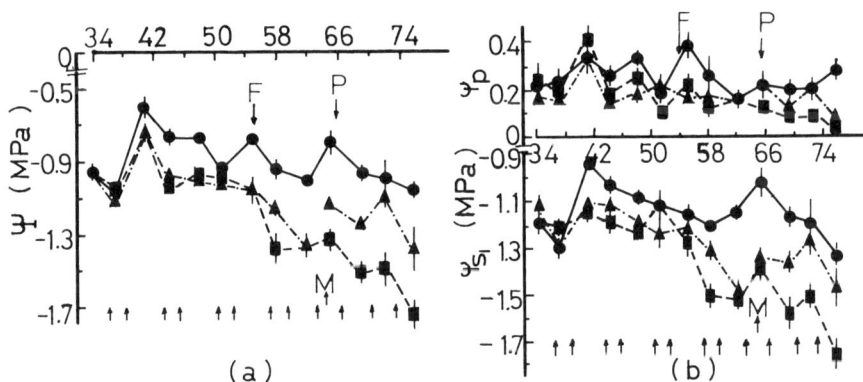

Figure 1. The time courses of some measures of plant water status in the three treatments. (a) Water potential (Ψ). (b) Solute (ψ_s) and pressure potentials (ψ_p). ●——● : wet, ▲—·—▲ : medium, ■---■ : dry treatment. The vertical bars indicate the standard errors of the means. The arrows show the beginnings of flowering (F) and podding (P), as well as the irrigation timing in the wet and medium (M) treatments.

The values of Ψ were clearly different in the wet and dry treatments, especially after day 52. There was no difference between the medium and dry treatments when the former was not irrigated. After the irrigation on day

64 the Ψ in the medium was shifted to less negative values. The general
pattern of Ψ in the wet during the observations was rather steady fluctua-
ting around -0.85 MPa, while that in the dry treatment was steadily declining to reach a value of -1.7 MPa at the final sampling. The pattern in the
medium was differentiated from that of the dry treatment after day 62, when
it became intermediate between the two extremes.

As regards the component potentials (fig. 1b), ψ_s was clearly differentiated in the wet and dry treatments for most of the observation period.
In the wet, the pattern was steady or slightly declining and reached eventually a value of -1.35 MPa. In the dry treatment it was steadily declining and reached a value of -1.77 MPa at the final sampling. The pattern in
the medium was similar to that observed for Ψ. The irrigation shifted ψ_s
to less negative values. It has been proved that the patterns of ψ_s in
field beans are dominated by osmoregulatory mechanisms (Kassam and Elston,
1976; Karamanos, 1978a). In contrast with ψ_s, the patterns of ψ_p were not
clearly separated between treatments, although the values in the wet were
higher than those in the other two treatments on most occasions. Despite
of the great day-to-day fluctuations probably arising from the variations
in the evaporative demand (Karamanos, 1978a), it can be said that ψ_p was
kept rather steady around a mean value of 0.28 MPa in the wet treatment. In
the dry, the pattern was declining from an initial value of 0.25 to a final
one of 0.04 MPa. Again, the medium treatment was temporarily differentiated
from the dry after day 64 reaching a final value of 0.09 MPa.

Growth in dry matter

The growth in dry matter of single plants (fig. 2) was virtually the
same in all treatments up to day 57. From then on, the plants in the wet
treatment exhibited significantly ($p < 0.05$) higher values than the ones
observed in the other two treatments. The temporary reduction in the rate
of dry matter increase between days 62-67 coincided with the beginning of
pod-setting. This reduction was common in all treatments, though more intense in the dry. The difference between the medium and dry treatments caused by the irrigation in the former never became statistically significant.
Similar conclusions can be drawn when considering the dry weights of leaves
and stems. As regards the pods, the well-watered plants showed consistently
the lowest values up to day 77, although the dry weights of the pods were
significantly higher than those in the medium and dry treatments at harvest

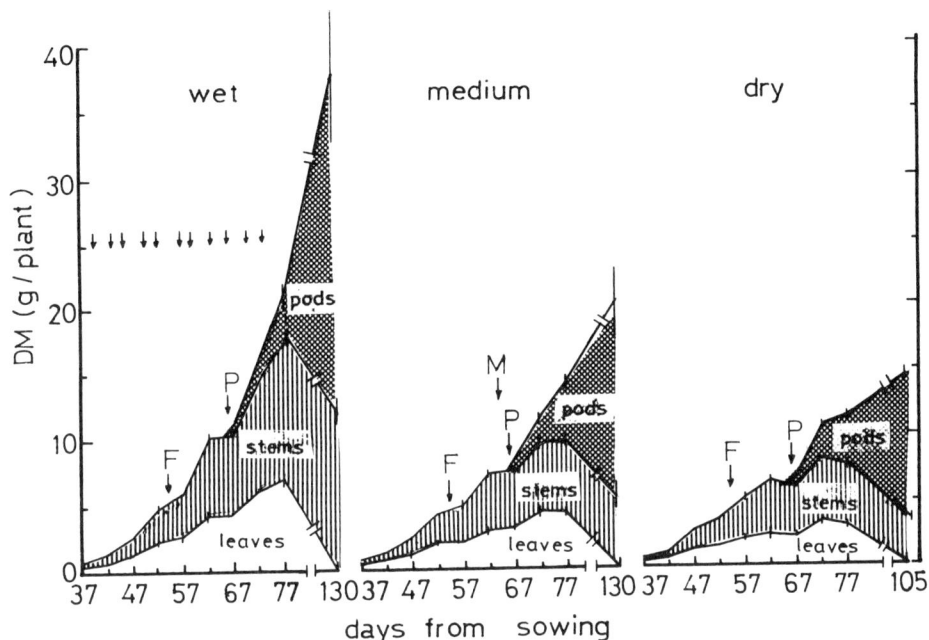

Figure 2   The time course of the growth in dry matter of whole plants and plant parts in the three irrigation treatments. The vertical bars indicate the standard errors of the means. The beginnings of flowering (F), podding (P) as well as the irrigation timing in the wet and medium (M) treatments are also shown .

time. Water stress was apparently associated with an earliness in pod growth and seed filling in the drier treatments which essentially resulted in an earlier maturity.

When considering the dry matter partition in the plant organs (fig. 3) it appears that the leaf weight ratio (LWR) showed a falling trend while the fruit weight ratio (FWR) an increasing one . The stem weight ratio (SWR) increased steadily up to day 62 and started falling thereafter. The LWR in the wet treatment was consistently the highest throughout the observation period. A marked decrease was observed in the dry treatment after day 62. SWR was the lowest in the wet treatment up to day 62, but it reached the highest values after pod-setting. FWR increased quicker in the dry and medium and slower in the wet treatment. Ishag (1973b) associated the fall in SWR after pod-setting with pod-filling. On this account, the higher values of FWR in the dry treatment can be related to the quicker drop in SWR. The irrigation in the medium treatment. shifted the LWR closer to the wet, but it left SWR close to the dry treatment.

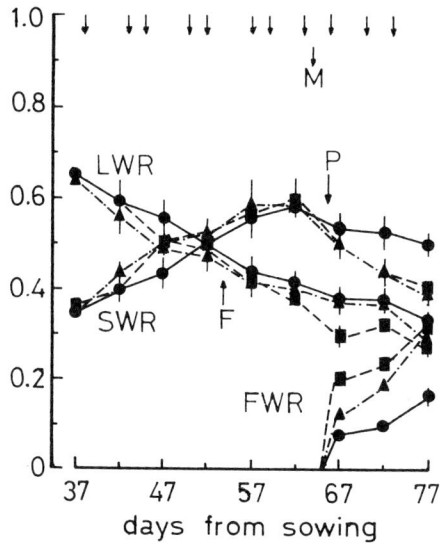

Figure 3   The time courses of the leaf (LWR), stem (SWR), and
fruit weight ratios (FWR) in the three irrigation treatments.
The symbols are the same as in fig. 1

The growth in dry matter of the crop for the various treatments
followed a pattern similar to that of fig. 2. On day 77, the dry weight of
the crop in the wet (391.7 g/m^2) was significantly higher than those in
the medium and dry treatments (287.5 and 270.6 g/m^2 respectively). The
same applies when considering the total dry weights at harvesting (Table 1)

Growth in leaf area

The time course of the leaf area index (LAI) of the crops in the
three treatments is shown in fig. 4. In the medium and dry treatments a
maximum value of about 2.1 was reached on day 72. In the wet, a value of
about 3.8 was observed on day 77. The LAI in the well-watered plots was
consistently the highest from day 52 onwards and the difference from the
other two treatments became statistically significant in the last four
samplings.A temporary cessation in leaf growth, similar to that observed
for dry matter, was apparent just before pod-setting.

The general pattern of leaf area growth for single plants did not

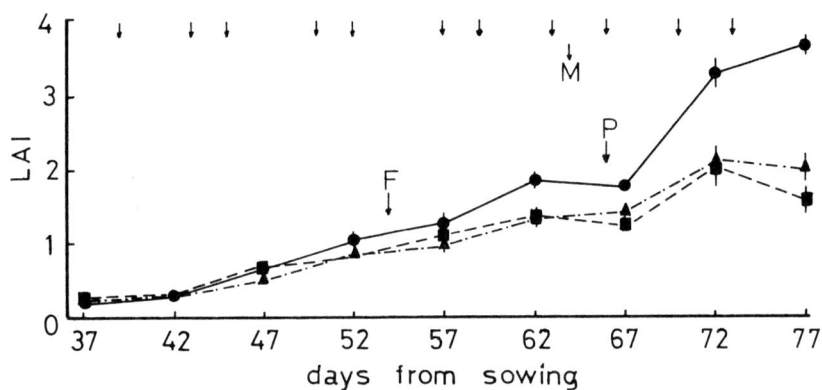

Figure 4   The time course of the leaf area index (LAI) in the three irrigation treatments. The symbols are the same as in fig.1

differ from that of fig. 4. On day 77, the total leaf area per plant in the wet (18.6 dm^2) was about twice as great as that of the medium (9.8 dm^2) and about three times greater than that in the dry treatment (6.7 dm^2). The differences between the wet and the two other treatments were highly significant (p< 0.01).

Growth rates

The time courses of the relative growth rates in dry matter (RGR) and leaf area (RLGR), net assimilation rate (NAR), and crop growth rate (CGR) are shown in fig. 5. NAR, RGR, and RLGR showed similar overall patterns. Thus, two well-defined minima were observed during the beginning of flowering and just before pod-setting, which were followed by high values. High values were also observed during the vegetative phase and especially between days 42-47. No consistent differences between the irrigation treatments were observed in the courses of NAR and RGR. However, in the course of RLGR there were many periods in which the wet treatment showed the highest and the dry the lowest values. The values of the wet were significantly higher (p < 0.05) than those of the dry treatment on the last sampling period. The negative values in the medium and dry treatments between days 72--77 were produced by the loss in leaf area due to drought-induced leaf senescence and death.

The pattern of CGR was slightly different from the previous ones. It showed an increasing trend in the wet and medium treatments. In the dry, it

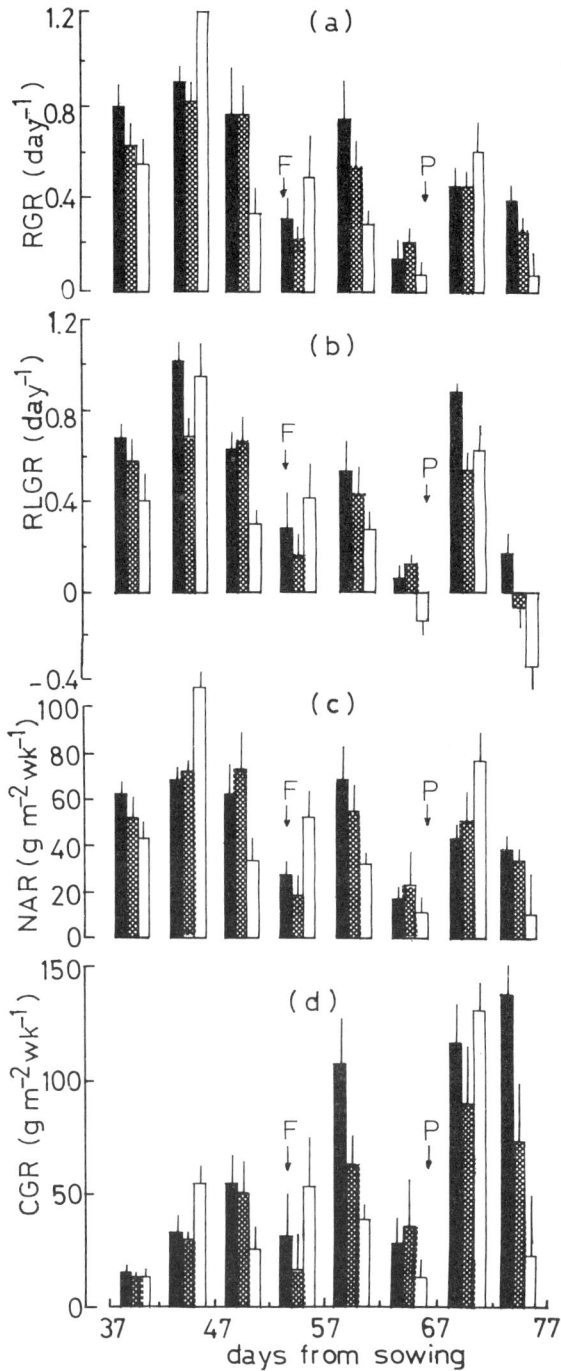

Figure 5   The time courses of RGR (a), RLGR (b), NAR (c), and CGR (d) in the different treatments. ■ : wet, ▨ : medium , □ : dry. The bars show the standard errors of the means.

fluctuated around 30 g m^{-2} week^{-1} up to the beginning of podding and then reached temporarily very high values. In all treatments the highest values were reached after podding, when the LAI was maximum. The peaks in CGR for the medium and dry treatments coincided with the peaks in their LAI between days 67-72. Alternatively, in the wet treatment CGR was still increasing after day 72, as also did the LAI (fig. 4). The wet, and to a minor extent the medium treatment, showed another peak immediately after flowering. The values of the wet were significantly higher (p<0.05) than those of the dry treatment between days 57-62 and 72-77. The dry showed significantly higher (p<0.05) values from the wet and dry treatments between days 42-47.

Final yields

The seed yields in the wet treatment were significantly higher (p<0.05) than those of the medium and dry treatments (Table 1) which did not differ significantly from each other. The same applies for the biological yields. The harvest index was significantly lower in the well-watered crop in comparison with the drier ones. The abundance of water in the wet treatment allowed a greater expansion of the vegetative organs at the expense of the produced seed.

TABLE 1 The seed and biological yields of crops and single plants as well as the harvest indices (HI) in the different treatments. The standard errors of the means are shown in parentheses

	Seed yield		Biological yield		HI
	t/ha	g/plant	t/ha	g/plant	
Wet	4.02 (0.61)	21.21 (3.07)	7.26 (1.02)	38.27 (4.89)	0.55 (0.02)
Medium	2.48 (0.37)	12.23 (1.76)	4.13 (0.60)	20.41 (2.83)	0.60 (0.01)
Dry	2.03 (0.11)	8.71 (0.23)	3.41 (0.22)	14.57 (0.39)	0.60 (0.01)

Growth and yield in respect to plant water status

Both the total dry matter and total leaf area of single plants on day 77 were highly correlated with the average Ψ between days 34-77 (Table 2). A high correlation (r=0.90) was also obtained between the total dry matter at harvest and the average Ψ over the same period.

The high correlations obtained between the total leaf area and Ψ were not surprising, since plant water status is known to affect a number of mechanisms associated with leaf growth in field beans. Thus, increasing

TABLE 2   The linear correlation coefficients between the total
dry matter (DM) and total leaf area (L) of plants on day 77 and
the average measures of plant water status (Ψ,ψ_s,ψ_p) during the
observation period. In addition, the correlation coefficients
between the various growth rates (RGR, RLGR, NAR, and CGR) during
plant development and the average measures of plant water status
during the corresponding periods are shown. *:p<0.05, **:p<0.01,
***:p<0.001

	Parameters of growth					
	DM	L	RGR	RLGR	NAR	CGR
Ψ	0.92***	0.94***	0.49*	0.56**	0.30	-0.16
ψ_s	0.89***	0.94***	0.51*	0.57**	0.34	-0.14
ψ_p	0.82***	0.76***	0.63**	0.62**	0.43	-0.19

water stress reduces drastically the laminar expansion of individual leaves
(Karamanos et al., 1982) and to a smaller extent the rates of leaf production and unfolding (Karamanos, 1978b; Farah, 1981a). Furthermore, increasing
water stress increased the rate of leaf senescence and death (Finch-Savage
and Elston, 1976). The paramount significance of leaf area on the productivity of different plant stands has been repeatedly emphasized (e.g. Watson,
1947, 1953; Thorne, 1971),although Ishag (1973b) found little or no correlation between the total dry weight at harvest and LAD in field beans.In
this work, LAD on day 77 was highly correlated with the total crop dry
matter both on day 77 (r=0.92) and at harvesting (r=0.82). On this account,
it can be postulated that the high correlations between total dry matter
and Ψ resulted mainly from the strong dependence of leaf area on Ψ. This
assumption is further supported from the fact that RGR showed worse correlations with all measures of plant water status than RLGR (Table 2). In
contrast with the conclusions drawn for leaf area, there was practically
no correlation between NAR and plant water status. The relative independence of NAR from environmental factors was postulated already by Watson (1947)
for other crops. No correlation was also found between CGR and plant water
status when the whole growth period was considered. However, significant
correlations between CGR and ψ_s were found in specific time intervals, as
for example between days 37-42 (r=0.61), 57-62 (r=0.77), and 72-77 (r=0.62).
    The use of the component potentials, especially of ψ_p, improved   the
correlations with the various rates in comparison with Ψ. However, ψ_s produced higher correlations with total dry matter and leaf area than ψ_p.
    Ishag (1973b) found that the seed yield of field beans depended to a

great extent on the leaf area before flowering. This was found not to apply
in our case, where the LAD after flowering exhibited the highest correlation
coefficients with seed yields (Table 3).

TABLE 3   The linear correlation coefficients between leaf area
duration (LAD) and plant water status (Ψ, ψ_s, ψ_p) against seed
yields for specific time intervals: up to the beginning of flowering (days 34–52), up to full bloom (34–57), up to pod-setting
(34–67), full bloom to full podding (57–77), and up to the end
of the observations (34–77). *:p<0.05,**:p<0.01,***:p<0.001

	Periods (days from sowing)				
	34–52	34–57	34–67	57–77	34–77
LAD	0.31	0.49	0.58*	0.77**	0.76**
Ψ	0.79**	0.79**	0.79**	0.78**	0.79**
ψ_s	0.52	0.53	0.61*	0.72**	0.69*
ψ_p	0.70*	0.81**	0.89**	0.89**	0.92***

In contrast with leaf area, plant water status seems to have affected
the seed yields already from the early growth stages. It is already known,
that the number of pods per plant is the most important yield component in
field beans (Kambal, 1969; Ishag, 1973a). In addition, variations in yield

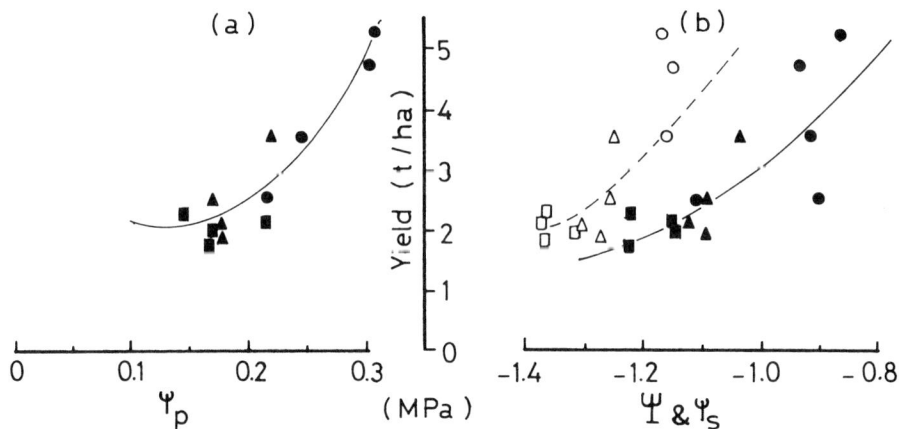

Figure 6   The relations between seed yield and the average measures
of plant water status during the observation period. Results from
the wet (circles), medium (triangles) and dry treatment (squares).
The fitted regression lines are also shown. (a) ψ_p, Y=3835.5 −
−28123.4 X + 105527.1 X^2 (r^2=0.90). (b) Ψ (filled symbols, Y=
=32376.3 exp 2.36 X, r^2=0.66) and ψ_s (open symbols, Y=103356.6
exp 2.91 X, r^2=0.52)

brought about by water shortage are mainly caused by a reduced number of pods per plant and, to a minor extent, by a reduced seed weight (Kogbe,1972; Farah, 1981a). It is, therefore, reasonable to suggest that water availability before flowering increased the number of flowers as well as the pod-retaining ability of the pod-bearing nodes. It is interesting that the contribution of ψ_p became more important as the age of the crop increased. In view of the importance of ψ_p on leaf growth rate (Table 2) and the late effects of LAD on seed yield, it is suggested that the high correlations of yields with ψ_p arise from both direct effects on the yield potential (number of flowers and pods per plant) and indirect effects via leaf area (weight of seeds).

The importance of ψ_p for seed yields can be further visualized when examining the plots relating seed yields with all measures of plant water status (fig. 6). It appears that curvilinear regressions explain better than linear ones these relations, since yields tend to reach a lower limit of 1.5-2.0 t/ha at high degrees of water stress. Thus, cell turgor throughout the growth period seems to be very important for the final yields of field beans.

ACKNOWLEDGEMENTS

The data of this work were collected while the author was at the Department of Agricultural Botany of the University of Reading. Many thanks are due to Mr R. Silver for technical assistance.

REFERENCES

Brouwer, W. 1949. Steigerung der Erträge der Hülsenfrüchte durch Beregnung sowie Fragen der Bodenuntersuchung und Düngung. Z. Acker- u. PflBau, 91, 319-346.
El Nadi, A.H. 1969. Water relations of beans. 1.Effects of water stress on growth and flowering. Exp. Agric. 6, 195-207.
El Nadi, A.H. 1970. Water relations of beans. 2. Effects of differential irrigation on yield and seed size of broad beans. Exp. Agric., 7, 107-111.
Elston, J., Karamanos, A.J., Kassam, A.H. and Wadsworth, R.M. 1976. The water relations of the field bean crop. Phil. Trans. R. Soc. Lond. B 273, 581-591.
Farah, S.M. 1979. An examination of the effects of water stress on leaf growth of crops of field beans *Vicia faba* L. Ph.D. Thesis, University of Reading.
Farah, S.M. 1981a. An examination of the effects of water stress on leaf growth of crops of field beans (*Vicia faba* L.). 1. Crop growth and yield. J. agric. Sci., Camb., 96, 327-336.
Farah, S.M. 1981b. An examination of the effects of water stress on leaf growth of crops of field beans (*Vicia faba* L.). 2. Mineral content.

J. agric. Sci., Camb., 96, 337–346.

Finch-Savage, W.E. and Elston, J. 1976. The death of leaves in crops of field beans. Ann. appl. Biol., 85, 463–465.

Ishag, H.M. 1973a. Physiology of seed yield in field beans. I. Yield and yield components. J. agric. Sci., Camb., 80, 181–189.

Ishag, H.M. 1973b.Physiology of seed yield in field beans. II. Dry matter production., J. agric. Sci., Camb., 80, 191–199.

Jones, L.H. 1963. The effect of soil moisture gradients on the growth and development of broad beans (*Vicia faba* L.). Hort. Res., 3, 13–26.

Kambal, A.E.1969. Components of yield in field beans, *Vicia faba* L.J. agric. Sci., Camb., 72, 359–363.

Karamanos, A.J. 1978a. Understanding the origin of the responses of plants to water stress by means of an equilibrium model. Praktika Acad. Athens 53, 308–341.

Karamanos, A.J. 1978b. Water stress and leaf area growth of field beans (*Vicia faba* L.) in the field. Leaf number and total leaf area. Ann. Bot., 42, 1393–1402.

Karamanos, A.J., Elston, J. and Wadsworth, R.M. 1982. Water stress and leaf growth of field beans (*Vicia faba* L.) in the field. Water potentials and laminar expansion. Ann. Bot. 49, 815–826.

Kassam, A.H. 1972. Determination of water potential and tissue characteristics of leaves of *Vicia faba* L. Hort. Res. 12, 13–23.

Kassam, A.H. and Elston, J. 1974. Seasonal changes in the status of water and tissue characteristics of leaves of *Vicia faba* L. Ann. Bot., 38, 419–429.

Kassam, A.H. and Elston, J. 1976. Changes with age in the status of water and tissue characteristics in individual leaves of *Vicia faba* L. Ann. Bot. 40, 669–679.

Keatinge, J.D.H. and Shaykevich, C.F. 1977. Effects of the physical environment on the growth and yield of field beans (*Vicia faba minor*) in the Canadian prairie. J. agric. Sci., Camb., 89, 349–353.

Kogbe, J.O.S. 1972. Factors influencing yield variation of field bean (*Vicia faba* L.). Ph.D. Thesis, University of Nottingham.

Mériaux, S. 1972. Influence de la sécheresse sur la croissance, le rendement et la composition de la féverole. Ann. agron., 23, 533–546.

Reisch, W. 1952. Variabilitätstudien an *Vicia faba* L. Z. Acker- u. Pflbau 94, 281–306.

Sprent, J.I. 1972. The effects of water stress on nitrogen fixing root nodules. 4. Effects of whole plants of *Vicia faba* and *Glycine Max*. New Phytologist, 71, 608–611.

Sprent, J.I., Bradford, A.M. and Norton, C. 1977. Seasonal growth patterns in field beans (*Vicia faba*) as affected by population density,shading and its relationship with soil moisture. J. agric. Sci., Camb., 88, 293–301.

Thorne, G.N. 1971. Physiological factors limiting the yield of arable crops In "Potential Crop Production. A Case Study" (Ed. P.F. Wareing and J.P. Cooper).(Heinemann, London). pp.143–158.

Waring, R.H. and Cleary, B.D. 1967. Plant moisture stress: evaluation by pressure bomb. Science, 155, 1248–1254.

Watson, D.J. 1947. Comparative physiological studies on the growth of field crops. II. The effect of nutrient supply on NAR and leaf area. Ann. Bot. 11, 375–407.

Watson, D.J. 1952. The physiological basis of variation in yield. Adv. Agron. 4, 101–145.

VEGETATIVE AND REPRODUCTIVE GROWTH OF FABA BEANS (VICIA FABA L.) AS
INFLUENCED BY WATER SUPPLY

G. Dantuma and C. Grashoff

Centre for Agrobiological Research, Wageningen, the Netherlands

ABSTRACT

Although drought stress is frequently a major yield limiting factor
in many faba bean growing regions, any avoidance of drought stress results
in enhanced vegetative growth at the expense of reproductive growth. Water
supply is a major factor in controlling the vigour of stem growth and LAI
and if non-limiting it often leads to an excessive vegetative growth in
terms of unacceptable heights of plants and LAI-values considerably above
optimum. Consequences are an increased pod abortion at lower nodes and a
decreased harvest index. Considerable variation exists between cultivars
and some appear to be more stable than others in relation to vegetative
growth in response to soil water status.

INTRODUCTION

As cell growth is quantitatively related to cell turgor, the
expansive growth of cells is a very sensitive response of the plant to
water supply. Leaf expansion and stem elongation is inhibited by drought
considerably earlier and more severely than processes as photosynthesis
and respiration. Leaf and canopy development are the determining factors
for the amount of solar radiation that is intercepted. Two main factors
that are determining the rate of development of the leaf area index (LAI)
are plant density and water supply. To maximize yields faba beans in many
regions are usually grown at relatively high plant densities of about
40 plants/m^2. The more favourable the growing conditions, the lower the
plant density can be and in the Netherlands in general an optimal yield of
most cultivars is already obtained with a plant density of about 20 plants/
m^2. The main reason for allowing such a lower plant density is most
probably a more regular water supply. For most of the existing cultivars,
however, growing conditions can be too good, as was shown by the results
of the joint faba bean trials in the Netherlands in 1980 and 1981. Whereas
in most years the highest yields are obtained in the N.O. Polder, in these
two years the trials near Wageningen gave better yields (table 1).

P.D. Hebblethwaite, T.C.K. Dawkins, M.C. Heath and G. Lockwood (eds.)
Vicia faba: Agronomy, Physiology and Breeding. ISBN 90-247-2964-5.

TABLE 1   Seed yield in t/ha (86% d.m.)
           Wageningen and N.O. Polder 1980 and 1981

	average	highest		lowest	
Minica	6,8	7,3	Wag. 81	5,9	N.O.P. 81
Strube's Ackb.	6,4	7,3	" 81	5,3	" 81
Wierboon	6,3	6,6	" 80	5,9	" 80
Kristall	5,3	6,1	" 81	4,5	" 80
Maris Bead	4,9	5,8	" 80	3,6	" 80

The main difference between the two locations is caused by the differences in soil type. The heavy clay soil near Wageningen has a considerable higher water retention capacity than the lighter clay soil in the N.O. Polder, resulting in certain restrictions in vegetative growth in the trials near Wageningen in most years. In the dry and sunny year 1982 the yields in the N.O. Polder were considerably higher than in Wageningen where vegetative growth was restricted too much by the drought (table 2).

TABLE 2   Seed yield in 1982 in t/ha (86% d.m.)

	Wageningen	N.O. Polder
Minica	5,3	9,4
Strube's Ackb.	5,0	7,3
Wierboon	4,3	8,1
Kristall	4,0	7,6
Maris Bead	3,6	6,6

The harvest index values of the six trials are given in table 3 and they clearly show the differences between cultivars and the variations caused by the environment. They also demonstrate the possibility and the need for breeding for a higher and more stable distribution of dry matter in faba beans.

TABLE 3   Harvest index

	Wageningen			N.O. Polder		
	'80	'81	'82	'80	'81	'82
Minica	63	65	61	60	54	60
Strube's Ackb.	54	58	55	49	47	54
Wierboon	44	55	50	43	47	50
Kristall	46	54	48	40	44	52
Maris Bead	48	54	48	39	48	47

The maximum plant lengths, measured in 1980, were 135, 190, 185, 200 and 195 cm for the cultivars Minica, Strube, Wierboon, Kristall and Maris Bead respectively. One of the aspects of such an excessive vegetative growth is the danger of lodging and this has to be avoided by breeding shorter cultivars with a higher stability of vegetative growth in response to soil water status.

EXPERIMENTS WITH AND WITHOUT IRRIGATION

A series of experiments was started in 1980 with the following aims:
a. as an attempt to get a better understanding of some of the results of the joint faba bean trials.
b. to study the plant – water relations of faba beans.
c. to investigate if the evaluation of breeding material can be improved and facilitated if the water supply in certain stages of development is regulated, in particular by avoiding drought stress.

1980:

The cultivar Minica was grown with and without irrigation. Water was supplied as soon as the water content of the soil decreased to 60% of field capacity. The weather conditions were such that long dry periods and severe drought stress did not occur and irrigation resulted only in a slight increase in total yield. Some short periods with mild to moderate water stress, however, were sufficient to cause considerable differences between the two crops. Whereas the unirrigated crop had a maximum LAI of 6,5 on July 9, a maximum LAI of 8,9 was measured on July 21 in the irrigated crop. Number of internodes did not differ between the two crops, so there were large differences in leaf area per plant, average leaf area and the SLW of the unirrigated crop was higher than that of the irrigated crop. Irrigation resulted in an increase of plant length, caused by an increased length of internodes, an increase in number of young pods, but also a higher degree of abortion of young pods, especially at the lower nodes. At harvest time the number of pods per plant between treatments was not significantly different. The final results of both crops are given in table 4.

TABLE 4   Final yield (15/9, 1980) cv. Minica, in kg d.m./ha

Wageningen 1980	V1	V2	
total yield	10270	10940	★
seed yield	6350	5820	★★
straw yield	3920	5120	★★
harvest index %	62	53	
V1 = unirrigated	★ s.d. at 5%		
V2 = irrigated	★★ s.d. at 0,5%		

1981:

The cultivars Minica and Kristall were grown with and without irrigation. As in 1980 water was supplied to keep the water content of the soil between 60 and 100% of field capacity. This involved irrigation on five occasions with 20 mm between middle and end of flowering and on one occasion, with 10 mm of water, 10 days before the beginning of flowering of Minica. Although final total dry matter yields of the irrigated crops were slightly higher than those of the unirrigated crops, seed yields and harvest index values of the unirrigated crops were higher (table 5).

TABLE 5   Wageningen 1981, final yields in kg/ha d.m.

	seed	straw	total	h.i.
Minica V1	6390	3680	10070	64
V2	5910	5030	10940	54
Kristall V1	5210	5090	10300	51
V2	4870	6130	11000	44

V1 and V2, see table 4.

Other characters such as plant height, number of pod bearing nodes and number of pods per node were markedly affected. As in 1980 avoidance of mild to moderate drought stress resulted in enhanced vegetative growth at the expense of reproductive growth, especially in terms of pod abortion at lower nodes. There is a critical stage for about two weeks after mid-flowering, when retention of young pods at the lower nodes is most important.

1982:

In this dry and sunny summer twelve cultivars were grown on two soil types near Wageningen, a heavy clay and a sandy soil and in each experiment five different water supply treatments were applied, ranging from dry to to freely available water supply, 60 - 100% of field capacity, throughout growth. Intermediate treatments included freely available water till end of flowering or after flowering with drought for the remainder of the growing period. As there are considerable differences in earliness between these cultivars, they were divided into three groups so that water could be applied in comparable stages of development. As a result of the unusual dry summer there was hardly any difference between the treatments V2 = "normal" growing conditions and V1 = dry by covering the soil between rows with plastic to drain off any rain water.

In the following the average results of two crops grown on the two soil types for seven of the cultivars are given for each of the treatments:

V2 = "normal" growing conditions

V3 = water freely available after flowering

V4 = water freely available till end of flowering

V5 = "optimal" water supply throughout growth.

The maximum plant heights, given as an indicator of the vigour of vegetative growth and hence canopy development and performance was 88, 89, 124, 178, 170, 195 and 196 cm for the cvs. Metissa, Optica, Minica, Wierboon, Alfred, Blaze and Kristall respectively.

The average seed yields are given in table 6 and the harvest index values in % in table 7.

TABLE 6   Average seed yields in t/ha d.m., 1982

treatments	V2	V3	V4	V5
Metissa	5.4	5.8	5.4	6.5
Optica	4.7	5.0	4.8	5.2
Minica	5.2	6.4	4.7	6.7
Wierboon	4.8	6.2	4.4	5.8
Alfred	4.2	6.5	4.2	6.4
Blaze	4.2	5.6	4.0	5.8
Kristall	4.0	5.9	4.0	6.0

TABLE 7   Harvest index, average values clay and sand

treatments	V2	V3	V4	V5
Metissa	67	69	63	64
Optica	62	64	60	60
Minica	63	64	51	57
Wierboon	50	50	41	43
Alfred	49	57	45	52
Blaze	45	49	41	44
Kristall	45	53	41	45

Clearly there was considerable variation between cultivars and it is interesting that some cultivars as Metissa and Optica appear to be more stable than others in relation to vegetative growth in response to soil water status. Not only does the harvest index differ considerably between the cultivars, but the response to irrigation at different times during the growth of the crop also differed. Some cultivars were relatively stable whereas others produced widely varying values.

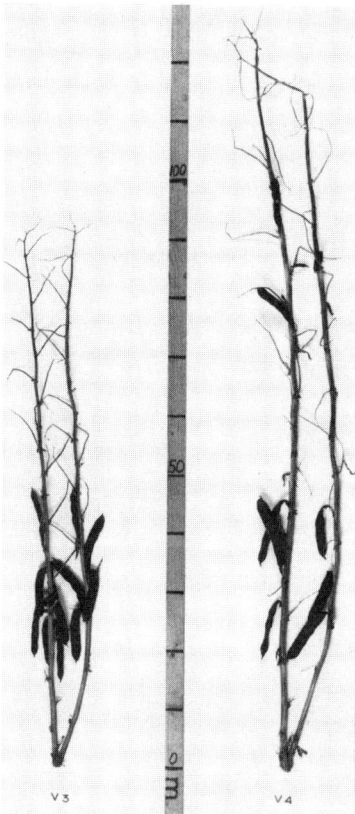

fig. 1 (left):
cv. Minica,
treatments V3 (l)
and V4 (r).

fig. 2 (right):
cv. Wierboon,
treatments V3 (l)
and V4 (r).

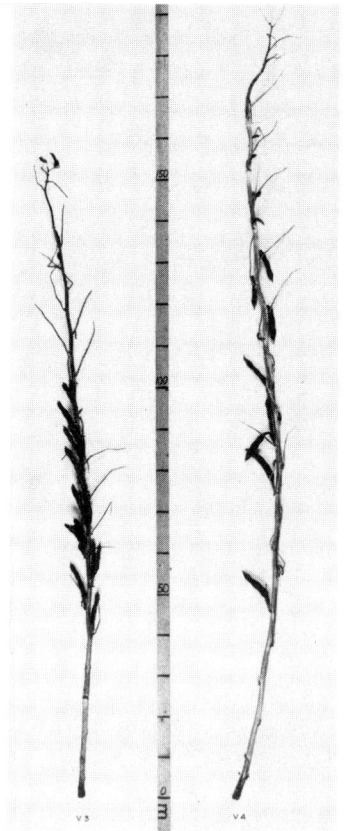

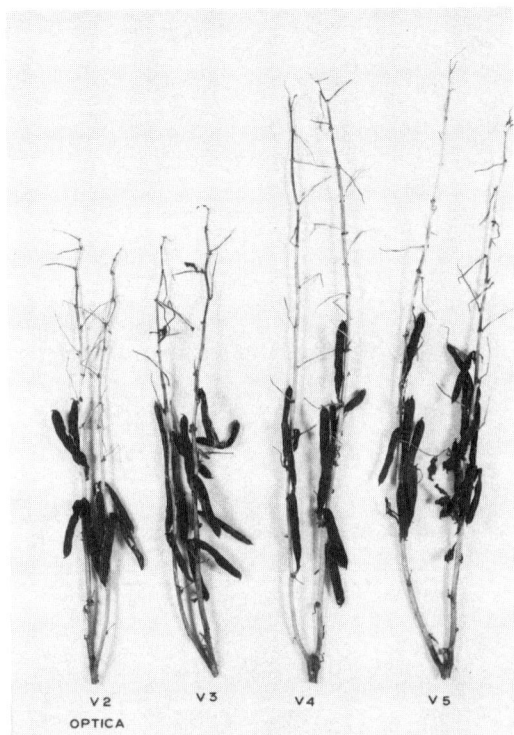

fig. 3: cv. Optica, treatments V2, V3, V4
and V5

As in the two preceding years
any avoidance of drought stress
resulted for most of the cultivars in excessive vegetative
growth and abortion of young
pods at the lower nodes (fig. 1
and 2). The treatment V5 not
only produced high total yields
but also despite rather low h.i.-
values high seed yields,
the latter as contrasted with the
results in 1980 and 1981.
Absence of adverse weather conditions and lodging in 1982 and
an amount of solar radiation
during formation and filling of
pods about 35% higher than in
the preceding years created the
possibilities for the formation
for sufficient pods at higher
nodes to compensate for the
aborted pods at lower nodes. The irrigated treatments in 1980 and 1981 were
more similar to the treatment V4 than V5 in 1982. The yield components
number of pods per plant and seed weight responded strongly to the various
irrigation regimes. Treatments resulting in limited vegetative growth,
but a high leaf area duration, such as were caused by post-flowering
irrigation, gave the highest values of average seed weight. The low yields
of Optica were caused by the plant density that was too low for this type
of plant. Even with an optimal water supply throughout the growing season
a closed canopy was not obtained (fig. 3).

PLANT - WATER RELATIONS

Leaf water potential (ψw) and osmotic potential (π) were measured
on several occasions during the growing seasons of 1980 and 1982 in
irrigated and non-irrigated treatments. Turgor potential (P) was calculated
from these measurements in order to investigate the relationship between
ψw, π,P and expansive growth.

68

During a relative short period in the wet summer of 1980 small, but
clear differences were found in diurnal changes in turgor between irrigated
and non-irrigated treatments of Minica. In the dry summer of 1982
measurements of plant water status could be obtained on many occasions and
it was possible to investigate the relationship between leaf water
potential and turgor during an important part of the period of expansive
growth (fig. 4).

fig. 4: The relation between leaf water
potential (ψw) and turgor (P) for
upper leaves of Minica (r=0,90). Data
obtained in the field on 7 days between
9/6 and 8/7, 1982 (normal dots). Non-
irrigated treatment 1/7:0 and
irrigated treatment 1/7:Δ.

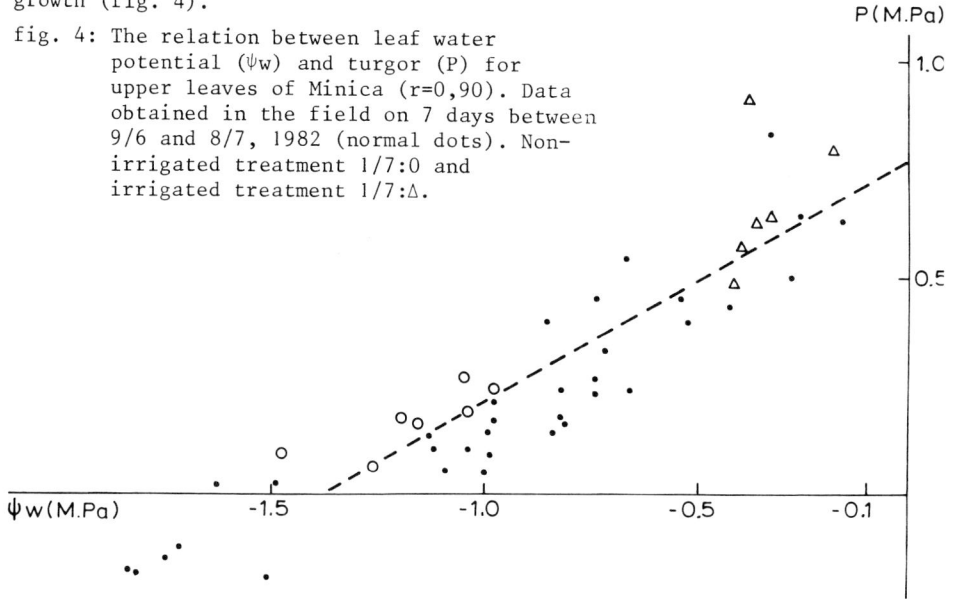

The relationship between ψw and P seemed to be nearly linear and
constant over a long period. Large differences in turgor were found between
irrigated and non-irrigated treatments, but important differences in the
ψw - P relationship could not be found, neither between different irrigatio
treatments nor between the cultivars Minica, Felix, Wierboon and Kristall.
Differences in turgor, as shown for the measurements on 1/7/'82 (fig. 4)
could be found on all the other measuring days, but there was a strong
influence of actual soil water content and weather conditions on the
absolute level of ψw and P. These differences were correlated with differences in relative water content of the leaves, stomatal closure and
daily measured expansive growth of young stem parts (Grashoff, to be
published). Differences in water stress, resulting in differences in turgor
have an important influence on stem elongation and leaf expansion and thus
on LAI and total vegetative growth, leading to different canopy
characteristics, which, with increasing amounts of evidence, are related

to pod set and pod retention.

SUMMARY AND CONCLUSIONS

a. Differences in water supply in various stages of growth affect to a
   large extent the ratio between vegetative and reproductive growth and
   the yield components number of pods and seed weight.

b. Results from investigations into plant – water relations have shown a
   strong correlation between turgor and canopy characteristics and the
   latter are closely related to factors as interception and penetration
   of light and stomatal behaviour. Mechanisms that regulate the ultimate
   productivity by means of sugar levels and source – sink relations,
   hormone balances, pod set and pod retention and leaf area duration need
   more thoroughly investigation. Attention has to be paid to the considerable variation between cultivars in reaction to growing conditions.

c. Optimal growing conditions, natural or artificially created, are helpful
   to select for increased and more stable harvest index and avoidance of
   excessive vegetative growth.

THE EFFECT OF IRRIGATION AND BEES ON THE YIELD AND YIELD
COMPONENTS OF VICIA FABA L.

P.D. Hebblethwaite, R.K. Scott and J.O.S. Kogbe
University of Nottingham
Department of Agriculture and Horticulture,
School of Agriculture,
Sutton Bonington, Loughborough, LE12 5RD. U.K.

ABSTRACT

The effects of irrigation and bees and their interaction on the
flowering, pod setting, yield components and seed yield of Vicia faba L.
were investigated in field experiments in 1969 and 1970 and the effects
of irrigation only in 1971.

No significant interactions were found between water availability
and bees. However, seed yield was substantially increased by irrigation
in 1969 and 1970 but not in 1971. In no years did bees have any
significant effect on seed yield, but a higher proportion of pods was set
on lower and middle nodes, a higher proportion of flowers set pods, and
seeds per pod were increased, though in the absence of bees this was
compensated for by larger seeds.

INTRODUCTION

Many workers have highlighted the enormous wastage of flowers and

pods before maturity in the field bean crop and estimates of flower and

pod shedding range from 72% to 97% (Soper, 1952; Rowland, 1955, 1961;

Hodgson and Blackman, 1956; Kambal, 1968, 1969; Ishag, 1969; Kogbe,

1972). Hodgson and Blackman (1956) considered that the phenomenon of

flower wastage might be associated with the restriction placed on the

internal supply of substrates by the light gradient within the canopy.

However, since it is the upper nodes that often lose most pods, it may

not be a direct effect of shading (Kogbe, 1972). Drayner (1959) and El

Nadi (1966) considered that pollination failure might account for some

flower shedding, and El Nadi (1966) found that flower shedding was more

extensive when water stress was imposed during flowering of broad beans.

The aim of these experiments was to assess more fully the potential

of using insect pollinators (honey bees) to advance and increase pod set,

and irrigation to increase both pod set and survival. The hypothesis was

that the combination of pollinators and irrigation might result in a

positive interaction with the full potential yield being realised:

irrigation might increase the numbers of flowers for the bees to

P.D. Hebblethwaite, T.C.K. Dawkins, M.C. Heath and G. Lockwood (eds.)
Vicia faba: Agronomy, Physiology and Breeding. ISBN 90-247-2964-5.
© 1984, Martinus Nijhoff/Dr W. Junk Publishers. Printed in The Netherlands.

pollinate and possibly nectar flow, the bees would subsequently increase pod set from the increased number of flowers, and by promoting the availability of water and nutrients, irrigation should enhance the survival of the increased number of pods set.

Bee treatments involve caging, and consequently shading and it was decided to examine the effects of shading by a treatment in which a cage was placed over the crop but open on the North side so that radiation would be comparable with closed cages. The open side gave access to free-flying wild bees to work the crop, albeit in a microclimate somewhat changed from that of the completely open crop.

EXPERIMENTAL PROCEDURE

The experiments were carried out in 1969, 1970 and 1971 at the School of Agriculture, Sutton Bonington on soil which was a fairly deep, coarse gravelly loam overlying Keuper Marl. The experiment in 1969 and 1970 consisted of eight treatments comprising the factorial combination of irrigated and non-irrigated crops with the following cage/bee treatments: (i) cage with bees, (ii) cage without bees, (iii) open-ended cage, (iv) open cage. In 1971 only irrigated and non-irrigated treatments were included. Details of experimental design, management treatments, harvesting and growth analysis are published elsewhere (Kogbe, 1972).

The varieties of field bean used were the small-seeded Tarvin in 1969, and Herz Freya in 1970 and 1971. Tarvin was chosen originally to represent the highest yielding of the recommended spring varieties at the time; Herz Freya was subsequently used because Tarvin's indeterminate growth habit, and late maturity made it necessary to prolong the caging treatment. With long straw, Tarvin was also prone to lodging especially where light intensity was decreased by the cages.

In all years the determination of irrigation use was calculated from potential transpiration (Penman, 1956) with corrections being made for crop cover by assuming a full crop cover at leaf area index of about 3. Data for accumulated potential soil water deficits for the three years are presented in Fig. 1.

No attempt was made to irrigate the crop at particular growth stages but water was applied from flowering onwards by trickle irrigation to maintain, as far as possible, a deficit less than 30 mm. This deficit was chosen as the probable maximum that spring field beans are able to withstand on this soil type without some yield loss occurring (French and Legg, 1979).

Cages similar to those used by Scriven, Cooper and Allen (1961) were placed on the crop on 13 June 1969 and 8 June 1970 and removed on 21 July and 14 July, respectively. Hives of honey bees were placed inside the cages and bees were fed sugar syrup which is said to increase pollen collection efficiency of the colony (Free and Spencer-Booth, 1961). Each hive consisted of a laying queen and workers. No bumble, robbers or long-tongued bees, were in any of the closed cages.

Cages reduced solar, net, and visible radiation by 32%, 39% and 43% respectively and wind speed by 44% (average 30 April to 2 May, 1971).

RESULTS

No interactions were found between irrigation and bees in 1969 or 1970. Consequently, results for irrigation and bees are presented separately. In 1969, caged plots were harvested prematurely because of severe storm damage to the plants at the end of July and consequently the irrigation section only includes data from plots that were not caged.

The effects of irrigation

Plant length and maturity

Irrigation increased plant length in all years and particularly in the very dry year of 1970 (Table 1) primarily because plants developed more nodes and longer internodes and as a result, plots receiving irrigation lodged in all years and this may have decreased the potential yield as lodging of irrigated crops has been shown to depress yields (McEwen et al., 1981). Irrigation delayed maturity in 1970 (Table 2).

74

TABLE 1   The effect of irrigation on plant length, number of
nodes per plant and internode length, 1969 and 1970

	Plant length (cm)	Number of nodes /plant	Internode length (cm)	Max. summer potential soil moisture deficits (mm)
1969				
Irrigated	171*	25*	6.5	30
Not-irrigated	144	21	7.0	125
S.E.	9.1	1.1	0.43	-
1970				
Irrigated	116***	23*	5.1**	30
Not-irrigated	64	16	3.9	225
S.E.	4.6	0.7	0.30	-

Significance of difference: *$0.01<P<0.05$; **$0.001<P<0.01$; ***$P<0.001$

TABLE 2   The effect of irrigation and caging on harvest date

Treatment	1969	1970	1971
Irrigated plots (outside cages)	9 Sept	2 Sept	9 Sept
Not-irrigated plots (outside cages)	9 Sept	17 Aug	6 Sept
Irrigated plots (inside cages)	4 Aug (prematurely)	2 Sept	-
Not-irrigated plots (inside cages)	4 Aug (prematurely)	17 Aug	-

Final harvest yield and yield components

Irrigation increased seed yield by 44% in 1969 and almost three fold
in 1970 but had no significant effect in 1971 (Table 3). Undoubtedly the
greater response in 1970 reflected a higher accumulated soil water
deficit (Fig. 1) and associated with this the application of 182 mm water
compared to 100 mm and 117 mm in 1969 and 1971 respectively.

Irrigation increased seed yield in 1969 and 1970 mainly because of
increased pod numbers (31% in 1969 and 120% in 1970), but in both years
all other yield components, except plant number, number of pods per
podding node and seeds per pod, were also increased (Table 3). In the
dry year of 1970 the average number of seeds per pod was also increased
because fewer ovules and seeds aborted (Table 3). In 1971 irrigation did
increase pod number per plant and seeds per pod. However, this was
compensated for by a decrease in the weight per seed.

TABLE 3   The effect of irrigation on the final seed yield components, ovule abortion and seed yield 1969 to 1971

	No. of plants/m²	No. of podding nodes/plant	No. of pods/ podding node	No. of pods/ plant	No. of calculated pods/m²	No. of seeds/ pod	Weight/ 1000-seeds (g at 85% D.M.)	Yield at 85% D.M. (t/ha)	Aborted (ovules or seeds) /pod
1969									
Irrigated	40	5.8***	1.8	10.4**	330***	3.8	448***	5.6***	0.26
Not-irrigated	43	4.0	1.7	6.9	252	3.8	412	3.9	0.28
S.E.	1.6	0.18	0.09	0.63	7.9	0.04	5.5	0.12	0.018
1970									
Irrigated	37	6.4***	1.9	12.2**	349***	3.6***	495***	6.1***	0.44
Not-irrigated	38	2.8	2.0	5.6	159	3.3	392	2.1	0.56***
S.E.	1.1	0.17	0.06	0.46	9.4	0.05	5.7	0.18	0.025
1971									
Irrigated	75	-	-	4.8*	356***	3.3**	354	4.2	-
Not-irrigated	74	-	-	4.1	303	3.1	401*	3.9	-
S.E.	1.8	-	-	0.17	3.8	0.02	14.1	0.19	-

Significance of difference: *0.01<P<0.05; **0.001<P<0.01; ***P<0.001

Dry matter production and distribution

In 1970, irrigation more than doubled total dry matter production at final harvest and the dry matter produced in different parts of the plant (Table 4).   In 1971 irrigation had no effect on any of the above measurements.   In both years it did not alter the proportion of dry matter in different parts of the crop (Table 5).

TABLE 4   The effect of irrigation on the partition of dry matter per unit area at final harvest, (g/m²), 1970 and 1971

	1970			1971		
	Irrigated	Not-irrigated	S.E.	Irrigated	Not-irrigated	S.E.
Total dry weight	1117***	429	37	997	958	31
Stem, petiole & leaf dry weight	315***	125	14	536	535	19
Pod + seed dry weight	803***	304	26	461	423	17
Seed dry weight	636***	236	20	357	328	16
Pod dry weight	166***	68	7	104	94	3

Significance of difference: ***P<0.001

Dry matter production in irrigated plots was similar in 1970 and 1971 but the proportion of the total dry matter represented by the seed (harvest index) was very much less in 1971 than 1970 (Table 5).   The 1971 observations showed that the crop was very tall (over 2 m high) possibly because of the very high population and wet cool conditions and consequently a large percentage of the dry matter was in the stem.

TABLE 5   The effect of irrigation on the partition of dry matter as a percentage of total dry matter at final harvest, 1970 and 1971

	1970			1971		
	Irrigated	Not-irrigated	S.E.	Irrigated	Not-irrigated	S.E.
Total	100	100	–	100	100	–
Stem, petiole and leaf	28	29	1	54	55	1
Pod + seed	72	71	1	46	44	1
Seed (harvest index)	57	55	1	36	34	1
Pod	15	16	1	10	9	1

Flowering and pod production

Flowering started on 8 June in 1969 but at the earlier date of 27 May in 1970 (Figs. 2 & 3) probably because of the earlier flowering of Herz Freya than Tarvin and the rapidly developing water stress in the latter year. By the time observations on labelled plants had started in 1969 some of the flowers had started to drop. Despite attempts to include the shed flowers in the counts by including the scars left by those shed, flower counts appeared to have been slightly under-estimated. In 1970, observations started at the bud stage.

Fewer flowers and immature pods were produced in 1970 than in 1969 (Figs. 2 & 3). This was varietal as well as environmental as Tarvin produced more flowers and immature pods than Herz Freya in adjacent variety experiments (Kogbe, 1972).

Irrigation increased flowers, immature and mature pods, primarily because they produced more flowering nodes and nodes bearing immature and mature pods rather than because of an increase in the productivity at individual nodes (Table 6). Irrigation increased the proportion of flowers forming mature pods; this resulted from an increase in flowers setting immature pods more than from an increase in the proportion of pods surviving (Figs. 2 & 3).

Additional water increased the number of maximum flowers per plant (Table 6). Flower shedding occurred mainly in early July 1969 and in late June and early July 1970 when pods were being set. Irrigation

TABLE 6   The effect of irrigation on flower and pod production
in labelled plants, 1969 and 1970

		Irrigated	Not irrigated	S.E.
Maximum flowers/plant	1969	61.5*	42.9	5.43
	1970	46.3***	30.4	2.15
Flowering nodes/plant	1969	11.8**	8.0	0.88
	1970	9.2***	6.8	0.29
Flowers per flowering node	1969	5.2	5.3	0.27
	1970	5.0**	4.4	0.13
Immature pods per plant	1969	36.1*	23.8	2.57
	1970	18.4***	8.5	0.62
Immature podding nodes/plant	1969	8.8*	6.9	0.57
	1970	7.8***	4.0	0.21
Immature pods/podding node	1969	4.1*	3.4	0.16
	1970	2.4	2.1	0.08
Mature pods/plant	1969	19.4**	11.5	1.85
	1970	13.4***	5.1	0.51
Podding nodes/plant	1969	8.1*	5.8	0.54
	1970	7.0***	2.7	0.18
Mature pods per podding node	1969	2.4	1.9	0.39
	1970	1.9	1.9	0.07
Percent flowers forming immature pods	1969	59.3	56.2	2.50
	1970	40.2***	28.0	1.10
Percent flowers forming mature pods	1969	31.4	26.0	1.94
	1970	29.3***	16.9	0.94
Percent immature pods forming mature pods	1969	53.3	46.7	3.43
	1970	72.0***	60.4	1.83

Significance of difference: *$0.01 < P < 0.05$; **$0.001 < P < 0.01$; ***$P < 0.001$

delayed flowering for 1 week in 1969 and 2 weeks in 1970. Although pods
were first formed on the non-irrigated plants, the more rapid increase in
pod set during the period when flowers were being shed, resulted in a
higher maximum pod production in the irrigated crop. In 1970, maximum
pod set was a week later in the irrigated crop because more pods were set
from flowers at upper nodes, but in 1969 it occurred at the same time in
both treatments (Figs. 4 & 5). Pods were shed mainly in late July 1969
and during the third week in July 1970, when accumulated soil moisture
deficits (Fig. 1) and average temperatures were very high, but irrigation
did not decrease the rate of pod shed (Figs. 2 & 3), which indicates that
other internal plant factors were also involved. The lack of any

interaction between irrigation and bees indicates that the abscission is not related to effects of water stress on fertilization. It has been suggested that high levels of ABA associated with water stress may be the agent for abscission (El Beltagy & Hall, 1975).

In 1969, pod shedding continued until early August, but in 1970 when a different variety was used, pod number remained almost constant after the third week in July in both treatments (Figs. 2 & 3).

The higher maximum flower production and pod set in the irrigated crop reflected the increase in the number of flowers and pods at the lower and middle nodes and a few extra nodes bearing flowers and pods at the uppermost part of the irrigated plants (Figs. 4 & 5). In 1970 the increased seed yield at individual nodes (Fig. 6) was due mainly to increased pod number (Fig. 5) and, to a lesser extent, to increased number of seeds per pod and weight per seed (Fig. 7). Node 9 produced the highest seed yield in the irrigated (3.3 g) and non-irrigated crops (2.0 g) (Fig. 6). These values constituted 23% and 28% of yield per plant in these treatments, respectively. Yield was also distributed more evenly and over more reproductive nodes in the irrigated crop (Fig. 6).

The effect of caging and bees

Final seed yield, yield components, flower and pod production and maturity

As there were no significant interactions all results are presented as means of combined irrigated and non-irrigated treatments (Table 7). When grown under open cages seed yield was decreased by 25% compared with the open crop in 1969, but was not affected in 1970 (Table 7). In 1969, plants produced fewer flowers, immature and mature pods when grown under open cages because they produced fewer flowering nodes, and fewer immature and mature pods per podding node than in the open field (Table 8). In 1970, plants produced more flowers under open cages than in the open crop; this resulted from slightly more flowering nodes, but the number of immature and mature pods was largely unaffected. Caging caused a smaller proportion of flowers to form mature pods, significantly so only in 1969. In this year, this was mainly due to the smaller proportion of flowers which set pods rather than the proportion of pods which survived (Table 8).

In both years, seed yield was further reduced in the crop grown in closed cages with or without bees than that grown under open cages. This was due to decreased average weight per seed under closed cages in 1969, and decreased pod number as a result of fewer pods per pod-bearing node in 1970 (Table 7).

TABLE 7   The effect of caging and bees on the final seed yield
components, ovule abortion and seed yield in 1969 and 1970

	No. of plants /m²	No. of podding nodes/plant	No. of pods/ podding node	No. of pods/ plant	No. of calculated pods/m²	No. of seeds/ pod	Weight/ 1000-seeds (g at 85% D.M.)	Yield at 85% D.M. (t/ha)	Aborted ovules and seeds/pod
1969									
Cage + bees	41	4.7	1.6	7.6***	246***	3.7	257**	2.3***	0.19**
Cage - bees	42	5.2	1.8	9.2	257***	3.4**	262**	2.3***	0.44**
Open cage	45	4.5	1.5*	6.7***	245***	3.7	303	2.7***	0.30
Open crop (Control)	43	5.7	1.9	10.9	313	3.7	314	3.6	0.28
S.E.	1.7	0.33	0.08	0.57	9.7	0.05	9.1	0.07	0.040
1970									
Cage + bees	38	4.5	1.7	7.2	212**	3.4	461	3.5	0.51*
Cage - bees	36	5.1*	1.6*	7.7	190**	3.2*	510***	3.2*	0.69*
Open cage	36	5.2*	1.8	9.0	239	3.5	476***	4.1	0.51*
Open crop (Control)	39	4.2	1.9	8.0	248	3.4	443	3.9	0.60
S.E.	1.7	0.23	0.07	0.46	10.3	0.05	7.6	0.20	0.039

Significance of difference: *0.01<P<0.05;   **0.001<P<0.01;   ***P<0.001

Pod numbers and seed yields were similar whether the crops were caged with or without bees and this applied in both years (Table 7). Similar results were obtained by Riedel and Wort (1960), and by Free (1966). However, caging plants without bees significantly decreased the average number of seeds per pod because more ovules aborted. Wafa and Ibrahim (1960) observed a similar effect in broad beans, and Free (1966) in broad and field beans. In 1970, the diminished number of seeds per pod in the absence of bees was compensated for by an increased average weight of individual seeds. There was no evidence of this in 1969, perhaps because the experiment was harvested prematurely.

Cages also appeared to delay ripening, more so in plants caged without bees. In plots caged without bees 55% of stems remained green at harvest. Percentages for cages plus bees, open cage and the open crop were 31%, 35% and 22%, respectively. The plants in cages without bees were slightly taller than those in other cage treatments at the end of the season. Bond, D.A.(personal communication) has observed similar responses in caged crops without bees.

TABLE 8   The effect of caging and bees on flower and
pod production in labelled plants, 1969
and 1970

		Cage + bees	Cage -bees	Open cage	Open crop	S.E.
Maximum flowers/plant	1969	38.7	43.8	37.4	52.2	3.84
	1970	41.2	46.4	42.8	38.5	2.19
Flowering nodes/plant	1969	7.5*	8.8	7.4*	9.9	0.62
	1970	9.2	9.6	9.2	8.3	0.47
Flowers/flowering node	1969	5.2	5.1	5.1	5.2	0.19
	1970	4.5	4.8	4.6	4.7	0.14
Immature pods/plant	1969	18.7**	18.8**	19.8**	29.9	1.82
	1970	12.2	11.4	12.9	12.7	1.06
Immature podding nodes/plant	1969	6.9	6.8	6.9	7.9	0.40
	1970	6.2	5.8	6.1	5.6	0.32
Immature pods/podding node	1969	2.7***	2.8***	2.8***	3.7	0.16
	1970	1.9	2.0	2.1	2.2	0.14
Mature pods/plant	1969	10.2*	9.4*	8.5*	15.5	1.31
	1970	8.3	7.1	8.0	8.2	0.52
Podding nodes/plant	1969	5.8	5.8	5.8	6.9	0.38
	1970	5.0	4.7	4.8	4.6	0.28
Mature pods per podding node	1969	1.7**	1.6**	1.5**	2.2	0.12
	1970	1.7	1.7	1.7	1.8	0.07
Percent flowers forming immature pods	1969	48.3	43.2	53.2	57.1	1.77
	1970	29.1	24.2	29.8	32.1	2.27
Percent flowers forming mature pods	1969	26.2	21.6*	23.1*	28.7	1.37
	1970	19.8	15.1**	18.0	20.8	1.06
Percent immature pods forming mature pods	1969	54.1	50.0	43.4*	50.0	2.43
	1970	67.9	64.8	60.5	64.2	4.56

Significance of difference: *0.01<P<0.05;   **0.001<P<0.01;   ***P<0.001

Plants produced more flowers when caged without bees (Table 8)
because they formed more flowering nodes in the upper part in 1969, and
more flowers at individual nodes in 1970 (Figs. 8 & 9).
Plants in cages without bees continued to open flowers for longer into
July, and flowers persisted in this treatment for a week longer in 1969
and two weeks in 1970 (Figs. 10 & 11).

Bees caused an earlier pod set at the lower nodes in 1969 and middle
nodes in 1970 (Figs. 8 & 9). This was reflected in the greater number of
pods in early July 1969, and late June to early July 1970 (Figs. 10 &
11). In late July, pod number became greater in plants caged without
bees as more pods were set at the upper nodes (Figs. 8 & 9). In 1969,
most of the later formed pods were shed, and the final pod number
remained similar in both treatments. In 1970, some of these pods on
plants caged without bees aborted, and remained green and attached to the
stem until final observation. The aborted pods contained no seeds. The
decreased number of flowers and the increased early pod set were

reflected in the greater proportion of flowers forming mature pods in plants caged with bees (Table 8). The increase in the average number of seeds per pod (in plants caged with bees in 1970) reflected the increased number of seeds at the lower and middle nodes (Figs. 8 & 9).

DISCUSSION

No significant interaction was found between irrigation and bees in the two years where tested (1969 and 1970). This indicates that the interaction hypothesis, that the combination of pollination and irrigation might result in a positive interaction with the full potential yield being realised (see Introduction), was not proven. However, irrigation during flowering and pod growth increased seed yield substantially under dry conditions. Other workers have also reported very substantial increases in field bean yields as a result of irrigation (Penman, 1962, 1970; Sprent, Bradford & Norton, 1977; French and Legg, 1979; Krogman et al., 1980; McEwen et al., 1981) and that these increases can be significantly correlated with increasing amounts of water applied (irrigation plus rain) (Krogman et al., 1980; Hebblethwaite, 1981; Day and Legg, 1983). The lack of response in 1971 compared to that in 1969 was possibly due to the higher maximum deficit (120 mm) in 1969 and this deficit coinciding with flowering and pod set in July. In 1971 maximum deficit (90 mm) only just exceeded 80 mm for brief periods in late July and August and well after pods were set. The lack of response to watering in 1971 indicates that the limiting deficit for field beans on this soil (loam) was similar to that on flinty silty clay loam over flinty clay at Rothamsted of 80 mm and well above that on loamy sand over sand at Woburn of 30 mm (French and Legg, 1979). The response of seed yield to irrigation was mainly through increases in pod number and to a lesser extent, increases in weight per seed, number of seeds per pod, and decrease in number of aborted ovules and seeds per pod. These data confirm that any or all seed yield components can be affected by water stress, and the time at which stress occurs and the degree of that stress determines which components are affected as reported by many workers (Jones, 1963; Myers, Corey, Lebaron and McMaster, 1957; El Nadi, 1970; Krogman et al., 1980; McEwen et al., 1981; Hebblethwaite, 1981). This supports the hypothesis that the crop is sensitive at most reproductive stages of growth and does not have any specific stress stage (French and Legg, 1979; Hebblethwaite, 1981).

The irrigated crop produced more flowers by increasing the number of flowering nodes at the top of the plant; that is the plant was more indeterminate in growth. Associated with more profuse flowering was an improved pod set (percentage of flowers producing mature pods) and it was due to improvement at this stage rather than at the survival stage (stage at which pods were rapidly aborting) that pod numbers and yield were improved. The pattern of pod loss in late July 1969 and mid to late July 1970 was similar whether or not the crop was irrigated. Consequently, it would seem that irrigation does improve pod set but does not increase proportionate pod survival. Addicott and Lynch (1955) suggest that the abscission of young fruit is due to the unsuccessful competition for limited metabolites, chiefly carbohydrates and nitrogen. A similar hypothesis was put forward by Meadley and Milbourn (1970, 1971) for pod abscission in vining peas. These authors state that yield may depend less on the ability of the crop to provide reproductive nodes than on its ability to fill a smaller number of pods and it is possible that field beans are affected in a similar way. El Nadi (1969, 1970) found that the imposition of water stress in broad beans at flowering resulted in a considerable increase in flower shedding and that the effects of water stress on seed yield was most severe at the time the seed was set. El Beltagy and Hall (1974, 1975) have shown that the abscission of reproductive organs in field beans is influenced by ethylene concentration which may vary with water stress. Furthermore, Sprent (1972) has shown that acetylene reduction activity in field beans is decreased by moisture stress. It has been suggested that high levels of ABA associated with water stress may be the agent for flower abscission (El-Beltagy and Hall, 1975).

Dry matter data indicated that the crop substantially increases amounts of dry matter produced when irrigated but irrigation does not increase the proportion of this dry matter going to the seed. Consequently, irrigation increases total dry matter and seed dry matter in the same proportion because of the greater production of carbohydrate. More carbohydrates are, therefore, available to fill more seeds to a greater weight and this is shown by an increase in average weight per seed, average number of seeds per pod and fewer ovules and seeds aborted in irrigated treatments.

Irrigation may also encourage an improvement in rhizobial activity, thus making more nitrogen available to the crop. Recent work (Bainbridge et al., 1977) has shown that irrigation increases the amount of nodular tissue but unexpectedly decreases the amount of nitrogen fixed per unit of nodule tissue. Despite this the total amount of nitrogen fixed in the season was more than doubled by irrigation. Sprent (1972) has shown a high degree of correlation between soil-water content in the field and nitrogen fixation.

Seed yield was decreased by open-ended caging (or the effects of shading) in 1969 only and by closed caging in both years. Radiation level during caging was similar in both years and would, therefore, not account for the differing responses in open-ended cages between years. Cages restricted air movement by about 50% and this could have increased the humidity in the caged crop, and consequently decreased transpiration. This was possibly a greater positive advantage to the non-irrigated crop in the drier year of 1970 as there was some suggestion that seed yield was increased by caging the non-irrigated crop, particularly as the estimated soil moisture deficit at the start of caging was in excess of 100 mm in 1970 and only about 40 mm in 1969 (Fig. 1). Furthermore, the interval between removing cages and harvest was longer in 1970 than in 1969 and the opportunity for compensatory growth correspondingly greater. It is also possible that wild bees were more effective in open cages in 1970 than in 1969. In the three years of their study, Scriven et al. (1961) were also able to measure a decrease in seed yield as a result of open-ended cages in one year. The consistent decrease in yield of the crop under closed cages in the irrigation treatment was probably due to poorer aeration, and less admission of solar radiation into the crop late in the evenings when the sun moved to the North-west before setting, during the long summer days. These two factors were likely to decrease growth to a greater extent in the crop under closed cages. It is also possible that honey bees were less efficient than wild bees when comparing caging plus bees with open-ended cages.

In two years of the experiment, caging the crop with bees did not increase seed yield, and similar results were obtained by Riedel and Wort (1960) and Free (1966) but not by Scriven et al. (1961) where significant increases were found in all three years of their work. The differing responses may have been due to the proportion of plants in each seed stock that were auto-fertile (Free and Williams, 1976). In these experiments and those of Riedel and Wort (1960) and Free (1966) spring beans were used while Scriven et al. (1961) used winter beans.

Confining honey bees to forage a crop may reduce their pollinating efficiency. The bees which worked the crop in closed cages were those from a new brood which had never been used to flying in the open. They seemed to settle down well to work the crop. The weather during flowering was dry and sunny, and suitable for bees to forage. Number of bees was non-limiting but it is doubtful whether they were all pollinating the crop since some robbed the flowers of nectar through holes pierced at the base. Some strains of honey bees are prone to robbing flowers (B. Cooper, personal communication) and this was obviously the case with the strain used in this experiment. It is also important to note that all bees (honey and bumble), had access to the flowers in open-ended cages, and this could have accounted for the higher yield from open-cages than from cages + bees.

About 30-40% of the plants in a crop of beans are reported to be from cross-fertilised seeds (Hua, 1943; Fyfe and Bailey, 1951; Rowland, 1958; Holden and Bond, 1960). These plants are known to be better able to set seed by selfing without tripping of the flower. Thus, in the absence of bees, deleterious effects would probably arise on the remaining 60-70% originating from self-fertilised seeds. The bees increased the number of seeds per pod, and the increase in this component of yield arose because fewer ovules and seeds aborted. The more aborted ovules and seeds in plants caged without bees probably arose from self-fertilisation without tripping in some of the 60-70% plants which originated from inbred seeds. Rowland (1958) and Holden and Bond (1960) observed that self-fertilisation was incomplete in untripped self-pollinated flowers especially in inbred plants and any ovules fertilised at all were likely to abort. That abortion of ovules may be due to zygotic infertility was suggested by Rowland (1960) who saw this as resulting from homozygosity of deleterious recessive genes which can accumulate in a naturally outbreeding species forced to inbreed.

The increase in number of seeds per pod did not result in increased seed yield in the crop caged with bees even though the number of pods was similar in the two treatments. This was because the larger seeds in the crop caged without bees compensated for the smaller number of seeds. There was no evidence for this in 1969, probably because the premature harvest prevented the filling of pods especially the later formed ones in the crop caged without bees. Free (1966) observed similar compensatory mechanisms in a field bean crop caged with and without bees, and suggested that the yield obtained was probably the limit of seed production the crop could attain under cages. This was because cages limit light intensity in the crop and thus the photosynthate available to the pods.

The bees tended to concentrate on pods in the lower and middle nodes of caged plants especially in 1969 which resulted in more even and earlier maturity. This may well be an advantage to growers in relation to ease of harvest.

Some of the pods formed in the crop caged without bees aborted, but remained green and attached to the plants, and only contained aborted ovules. Rowland (1960) considered this phenomenon as evidence of infertility at zygotic level but Chapman, Fagg and Peat (1979) showed that such pods were parthenocarpic and not parthenogenetic. These pods may be an evidence of apomixis in field beans. Further work is necessary to establish whether field beans can set pods without fertilisation.

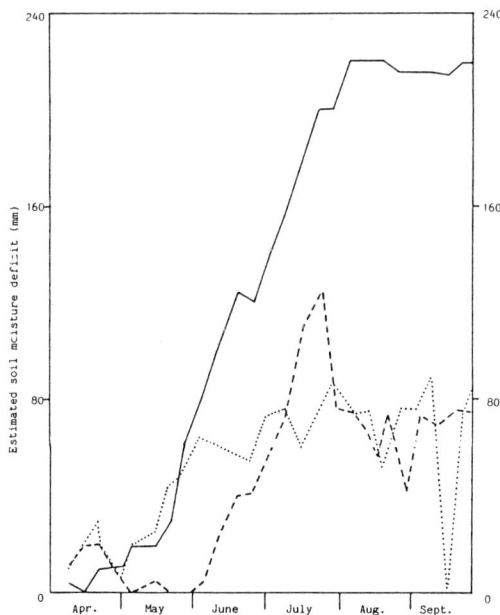

Fig. 1 Weekly estimated soil moisture deficit (mm),
1969, - - - - ;   1970, ———— ;   1971, ● ● ● ●.

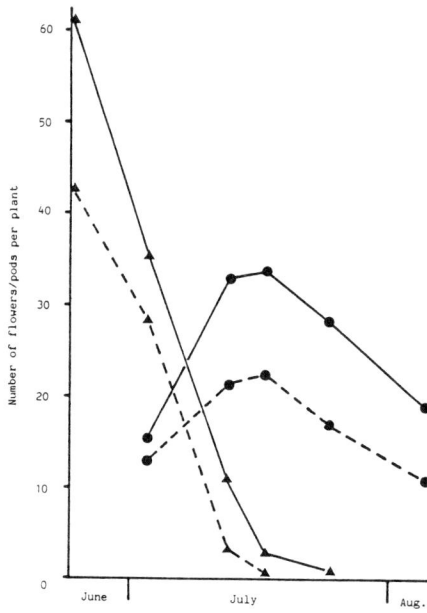

Fig. 2   Changes in the number of flowers and pods in
irrigated and non-irrigated plots in 1969.
Irrigated flowers, ▲——▲ ; irrigated pods, ●——● ;
non-irrigated flowers, ▲– – –▲ ; non-irrigated pods, ●----●

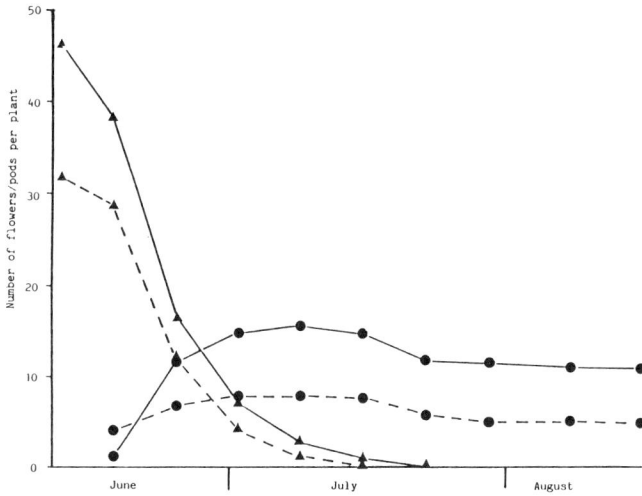

Fig. 3   Changes in number of flowers and pods in irrigated and non-irrigated plots in 1970.
Irrigated flowers, ▲——▲ ; irrigated pods, ●——● ; non-irrigated flowers, ▲---▲ ;
non-irrigated pods, ●---● .

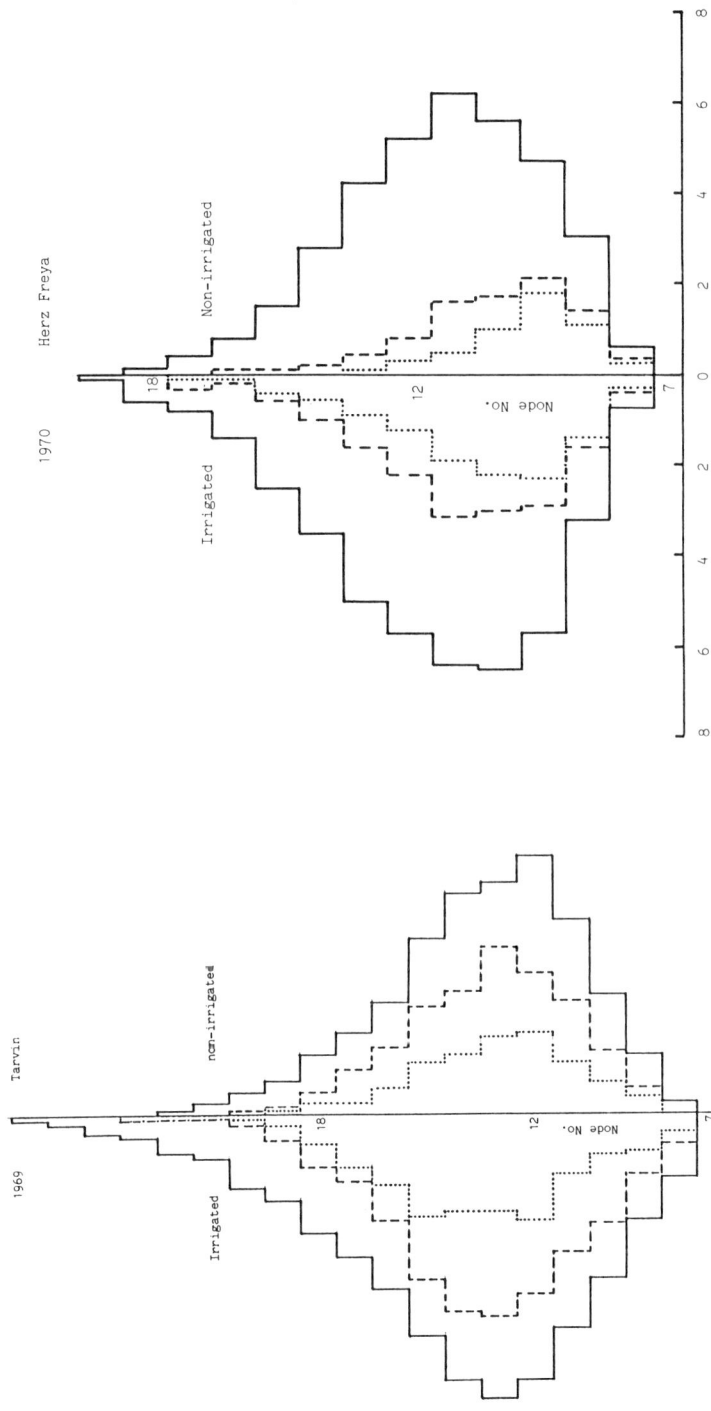

Fig. 5 Node by node distribution of flowers, immature and mature pods of irrigated and non-irrigated plants in 1970, (variety, Herz Freya). Flowers and flower buds per node, ——— ; immature pods per node, ----- ; mature pods per node, ●●●.

Flowers and flower buds, immature and mature pods per node

Fig. 4 Node by node distribution of flowers, immature and mature pods of irrigated and non-irrigated plants in 1969, (variety, Tarvir). Flowers and flower buds per node, ——— ; immature pods per node, ----- ; mature pods per node, ●●●●.

Flowers and flower buds, immature and mature pods per node

88

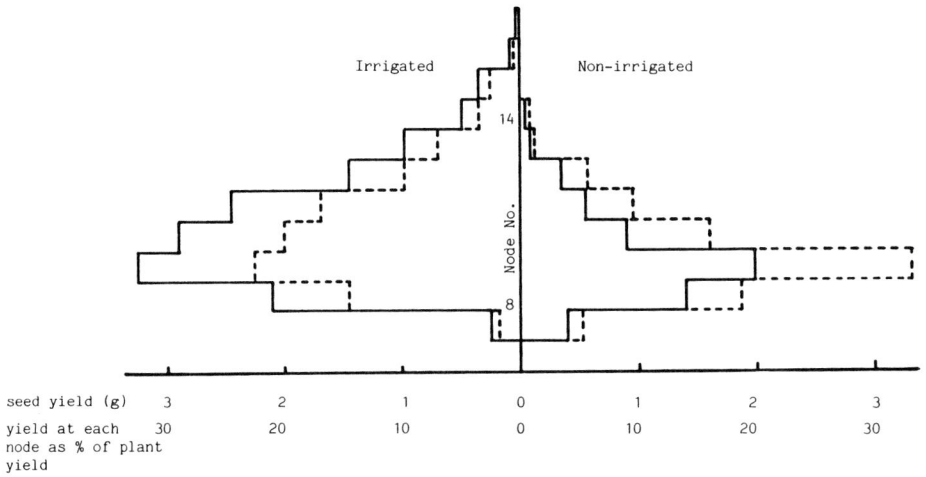

Fig. 6 Node by node distribution of seed yield (g) and yield at each node as a percentage of yield per plant in irrigated and non-irrigated crops in 1970. Seed dry weight (g) per node, ——— ; seed dry weight per node as a percentage of total plant yield, —————— .

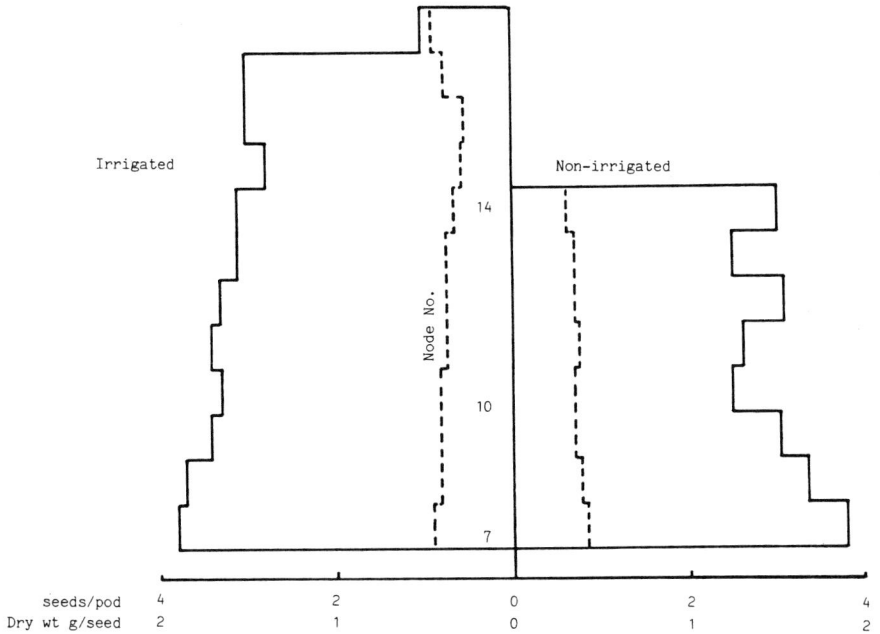

Fig. 7 Node by node distribution of seeds per pod and average weight (g) per seed in irrigated and non-irrigated crops in 1970. Seeds per pod, ——— ; average weight per seed (g), ———— .

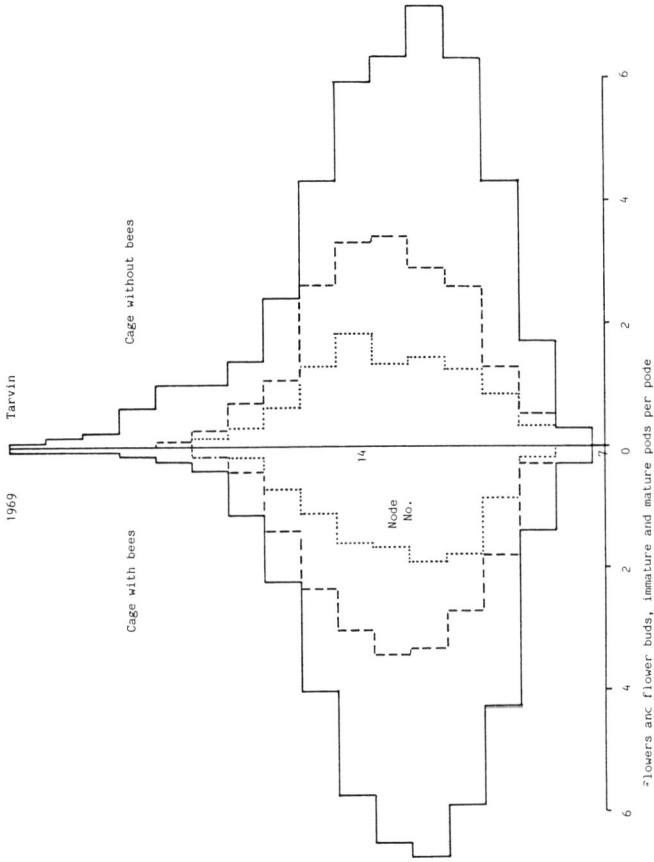

Fig 8 Node by node distribution of flowers, immature and mature pods of plants caged with and without bees in 1969, (variety, Tarvin). Flowers and flower buds per node, ⎯⎯⎯ ; immature pods per node, ‒‒‒ ; mature pods per node, ●●● .

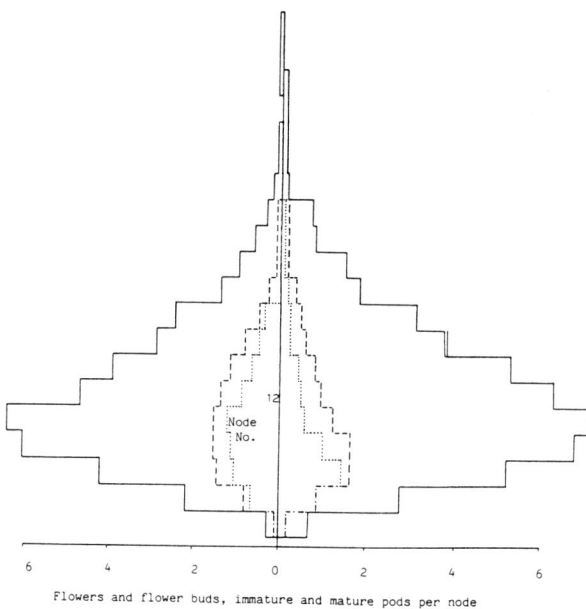

Flowers and flower buds, immature and mature pods per node

Fig. 9 Node by node distribution of flowers, immature and mature pods
of plants caged with and without bees in 1970, (variety, Herz Freya).
Flowers and flower buds per node, _____ ; immature pods per node, ---- ;
mature pods per node, ● ● ● .

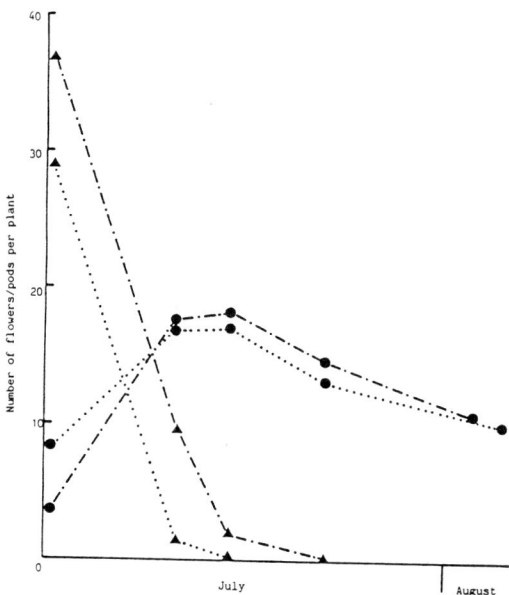

Fig. 10 Changes in number of flowers and pods in plants caged with and
without bees in 1969 (variety, Tarvin). With bees, flowers, ▲....▲ ;
with bees, pods, ● ...●; without bees, flowers, ▲-·-·- ▲;
without bees, pods, ●-·-·- ● .

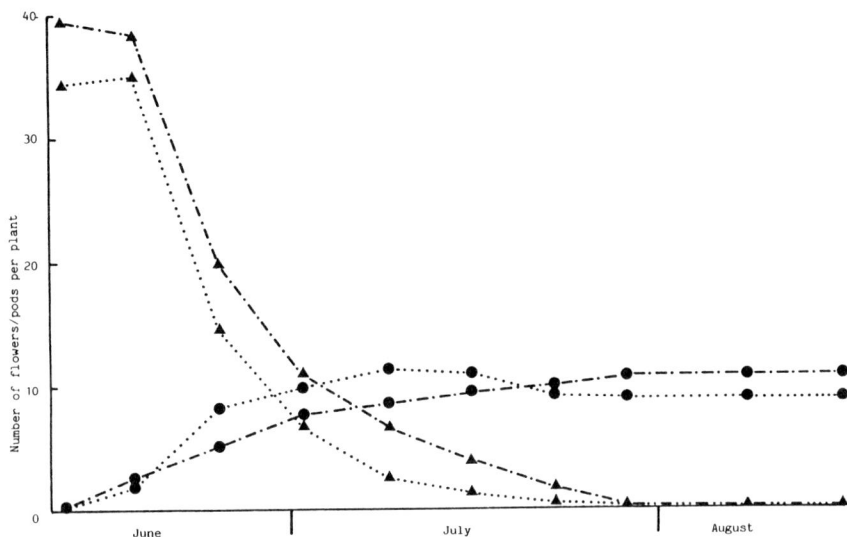

Fig. 11 Changes in number of flowers and pods in plants caged with and without bees in 1970 (variety, Herz Freya). With bees, flowers,▲ ..... ▲; with bees, pods,● ······ ●; without bees, flowers,▲ ‾·‾·‾·‾ ▲; without bees, pods,● ‑·‑·‑ ● .

REFERENCES

Addicott, F.T. and Lynch, R.S. 1955. Physiology of abscission. Ann. Rev. Pl. Physiol. 6, 211-238.

Bainbridge, A., Bardner, R., Cockbain, A.J., Day, J.M., Fletcher, K.E., Hooper, D.J., Legg, B.J., McEwen, J., Salt, G.A., Webb, R.M. and Wilding, N. 1977. Recent work at Rothamsted on factors limiting yields of field beans (Vicia faba L.). Proc. Symp. on the Production, Processing and Utilisation of the Field Bean (Vicia faba L.). (Ed. R. Thompson). Bulletin No.15 SHRI.

Chapman, G.P., Fagg, C.W. and Peat, W.E. 1979. Parthenocarpy and internal competition in Vicia faba L. Z. Pflanzenphysiol. 94, 247-255.

Day, W. and Legg, B.J. 1983. Water relations and irrigation response. In: "The Faba Bean - A basis for improvement". (Ed. P.D. Hebblethwaite). (Butterworths, London). pp. 217-231.

Drayner, J.M. 1959. Cross and self-fertility in field beans. J.agric. Sci. Camb. 53, 385-403.

El-Beltagy, A.S. and Hall, M.A. 1974. Effect of water stress upon endogenous ethylene levels in Vicia faba. New Phytology, 73, 47-60.

El-Beltagy, A.S. and Hall, M.A. 1975. Studies on endogenous levels of ethylene and auxin in Vicia faba during growth and development. New Phytology, 75, 215-244.

El Nadi, A.H. 1966. Effects of water stress on the growth and flowering of broad beans. Ph.d. Thesis, University of Nottingham.

El Nadi, A.H. 1969. Water relations of beans. Effects of water stress on growth and flowering. Expt. Agric., 5, 195-207.

El Nadi, A.H. 1970. Water relations of beans. II. Effects of differential irrigation on yield and seed size of broad beans. Exp. Agric. 6, (No.2), 107-111.

Free, J.B. 1966. The pollination requirements of broad beans and field beans (Vicia faba). J.agric. Sci. Camb., 66, 395-397.

Free, J.B. and Spencer-Booth, Y. 1961. The effect of feeding sugar syrup to honeybee colonies. J.agric. Sci. Camb., 57, 147-151.

Free, J.B. and Williams, I.H. 1976. Pollination as a factor limiting the yield of field beans (Vicia faba L.). J.agric. Sci. Camb., 87, 395-399.

French, B.K. and Legg, B.J. 1979. Rothamsted irrigation 1964-76. J.agric. Sci. Camb., 92, 15-37.

Fyfe, J.L. and Bailey, N.J.T. 1951. Plant breeding studies in leguminous forage crops. I. Natural crossing in winter beans. J.agric. Sci. Camb., 41, 371-378.

Hebblethwaite, P.D. 1981. The effects of water stress on the growth of Vicia faba L. In "Faba Bean Improvement: Proceedings of the International Faba Bean Conference, Cairo, March 7-11, 1981". (Ed. G. Hawtin and C. Webb). (Martinus Nijhoff/Dr. W. Junk, Netherlands). pp. 161-175.

Hodgson, G.L. and Blackman, G.E. 1956. An analysis of the influence of plant density on the growth of Vicia faba. I. The influence of density on the pattern of development. J.Expt. Bot. 7, 147-165.

Holden, J.H.W. and Bond, D.A. 1960. Studies on the breeding system of field bean Vicia faba (L). Heredity, 15, 175-192.

Hua, H. 1943. (Natural crossing in Vicia faba). Chin. J. Scient. Agric., 1, 63-65.

Ishag, H.M.H. 1969. Physiology of seed yield in Vicia faba L. Ph.D. Thesis, Univ. of Reading.

Jones, L.H. 1963. The effect of soil water gradients on the growth and development of broad beans (Vicia faba L.). Hort. Res., 3, 13-26.

Kambal, A.E. 1968. A study of the agronomic characters of some varieties of Vicia faba. Sudan Agric.J., 3, 1-10.

Kambal, A.E. 1969. Flower drop and fruit set in field beans, Vicia faba L. J.agric. Sci. Camb., 72, 131-138.

Krogman, K.K., McKenzie, R.G. and Hobbs, E.H. 1980. Response of faba bean yield, protein production, and water use to irrigation. Can. J. Pl. Sci., 60, 91-96.

Kogbe, J.O.S. 1972. Factors influencing yield variation of field beans (Vicia faba L.). Ph.D. Thesis, University of Nottingham.

McEwen, J., Bardner, R., Briggs, G.G., Bromilow, R.H., Cockbain, A.J., Day, J.M., Fletcher, K.E., Legg, B.J., Roughley, R.J., Salt, G.A., Simpson, H.R., Webb, R.M., Witty, J.F. and Yeoman, D.P. 1981. The effects of irrigation, nitrogen fertiliser and the control of pests and pathogens on spring-sown field beans (Vicia faba L.) and residual effects on two following winter wheat crops. J.agric. Sci. Camb., 96, 129-150.

Meadley, J.T. and Milbourn, G.M. 1970. The growth of vining peas. II. The effect of density of planting. J.agric. Sci. Camb., 74, 273-279.

Meadley, J.T. and Milbourn, G.M. 1971. The growth of vining peas. III. The effect of shading on abscission of flowers and pods. J.agric. Sci. Camb., 77, 103-108.

Myers, V.I., Corey, G.L., Lebaron, M. and McMaster, G. 1957. In "Idaho Agricultural Research Bulletin No.37". pp 1-16.

Penman, H.L. 1956. Evaporation: An introductory survey. Netherlands Journal of Agricultural Science, 4, 9-29.

Penman, H.L. 1962. Woburn Irrigation, 1951-1959. III. Results for
rotation crops. J.agric. Sci. Camb., 58, 365-379.
Penman, H.L. 1970. Woburn Irrigation, 1960-1968. VI. Results for
rotation crops. J.agric. Sci. Camb., 75, 89-102.
Riedel, I.B.M. and Wort, D.A. 1960. The pollination requirements of
field beans (Vicia faba L.). Ann. appl. Biol., 48 (1), 121-124.
Rowland, D.G. 1955. The problems of yield in field beans. Agric. Progr.
30, 137.
Rowland, D.G. 1958. The nature of the breeding system in the field bean
(V. faba L.) and its relationship to breeding for yield. Heredity,
12, 113-125.
Rowland, D.G. 1960. Fertility studies in the field bean (Vicia faba
L.). I. cross- and self-fertility. Heredity, 15, 161-173.
Rowland, D.G. 1961. Fertility studies in the field bean (Vicia faba
L.). II. Inbreeding. Heredity, 16, 497-508.
Scrivin, W.A., Cooper, B.A. and Allen, H. 1961. Pollination of field
beans. Outlook on Agriculture, 3, 69-75.
Soper, M.H.R. 1952. A study of the principal factors affecting the
establishment and development of field beans. J.agric. Sci. Camb.,
42, 335.
Sprent, J.I. 1972. The effects of water stress on nitrogen-fixing root
nodules. IV. Effects on whole plants of Vicia faba and Glycine max.
New Phytology, 71, 603-611.
Sprent, J.I., Bradford, A.M. and Norton, C. 1977. Seasonal growth
patterns in field beans (Vicia faba) as affected by population
density, shading and its relationship with soil moisture. J.agric.
Sci. Camb., 88, 292-301.
Wafa, A.K. and Ibrahim, S.H. 1960. The effect of the honeybee as a
pollinating agent on the yield of broad bean. Bull. Fac. Agric. Ain
Shams Univ. No.205, 17-28.

BEHAVIOUR OF FIELD BEAN LINES WITH A WATER TABLE MAINTAINED AT DIFFERENT LEVELS

A. Alvino[*], G. Zerbi[*], L. Frusciante[**], L.M. Monti[**]

[*] Irrigation Institute of National Council of Research
1085 Via Argine, Naples (Italy)
[**] University of Naples, Institute of Plant Breeding
Portici, Naples (Italy)

ABSTRACT

Fourteen accessions of field bean obtained from a previous screening were grown on different shallow water tables mantained from almost water-logging to optimal water table level. The performance of the lines generally confirmed the results obtained in the previous year. Manfredini, Chiaro T. L. and P.11 lines showed good resistence to high water table. Pod setting seemed to be the main trait linked to yield reduction in excess water conditions.

INTRODUCTION

Yields and biometric characteristics of 26 Vicia faba L. minor acces-sions in relation to four shallow water table levels have been reported previously (Alvino et al., 1983) with the aim of investigating the effect of waterlogging on the behaviour of the crop and on the differences among lines.

The results suggested a continuation of this study testing the more interesting varieties and lines. The present work reports the results obtained during a second year of study.

MATERIALS AND METHODS

This study was performed on the Experimental Farm of the CNR Irrigation Institute in the Volturno river plain (Southern Italy).

The field layout (Fig. 1) to provide a double inclined water table, described in the previous work, was modified by deepening the drainage ring from 1.0 m to 1.5 m, to provide a more inclined water table (Fig. 2). Four-teen lines of Vicia faba were screened from the previous year's experimen-tal material.

P.D. Hebblethwaite, T.C.K. Dawkins, M.C. Heath and G. Lockwood (eds.)
Vicia faba: Agronomy, Physiology and Breeding. ISBN 90-247-2964-5.
© 1984, Martinus Nijhoff/Dr W. Junk Publishers. Printed in The Netherlands.

96

Four lines (Manfredini, Chiaro T.L., P.11 and P.41) were chosen on the basis of overall good yielding ability, five lines (P.23, P.26, P.28, P.30, Scuro T.L.) were chosen on the basis of good yielding ability at deep water table level, two lines (P.85, P.21) were chosen on the basis of good yielding ability at high water table, two lines (Maris Bead, Maris Blaze)were chosen on the basis of their good total plant yield while grain production was very low, and one line (P.98) was chosen because of its steep response to decreasing water table level.

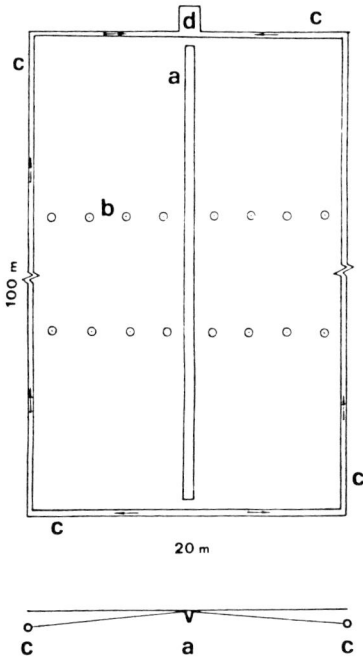

Fig.1. Experimental field: a) ditch; b)observation well;c)drainage ring; d)pumping well.

Field plots were arranged to test four water table levels of increasing depth in a split-plot design with four replicates. An individued plot was 3 m^2. Sowing was performed in December 1979 and the inclined water table was set up on January 25, 1980 when plant emergence was complete and was maintained until May 15. The ditch was then emptied and the water table level was left to equilibrate with the remainder of the field.

Plant height was measured at the following stages and dates: vegetative stage I, March 13; vegetative stage II, March 26; vegetative stage III, April 4; beginning of flowering, April 11; flowering, April 18; end of flowering, April 29; complete pod setting, May 30; and harvesting, June 23.

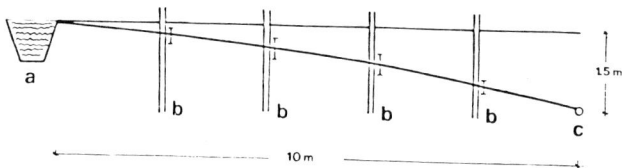

Fig. 2. Section of the experimental field and mean depth of water table.

At harvest, the dry weight of grain and total plant (plant + pods) were determined. Harvest index was calculated as the ratio between grain and total plant weight. Furthermore, average seed weight, number of pods per plant, number of ovules per pod, number of seeds per pod, and number of branches per plant were recorded. Crude protein percentage of seeds was determined from the plots with the highest and lowest water table.

RESULTS

The artificial water table level fluctuated during the trial because of rainfall. The values measured in the plots corresponding to the four water table levels are given in Fig. 3, together with the daily rainfall values. The average values of the water table levels over the period, expressed as distance from the surface, were 106, 65, 43 and 25 cm. These have been referred to in the text as levels 1, 2, 3 and 4 respectively.

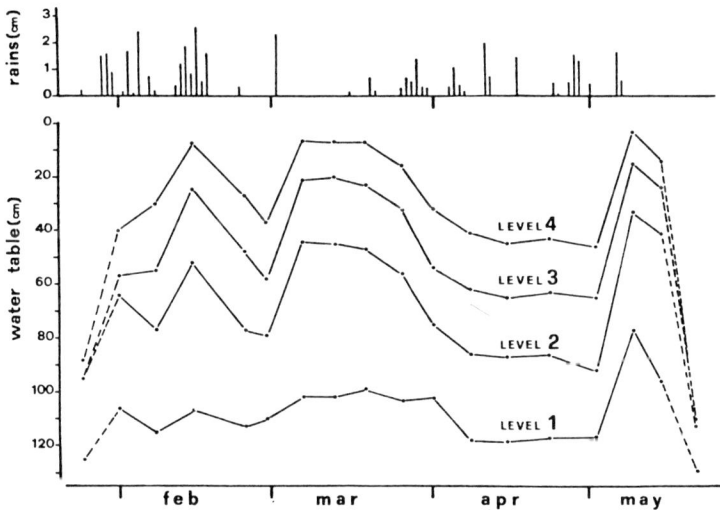

Fig. 3. Fluctuations in the level of the water table during the growing period due to the rainfall.

Growth during the crop cycle, expressed as plant height averaged over all lines, is reported in Fig. 4; only the highest water table tested (level 4) had a significant effect (P = 0.01) on plant height over the duration of crop growth. Table 1 reports data about yield and its components in relation to water table level averaged over all lines. Grain yield and

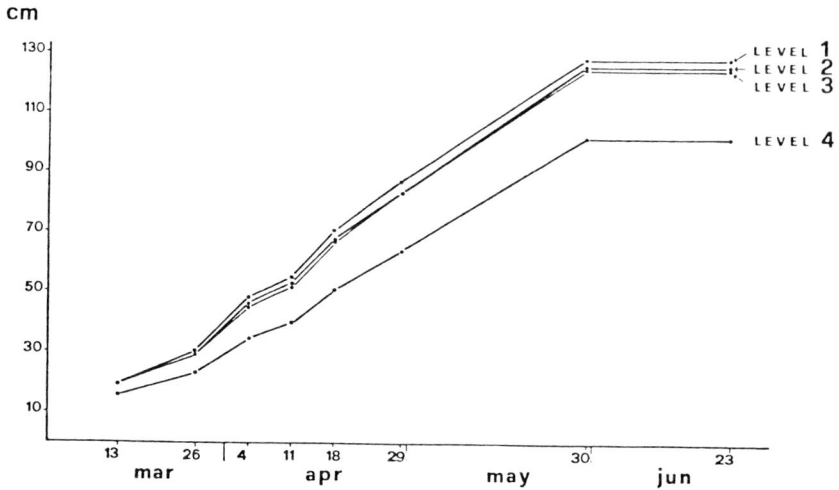

cm

Fig.4. Plant growth(height)during the crop cycle in response to the levels of the four water table tested

total dry matter were lower at the highest water table; harvest index was not influenced. Number of pods per plant was the yield component that had the largest effect on yield. Number of branches per plant was also influential. No effect was noted on average seed weight and on number of seeds per pod. A slight effect was remarked on the number of ovules per pod but not on pod fertility. The percentage of crude protein of grain was higher at water table level 1.

TABLE 1 . Yield and other parameters with relation to different levels of water table. Values are averaged over all the lines. Values followed by the same letter do not differ significantly at the 0.01 level according to Duncan's MRT.

	Water table			
	Level 1	Level 2	Level 3	Level 4
Grain yield (t . ha^{-1})	2.36 b	2.14 b	2.12 b	1.35 a
Total dry matter (t . ha^{-1})	8.59 b	7.70 b	7.80 b	4.74 a
Harvest index %	27.5	27.8	27.2	28.5
N° of pods per plant	8.33 B	6.55 AB	5.83 A	5.50 A
100 Seeds weight (g)	38.1	38.2	40.2	38.7
N° of branches per plant	1.65 B	1.65 B	1.65 B	1.45 A
N° of seed per pod	3.27	3.20	3.19	3.18
N° of ovules per pod	3.59 B	3.46 AB	3.36 A	3.57 AB
Seeds/ovules %	91.1	92.5	94.8	89.3
Crude protein (N x 6.25%) content of grain	33.6 b	–	–	28.5 a

Table 2 gives the values of correlation coefficients calculated between the yield and other characteristics for each of the four water table treatments. The only traits correlated with grain yield were the number of pods per plant and plant height at flowering. When the water table was highest (level 4) grain yield was correlated with total dry matter, average seed

TABLE 2. Correlation coefficients between grain yield and other characteristics of field beans grown with four different levels of water table.

	Water table			
	Level 1	Level 2	Level 3	Level 4
Grain yield – total dry matter	n.s.	n.s.	n.s.	.896 ***
" " – 100 seeds weight	n.s.	n.s.	.534 *	.863 ***
" " – N° of pods per plant	.639 *	.815 ***	.718 **	.742 **
" " – N° of branches per plant	n.s.	n.s.	n.s.	.814 ***
" " – N° of seeds per pod	n.s.	n.s.	n.s.	n.s.
" " – N° of ovules per pod	o n.s.	n.s.	n.s.	n.s.
" " – Height at march 13	n.s.	n.s.	n.s.	n.s.
" " – " " march 26	n.s.	.569 *	n.s.	n.s.
" " – " " april 4	n.s.	.726 **	.693 **	n.s.
" " – " " april 11	.659 *	.813 ***	.709 **	.555 *
" " – " " april 18	.718 **	.852 ***	.716 **	.751 **
" " – " " april 29	.560 *	n.s.	.559 *	.848 ***
" " – " " may 30	n.s.	n.s.	n.s.	.850 ***
" " – Crude protein content of grain	-.791 ***	–	–	.560 *

weight, number of pods per plant, number of branches per plant and final plant height. Crude protein content was negatively correlated in the case of the lowest water table and positively correlated in the case of the highest.

Yield performance of the different accessions differed over the four water table levels tested. This interaction is given in Table 3 where the accessions are presented in decreasing order using their average value as a reference. The lines Chiaro T.L., Scuro T.L., Manfredini and P.11 gave the highest yields at all water table levels tested. No significant interactions were noted in the analysis of the variance for the other traits considered; Table 4, therefore, reports these values averaged over the four water table treatments. Only grain protein content showed a different response at water table levels 1 and 4 (Table 5).

DISCUSSION

Results obtained confirm the data attained in previous tests (Williamson and Kriz, 1970; Alvino and Zerbi, 1982; Alvino et al., 1983). Reduced
growth and low production are related to an excess of soil water; this was

TABLE 3.Grain yield of different lines on the four water table levels.
The last column reports their average values.Differences
among w.t. treatments were tested by means of Duncan's MRT.

	Water table				$\bar{X}$
	Level 1	Level 2	Level 3	Level 4	
Chiaro T.L.	3.42 b	2.81 ab	2.60 ab	2.22 a	2.76
Scuro T.L.	3.20 b	2.62 ab	2.53 ab	1.87 a	2.56
Manfredini	2.89 b	2.76 ab	2.62 ab	1.94 a	2.55
P.11	2.87 a	2.67 a	2.60 a	2.04 a	2.54
P.41	2.53 b	2.20 ab	2.11 ab	1.36 a	2.05
P.85	2.42 b	2.33 b	2.04 b	1.03 a	1.95
P.98	2.18 a	2.09 a	1.98 a	1.30 a	1.89
P.23	2.36 b	2.27 b	1.73 b	0.81 a	1.79
P.30	2.40 b	2.11 b	1.69 ab	0.83 a	1.76
P.28	2.22 b	2.07 b	1.91 b	0.79 a	1.75
P.26	2.76 c	2.00 bc	1.58 b	0.51 a	1.71
P.21	2.04 a	1.56 a	1.47 a	1.17 a	1.56
Maris Bead	1.78 a	1.78 a	1.41 a	1.27 a	1.56
Maris Blaze	1.82 b	1.64 ab	1.60 ab	0.78 a	1.46

TABLE 4. Measured characteristics averaged over the four water table levels.
Means within the columns were tested following the Duncan's MRT.

Lines	Whole plant (d . m) (t.ha^{-1})	Harvest index (%)	100 seeds weight (g)	N° of pods per plant (n)	N° of branches per plant (n)	N° of seeds per pod (n)	N° of ovules per pod (n)	Seeds ovules %
Chiaro T.L.	10.45 d	24.0 b	42.5 de	8.26 d	1.74 ef	3.52 de	3.78 c	93.6 bc
Scuro T.L.	9.53 bc	24.1 bc	39.6 cd	7.29 cd	1.72 def	3.40 cde	3.58 abc	95.1 c
Manfredini	8.60 b	26.7 bcd	49.9 f	8.11 d	1.81 f	3.08 abc	3.75 c	82.1 a
P.11	10.23 cd	22.6 b	44.9 e	6.49 bcd	1.78 ef	3.33 bcde	3.56 abc	93.7 bc
P.41	5.95 a	31.4 de	39.5 bcd	7.17 cd	1.58 abcde	3.12 abc	3.43 ab	90.9 bc
P.85	5.29 a	33.7 e	34.0 a	7.20 cd	1.40 a	3.15 abc	3.37 a	93.8 bc
P.98	5.56 a	30.5 de	39.6 cd	7.08 cd	1.44 ab	3.01 ab	3.36 a	89.7 bc
P.23	5.43 a	30.5 de	34.3 ab	6.80 cd	1.49 abcd	3.15 abc	3.35 a	94.0 bc
P.30	5.03 a	30.9 de	32.9 a	6.87 cd	1.58 abcde	3.19 abcd	3.41 a	93.9 bc
P.28	5.34 a	28.8 cd	35.7 abc	6.12 bcd	1.47 abc	2.92 a	3.30 a	88.5 ab
P.26	5.43 a	27.2 bcd	32.0 a	6.96 cd	1.57 abcde	3.08 abc	3.40 a	90.4 bc
P.21	5.45 a	26.7 bcd	32.5 a	5.39 abc	1.50 abcd	3.13 abc	3.43 ab	91.4 bc
Maris Bead	9.20 b	15.0 a	40.6 cde	4.30 ab	1.65 bcdef	3.57 e	3.72 bc	95.9 c
Maris Blaze	9.50 bc	14.6 a	45.2 e	3.69 a	1.68 cdef	3.32 bcde	3.53 abc	93.8 bc

TABLE 5. Protein content of different lines at levels 1 and 4. Differences between w.t. treatments were tested by means of Duncan's MRT

| | Water table | |
	Level 1	Level 4
Chiaro T.L.	32.1 b	29.3 a
Scuro T.L.	31.4 a	30.1 a
Manfredini	31.9 b	28.0 a
P.11	31.9 b	28.5 a
P.41	33.7 b	27.2 a
P.85	34.0 b	28.5 a
P.98	34.5 b	30.1 a
P.23	36.5 b	26.4 a
P.30	34.1 b	27.0 a
P.28	34.3 b	26.6 a
P.26	33.3 b	25.4 a
P.21	34.5 b	27.0 a
Maris Bead	34.4 a	32.5 a
Maris Blaze	33.5 a	31.9 a

the case when the average water table was at 25 cm. When the water table was deeper (from 43 to 106 cm) the differences in yields and their components between treatments were slight. Within the critical water table level (level 4) the different accessions demonstrated a range of behaviour with high yielding (Chiaro T.L., P.11, Manfredini, Scuro T.L.) and low yielding accessions (P.26, Maris Blaze, P.28, P.23, P.30). Under opposite conditions (level 1) the range was reduced and the ratio between the lowest and the highest yielding accessions (Maris Bead and Chiaro T.L., respectively) was 1:1.9 compared to the ratio of 1:4.4 in the first case. The existence of lines capable of producing good yields at high water table level explains the significant correlations between yield and yield components reported in Table 2. At levels 1 and 2 yield depended on number of pods per plant but seemed to be independent from vegetative growth, average seed weight, number of branches per plant and number of seeds per pod. This confirms the hypothesis stated in the first study; that the yield reduction is due to decreased pod setting.

The behaviour of the lines in comparison with the results of the previous studies confirmed that (i) Chiaro T.L., Manfredini and P.11 demonstrated good yielding ability over the four water table levels, (ii) P.23, P.26, P.28, P.30 and Scuro T.L. demonstrated their high productivity when grown at lower water tables, (iii) Maris Bead and Maris Blaze produced low grain yields notwithstanding good vegetative development (Harvest indices of around 15% were noted for these two varieties and this value is very low compared with values found in the literature (Hebblethwaite, 1981)), and finally that (iv) the other four accessions demonstrated a different behaviour with respect to the previous study which could be attributed to different conditions of water table and weather between the two years.

REFERENCES

Alvino, A. and Zerbi, G. 1982. Esperienze in lisimetri per la valutazione
    dell'influenza del livello di falda sull'accrescimento e sulla resa
    di tre cultivar di Vicia faba L. Riv. di Agronomia, 3, 309-316.
Alvino, A. et al. 1983. Evaluation of field bean lines grown with a shallow
    water table maintained at different levels. Field Crops Res., 6, 179-
    188.
Hebblethwaite, P. 1981. The effect of water stress on the growth, develop-
    ment and yield of Vicia faba L. In "Faba bean improvement" (Ed. Hawtin
    G. and Webb C.) Proceedings of the Faba bean Conference. Cairo, March
    7-11, 1981.
Williamson, R.E. and Kriz, G.J. 1970. Response of agricultural crops to
    flooding, depth of water table, and soil gaseous composition. Amer.
    Soc. Agr. Eng., Trans.,13,216-220.

EVALUATION OF SEED METERING DEVICES

FOR SOWING GRAIN LEGUMES

G.G. Rowland and R.M. Kehrig

Crop Development Centre
University of Saskatchewan
Saskatoon   S7N 0W0   CANADA

ABSTRACT

Fifteen seed metering devices (seedcups) were tested on lentils (Lens culinaris), peas (Pisum sativum), faba beans (Vicia faba), chickpeas (Cicer arietinum) and a small-seeded broad bean (Vicia faba) in laboratory and field tests. Data on seed damage, seed germination and field stands indicate that seed metering systems which allow for modification of the metering device gave the greatest flexibility in types of crops they could handle. Furthermore, seedcups which allowed some adjustments proved more flexible than unadjustable systems that generally would work well only on one or two crops.

INTRODUCTION

Ever since the western plains of Canada were first cultivated in the late nineteenth century cereal grains have been the main crops produced. Grain legume production has been limited to field peas (Pisum sativum L.), soybeans (Glycine max L.) and Phaseolus beans. However, since 1970 lentils (Lens culinaris Medic.), faba beans (Vicia faba L.) and broad beans (V. faba L. Major) have been grown. Furthermore, research is underway at the Univ. of Saskatchewan that may add chickpeas (Cicer arietinum L.) to this list.

Sowing equipment used by farmers in western Canada is well suited to small cereal grains but difficulties can arise when sowing larger-seeded grain legumes. While problems with lentils are minimal, farmers have complained that their drills or discer seeders badly crack faba bean seed and do not distribute the seed evenly. Since chickpeas and small-seeded broad beans are even larger and more irregular in their shape, it is anticipated farmers will have similar problems in sowing these crops.

The types of drills and discer seeders used in western Canada are quite varied in type and varieties of seed metering devices (seedcups) used. The problems of seed cracking and poor seed distribution are primarily due to the seedcups used in these drills and discer seeders. Recommendations are not available to farmers as to what seed metering system might work best for a particular grain legume and it was for this

P.D. Hebblethwaite, T.C.K. Dawkins, M.C. Heath and G. Lockwood (eds.)
Vicia faba: Agronomy, Physiology and Breeding. ISBN 90-247-2964-5.
© 1984, Martinus Nijhoff/Dr W. Junk Publishers. Printed in The Netherlands.

reason that this study was undertaken.

MATERIALS AND METHODS

The majority of the seedcups tested were from drills or discer seeders commonly used in western Canada, but others were included to see if better systems exist elsewhere in the world. Twelve companies supplied 15 types of seed metering devices and many of these could be set up in a number of different ways. A list of seedcups and test-crops is found in Table 1.

Seed metering systems are of generally two basic kinds; the internal force feed and the external force feed. The internal force feed system has a rotating feed wheel containing seed cells or flutes on its interior surface which meters the seed. External force feed systems are of two types; the Hoozier (fluted roll) and the Siedersleben (studded roll). In both types the seed is metered on the exterior of the feed roller.

A single row seed drill with a flexible drive system was designed and built for testing the seedcups both in the laboratory and the field. The seedcups were mounted on individual metal plates which could then be fastened to a small seedbox on the drill. The drill had a variable speed unit which allowed testing at different shaft speeds (r.p.m.).

For the laboratory tests the drill was driven by a 0.5 horsepower electrical motor. The shaft speeds tested on each seedcup and crop were 15, 20, 25 and 30 r.p.m. The crops tested were Laird lentils, Century peas, an experimental line of faba beans - 77RM1316, a bulk selection of kabuli chickpeas and a Chinese broad bean.

In the laboratory three repetitions (runs) of each seedcup-crop-shaft speed combination were performed. Seed was collected from each run and evaluated for the amount of seed-splitting and the number of damaged seeds were counted. Thirty-three captan (N-trichloromethylmercapto-4-cyclohexene-1, -2-dicarboximide) treated seeds of each combination were sown in vermiculite, grown in a controlled environment chamber for 15 days at 21°C after which germination counts were taken.

If a particular seedcup produced an excessive amount of splitting and damage in one crop, it was not tested on crops of a larger seed size. Similarly, if a seedcup did not meter a crop (e.g., because of small seed) it could not be tested.

In the field, each seedcup was tested with each crop (provided it

TABLE 1   Equipment manufacturers and their seed metering devices
classified as to type and crops tested on.

Manufacturer	No.	Seed metering device[+]	Type	Crops tested[*]
Aitchison	1	SFS	smooth pad	l,p,f,c,b
	2		serrated pad	l,p,f,c,b
Canadian Co-op	3	IFF	double run small	l,p
Implements	4		double run large	l,p,f
	5	EFF(H)	fluted roller)	l,p,f,c,b
Crustbuster	6	EFF	slot feed	l,p,f
Duncan	7	EFF(H)	fluted roller	l,p,f,c,b
International Harvester	8	EFF(H)	fluted roller	l,p,f,c,b
John Deere	9	IFF	double run small	l
	10		double run large	l,p,f,c,b
	11	IFF	single run	l
	12	EFF(H)	fluted roller	l,p,f,c,b
Lilliston	13	EFF(S)	medium wheel	l,p
	14		intermediate wheel	f,c,b
	15		coarse wheel	c,b
Massey-Ferguson	16	IFF	double run small	l,p,f
	17		double run large	l,p,f
Midland	18	EFF(S)	single run	p,f,c,b
Morris	19	EFF(S)	normal grain wheel	l,p,f,c,b
	20		bean wheel	l,p,f,c,b
	21		custom bean wheel	f,c,b
	22		fertilizer wheel	l,p,f,c,b
Tye	23	IFF	single run	l,p,f,c,b
Versatile-Noble	24	EFF(H)	fluted roller	l,p,f

[+]SFS=sponge feed system, IFF=internal force feed, EFF=external force feed,
(H)=Hoozier type, (S)=Siedersleben type.

[*]l=lentil, p=peas, f=faba beans, c=chickpeas, b=broad beans

metered seed without severe damage in the laboratory tests) at what was
ascertained to be the optimum shaft speed.   The single, 20 meter rows of
each crop-seedcup combination were replicated three times.   Counts were
taken in each row of the numbers of seedlings emerging and the stand
uniformity was rated on a scale of 1-5 with 1 being uniform and 5 being
irregular with large gaps and clumping of the seed.

An analysis of variance was performed on the data from within each
crop.   Where significant differences occurred, means were ranked by
Duncan's multiple range test.

RESULTS

The various shaft speeds used in the laboratory tests had, with one exception, no effect on any of the seed characters studied. This is in agreement with a Prairie Agriculture Machinery Institute of Canada study (unpublished results) on faba beans which found that speed of the fluted rollers or feed wheels in three seedcups tested had relatively little effect on seed damage. The one exception was for broad beans where shaft speed was significant (P=0.05), apparently due to the Duncan seedcup. The seedcup by shaft speed interaction was not significant.

While chickpeas were included in the field study their emergence was so poor that no data were taken. In contrast to the laboratory tests, seeds sown in the field were not treated with captan as it was assumed that soil temperatures were too warm to warrant the treatment.

Significant differences among seedcups in each crop occurred for some of the characters examined. Results for the significant characters are presented in Tables 2-4. Differences in seed damage for broad beans and germination percentage for lentils and chickpeas were found to be not significant.

Aitchison smooth pad and serrated pad

This foam-rubber disc system minimized the amount of seed damage within each crop, but apart from faba beans seed distribution was very uneven. With the exception of broad beans, the uniformity of seed distribution was better with the serrated disc. For broad beans the serrated disc decreased seed germination and uniformity of feeding was very poor.

Canadian Co-operative Implements internal force feed

This seedcup was only tested with the smaller seeded crops. The small double run was not suited to any of the crops, while the large double run was acceptable with lentils and faba beans.

Canadian Co-operative Implements external force feed

While seed damage was acceptable on lentils and peas and germination was reasonable, the stands were uneven for all crops.

Crustbuster external force feed

This seedcup produced acceptable results only with lentils. In the

other crops the amount of seed damage and irregular spacing of the seed
was high.

Duncan external force feed

The Duncan seedcup performed best with peas resulting in minimal
seed damage, little reduction in germination and a uniform stand. It was
marginally acceptable on faba beans, but performed poorly with other
crops.

TABLE 2   The percentage of damaged[+] seeds in each crop after feeding
through the various seed metering devices in the laboratory.

Seed metering device	No.	Test-crop[*]			
		Lentils	Peas	Faba beans	Chickpeas
Aitchison	1	11ab	6ab	9bc	4bc
	2	12bcd	3a	9bc	4bc
Canadian Co-op	3	17g	15f	–	–
Implements	4	12bcd	10e	12d	–
	5	12bcd	7bc	12d	7ef
Crustbuster	6	12bcd	8cd	19f	–
Duncan	7	14cdef	6ab	11cd	6de
International Harvester	8	14cdef	7bc	12d	7ef
John Deere	9	13cde	–	–	–
	10	14cdef	6ab	10bcd	8f
	11	17g	–	–	–
	12	12bcd	7bc	12d	7ef
Lilliston	13	10a	14f	–	–
	14	–	–	6a	3a
	15	–	–	–	4bc
Massey-Ferguson	16	13cde	8cd	16e	–
	17	15ef	10e	11cd	–
Midland	18		8cd	11cd	5cd
Morris	19	16fg	9d	10bcd	4bc
	20	12bcd	7bc	10bcd	3ab
	21	–	–	10bcd	4bc
	22	14cdef	6ab	9b	4bc
Tye	23	13cde	7b	11cd	6de
Versatile-Noble	24	20h	7bc	11cd	–

[+]Split seeds were considered separately.
[*]Means within each column followed by the same letter are not significantly different at the 0.05 level as determined by Duncan's multiple range
test.

International Harvester external force feed

For all crops, this seedcup produced the most uniform stand of those

seedcups with a Hoozier type system. While it severely reduced germination of peas and increased damage to the broad beans, field stands were
quite uniform.

John Deere internal force feed (double run)

The small side was better than the large side for lentils. The
large run was acceptable for peas and faba beans, but not for chickpeas
or broad beans.

John Deere internal force feed (single run)

While seed damage to lentils was high with this seedcup, stand
uniformity was good.

TABLE 3   The percentage of seeds germinating in each crop after feeding
through the various seed metering devices in the laboratory.

Seed metering device	No.	Test-crop *		
		Peas	Faba beans	Broad beans
Aitchison	1	90ab	95b	89b
	2	92a	93bcd	84cdef
Canadian Co-op Implements	3	74g	–	–
	4	79f	92cd	–
	5	88bcd	93bcd	92a
Crustbuster	6	89abcd	87e	–
Duncan	7	89abcd	92cd	87bcd
International Harvester	8	84e	92cd	86bcde
John Deere	10	86cde	93bcd	84def
	12	86cde	92cd	87bcd
Lilliston	13	72g	–	–
	14	–	92cd	83ef
	15	–	–	86bcde
Massey-Ferguson	16	86cde	88e	–
	17	86cde	93bcd	–
Midland	18	89abcd	93bcd	84cdef
Morris	19	86cde	92cd	89b
	20	90ab	91d	84cdef
	21	–	94bc	89b
	22	86cde	97a	89b
Tye	23	89abcd	93bcd	83ef
Versatile-Noble	24	81f	93bcd	–

*
Means within each column followed by the same letter are not significantly different at the 0.05 level as determined by Duncan's multiple range
test.

John Deere external force feed

While this seedcup significantly increased seed damage in faba beans and chickpeas, and reduced seed germination in peas, faba beans and broad beans, it produced uniform field stands for all crops.

Lilliston external force feed

The medium wheel was good for lentils, but severely increased seed damage, decreased germination and produced irregular stands of peas. The intermediate wheel was not suitable for lentils or peas but was suitable for faba beans. It also produced little damage on chickpeas. However, it produced an irregular stand of broad beans and the coarse wheel performed only slightly better.

TABLE 4   Stand uniformity ratings[+] of field plots sown with the various seed metering devices.

| Seed metering device | No. | Test crop[*] | | | |
		Lentils	Peas	Faba beans	Broad beans
Aitchison	1	4.7g	4.0e	3.0e	3.0cde
	2	2.3cd	3.0d	1.7bc	4.3g
Canadian Co-op Implements	3	3.0ef	4.3e	–	–
	4	1.7b	4.0e	2.0bcd	–
	5	3.0ef	2.7cd	2.7de	4.0fg
Crustbuster	6	2.3cd	4.7e	5.0f	–
Duncan	7	3.3f	2.0bc	2.0bcd	4.0fg
International Harvester	8	2.7de	2.0bc	2.0bcd	2.3c
John Deere	9	2.3cd	–	–	–
	10	2.7de	2.3bcd	2.3cde	4.3g
	11	2.0bc	–	–	–
	12	2.0bc	2.3bcd	1.3ab	2.3c
Lilliston	13	1.7b	4.3e	–	–
	14	–	–	2.3cde	4.3g
	15	–	–	–	4.0fg
Massey-Ferguson	16	2.7de	4.0e	–	–
	17	3.3f	3.0d	3.0e	–
Midland	18	–	2.3bcd	2.3cde	4.0fg
Morris	19	3.3f	4.3e	2.3cde	2.7bcd
	20	2.0bc	2.0bc	2.7de	2.7bcd
	21	–	–	2.0bcd	2.0b
	22	2.7de	2.0bc	1.7bc	3.3def
Tye	23	2.0bc	1.7b	1.7bc	2.7bcd
Versatile-Noble	24	2.0bc	2.7cd	2.3cde	–
Control	25	1.0a	1.0a	1.0a	1.0a

[+]Ratings were 1 to 5; 1=uniform spacing, 5=irregular spacing.
[*]Means within each column followed by the same letter are not significantly different at the 0.05 level as determined by Duncan's multiple range test.

Massey-Ferguson internal force feed (double run)

Neither the small or large run could be tested on chickpeas and broad beans due to severe seed cracking. Generally seed damage was high, germination was reduced and stands were irregular with this seedcup.

Midland external force feed

Lentils were not metered properly with this seedcup and could not, therefore, be tested. The amount of seed damage was acceptable with peas, faba beans and chickpeas and germination was reasonable for peas and faba beans. This seedcup also handled peas and faba beans quite well in the field, but stands of broad beans were very uneven.

Morris external force feed

This Siedersleben type system was tested with four different wheels. The normal grain wheel did quite well with chickpeas and broad beans as far as seed damage and germination were concerned and stand uniformity was acceptable in the field for broad beans. The bean wheel produced a minimum of damage on all crops and stand uniformity in the field was quite good. This was the best wheel for lentils. The custom bean wheel worked very well for faba beans and broad beans and caused very little damage with chickpeas. The fertilizer wheel was very good for all characters on peas and faba beans and there was little damage to chickpeas.

Tye internal force feed

This seedcup was quite good with lentils, peas and faba beans and slightly poorer with broad beans, where seed damage was the greatest problem.

Versatile-Noble external force feed

This seedcup produced quite severe visible damage on lentils but field stands were acceptable. Field stands were fairly uniform with peas and faba beans, but germination was significantly decreased in peas.

DISCUSSION

While some of the seedcups which could not be modified by adding different wheels or rollers, worked quite well on a range of crops, those systems which could be modified performed the best. The Siedersleben system used by Morris, was the most flexible on the five crops tested and a wheel could always be found that would do an acceptable job. The same applied for the Siedersleben system used by Lilliston with the exception that peas could not be handled by any of the wheels. The Aitchison system, while producing a minimum of seed damage, gave problems in metering peas, chick peas and broad beans.

Of the non-modifiable seedcups the International Harvester, John Deere external force feed and the Tye performed the best. The Versatile-Noble was suitable for lentils and faba beans and the Midland, Duncan and John Deere double large run internal force feed seedcups were suitable for peas and faba beans. Generally, seedcups with adjustable feed gates performed better than those without adjustable gates.

The field portion of the study needs to be repeated and modifications made to the drill so that actual seed numbers going into the ground may be counted. The seed must also be treated with a fungicide to prevent seed rotting.

The support of the Saskatchewan Agricultural Research Fund is acknowledged with thanks.

THE EFFECTS OF POOR SOIL PHYSICAL CONDITIONS ON THE GROWTH AND YIELD OF <u>VICIA FABA</u>

T.C.K. Dawkins, J.C. Brereton

University of Nottingham, School of Agriculture,
Sutton Bonington, Loughborough, Leics. LE12 5RD, U.K.

ABSTRACT

The effects of poor soil physical conditions on a number of crops including beans is under investigation at the University of Nottingham. A preliminary trial conducted in 1982 showed that the passage of a tractor wheelings across the entire plot post-sowing could reduce the yield of beans by up to 48% depending upon sowing date.

Further experiments were conducted in 1983 to investigate the causes of such yield reductions. Significant reductions in crop height, dry matter and leaf area index occurred as a result of soil compaction but there was no difference in either plant population or combine yields at final harvest. Nevertheless, hand harvesting indicated a significant yield depression as a result of compaction.

INTRODUCTION

Faba beans (<u>Vicia faba</u> L.) are a well established crop in the U.K. although the area under cultivation has always been a small fraction of the total arable area.

Traditionally, faba beans have been regarded as a crop most suited to heavy soils but in practice a wide range of soil types are used. The crop is not considered to be very demanding in terms of seed bed preparation and satisfactory crops have frequently been produced by ploughing in the seed. However, with increasing pressure on the land and falling soil organic matter levels combined with increasing machinery size the problems of soil compaction cannot be ignored.

A preliminary study conducted at the University of Nottingham School of Agriculture in 1982 showed that the yield of field beans could be reduced by up to 48% as a result of topsoil compaction induced by one passage of a tractor wheeling over the entire plot post sowing. The loss of yield could not be accounted for by loss of plants but rather to a fundamental modification of plant growth habit. Studies by Kays, Nicklow and Simons (1974) demonstrated that impeded bean roots produced ethylene which is known to be a regulator of cell growth. Wilkins, Wilkins and Wain (1976) demonstrated that the growth of pea roots could be

P.D. Hebblethwaite, T.C.K. Dawkins, M.C. Heath and G. Lockwood (eds.)
Vicia faba: Agronomy, Physiology and Breeding. ISBN 90-247-2964-5.
© 1984, Martinus Nijhoff/Dr W. Junk Publishers. Printed in The Netherlands.

significantly improved in compact soil by the incorporation of D.I.H.B. (3.5. Di-iodo hydroxy benzoic acid) and subsequent studies have revealed that D.I.H.B. was a non-specific inhibitor of ethylene bio-synthesis.

There are few reports of the effects of chemical alleviation of soil compaction under field conditions and so attempts were made to assess the potential of D.I.H.B. as a means of reducing the severity of soil compaction in faba beans.

This study demonstrated the effect of topsoil compaction with and without D.I.H.B. incorporation on the growth and yield of faba beans.

MATERIALS AND METHODS

The trial was conducted on a coarse gravelly sandy loam (Arrow series) overlying Keuper Clay at 0.3 - 1.0 m depth. The trial consisted of three treatments; a surface compacted treatment (SC) achieved by progressively wheeling the entire plot with one pass of a tractor post sowing, and identical treatment (SCD) but sprayed with 50g ai ha^{-1} D.I.H.B. prior to emergence of the crop and a rolled treatment (RL) achieved by rolling the plot three times post sowing with a two-tonne roller. A control treatment (C) equivalent to standard farm practice was also included. Details of experimental management techniques are given in Table 1.

TABLE 1   Experimental and management techniques 1983

Sowing date	11 March 1983
Sowing method	Nordsten drill
Variety	Maris Bead
Experimental design	Randomised block
Plot size	2.5m x 18m
Number of replicates	4
Final harvest area	2.5m x 6m
Inter row spacing	11.9cm
Previous crop	Potatoes
Seed bed fertilizer	None
Weed control	Trietazine + Simazine (Remtal SC) 2.4l ha^{-1}
Disease control	Benomyl (Benlate) 1kg ha^{-1}
Insect control	Pirimicarb (Aphox) 0.02kg ha^{-1}
Date of final harvest	16 August 1983

Detailed measurements of crop growth were made using destructive growth analysis techniques. Final yield and yield components were estimated from a ten plant subsample and from a 2.5m x 6m harvest area.

Soil conditions were measured using a Bush penetrometer (Anderson, Pidgeon, Spencer and Parks, 1980) at 3.5 cm increments to a depth of 22 cm. Effective extraction depths, indicative of zones of root activity were determined from neutron probe records (Bell, 1969; McGowan, 1974). At peak anthesis profile pits were dug to a depth of 90 cm on the SC and C treatments and root distribution was mapped using a modification of the technique of Böhm (1976). Soil cores were also extracted using a Jarrat auger and the roots from selected horizons washed onto a 1 mm sieve and root length determined.

RESULTS

Soil conditions

Soil compaction from tractor wheelings inreased soil strength between 7.0 cm and 10.5 cm in the SC and SCD treatments (Figure 1). The RL treatment did not significantly increase soil strength over the control (C).

Emergence

Plants m^{-2} at establishment and final harvest were unaffected by any of the treatments although plant population declined between sowing and final harvest (Table 2)

TABLE 2   Plants per m^2 at full establishment and final harvest

	Plants per m^2				
	SC	C	SCD	RL	LSD P < 0.05
11.5.83	58	50	44	44	NS
16.8.83	38	37	34	40	NS

The effects of compaction on crop growth and yield

Plant height

Plant height was significantly reduced throughout the growing season in the SC and SCD treatments (Figure 2).

The SCD treatment did not increase crop height over the SC treatment. The RL treatment had little effect of crop height and so measurement was discontinued. A severe thunderstorm damaged some of the plants two weeks before harvest thus preventing crop height measurement at final harvest.

Leaf area index (L.A.I.) and dry weight per m^2

Significant decreases in L.A.I. (Figure 3) and dry matter production per m^2 (Figure 4) had occurred in the SC and SCD treatments by the end of May. It seems probable that the effects of compaction on L.A.I. and dry matter production per m^2 were exacerbated by drying conditions combined with a restricted root development from late May onwards (Figure 4). D.I.H.B. did not improve either L.A.I. or dry matter production per m^2 over the SC treatment.

Leaf area (Table 3) and dry matter per plant (Table 4) were significantly reduced by the SC and SCD treatments from late May onwards indicating a fundamental change in plant growth habit induced by compaction.

TABLE 3   Leaf area per plant as influenced by compaction

	Leaf area per plant (cm^2)					
	SC	C	SCD	RL	LSD	Sig
11.5.83	70	85	60	70	NS	
27.5.83	179	223	151	211	57	***
3.6.83	332	460	345	431	63	***
20.6.83	848	990	769	-	NS	NS

TABLE 4   Dry weight per plant as influenced by compaction

	Dry weight per plant (g)					
	SC	C	SCD	RL	LSD	Sig
11.5.83	0.5	0.6	0.4	0.5	NS	
27.5.83	1.2	1.5	1.0	1.3	0.2	*
3.6.83	2.2	2.9	2.0	2.6	0.6	***
13.6.83	5.1	7.6	6.0	8.5	2.9	**
17.6.83	6.9	10.2	8.1	-	1.7	*
20.6.83	10.2	11.6	9.7	-	NS	
16.8.83	20.7	28.7	23.5	26.5	5.8	**

The RL treatment had no effect on these two components.

Root growth

Analysis of effective extraction depth by roots from neutron probe data revealed that compaction delayed root development and reduced the extent of rooting (Figure 5).

The period of rapid root development (80-110 days after sowing) coincided with significant decreases in LAI and dry matter production per m^2 and this may have resulted from diversion of assimilates from the above ground components into root growth.

Inspection of the profile pits (Figure 6) demontrated a less extensive root system in the SC treatment but this technique must be viewed with caution as it is difficult to replicate.

The root length distribution analysis revealed that there was a reduction in root length in the SC and SCD treatments in the 30cm - 70cm zone (Figure 7).

Yield and yield components

Combine harvested yields were reduced by the SC treatment in 1982 but not in 1983 (Table 5). Yield estimates from growth analysis showed the converse (Table 5). Harvest index and mean seed weight were unaffected by the SC treatment in either year. The D.I.H.B. treatment did not increase yields over the SC treatment and had no effect on harvest index or mean seed weight.

Analysis of yield components in 1983 from the SC treatment suggest that the reduced yield obtained by subsample growth analysis might be accounted for in terms of a reduced number of podding nodes per plant resulting in fewer pods and seeds per plant. Pods per podding node and seeds per pod were unaffected (Table 6).

TABLE 5   Yields of beans 1982-83

		SC	C	SCD	RL	LSD	Sig
Combine yields	1982	2.23	3.13	–	–	0.50	*
t ha^{-1} @ 10% MC	1983	2.8	2.5	2.0	2.6	NS	
Growth analysis	1982	234	273	–	–	NS	
gm^{-2} @ 10%	1983	471	637	494	637	165.3	**
Yield per plant	1982	5.3	6.3	–	–	NS	
@ 10%	1983	12.6	17.5	14.6	15.9	0.94	**
HI	1982	0.51	0.46	–	–	NS	
	1983	0.56	0.56	0.57	0.55	NS	
Mean seed wt. (g)	1982	0.33	0.35	–	–	NS	
	1983	0.29	0.30	0.30	0.29	NS	

The RL treatment had few consistent effects on any component.

TABLE 6   Yield components 1983

	SC	C	SCD	RL	LSD	Sig
Pod nodes m^{-2}	147.1	183.4	147.0	168.1	NS	
Pod nodes plant^{-2}	3.90	4.97	4.32	4.15	0.81	*
Pod number m^{-2}	375	462	365	469	793	*
Pods per plant	9.92	12.65	10.70	11.72	0.68	*
Seeds m^{-2}	1418	1849	1445	1917	461	**
Seeds plant^{-1}	38.1	51.1	42.6	48.2	45.1	NS
Pods per pod node	2.5	2.5	2.5	2.9	NS	
Seeds pod^{-1}	3.9	4.0	4.1	4.1	NS	

DISCUSSION AND CONCLUSIONS

Top soil compaction can reduce the yield of Vicia faba in the U.K.
This type of compaction may result from post-sowing operations such as
herbicide application and may be most apparent where tractors have turned
on headlands.

Reductions in yield occurred as a result of fewer podding nodes per plant and hence pods and seeds per plant. Harvest index, seeds per pod, plants per m^2 and pods per podding node were unaffected by compaction. Compaction also reduced leaf area index, dry matter accumulation per m^2 and root exploration of the soil.

The use of D.I.H.B. did not appear to improve yield over a non-treated compacted plot and surface rolling was found to have no effect on yield.

Work is continuing at Sutton Bonington to assess the effect of compaction on field beans for a further year and attempts will be made to improve the sampling techniques.

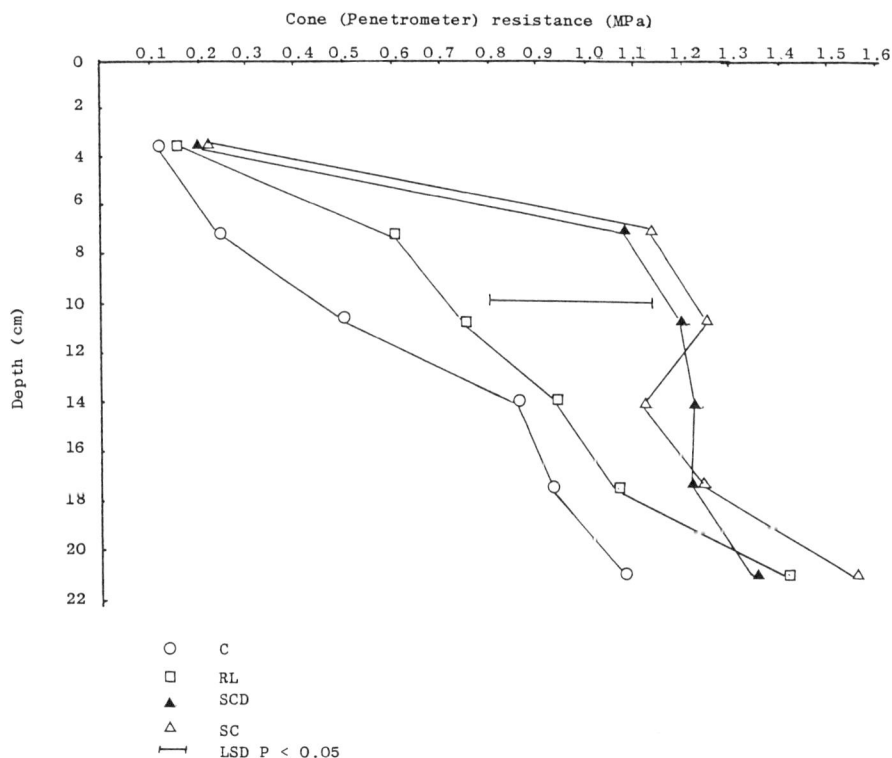

Fig. 1. Variation in cone resistance with depth as influenced by compaction

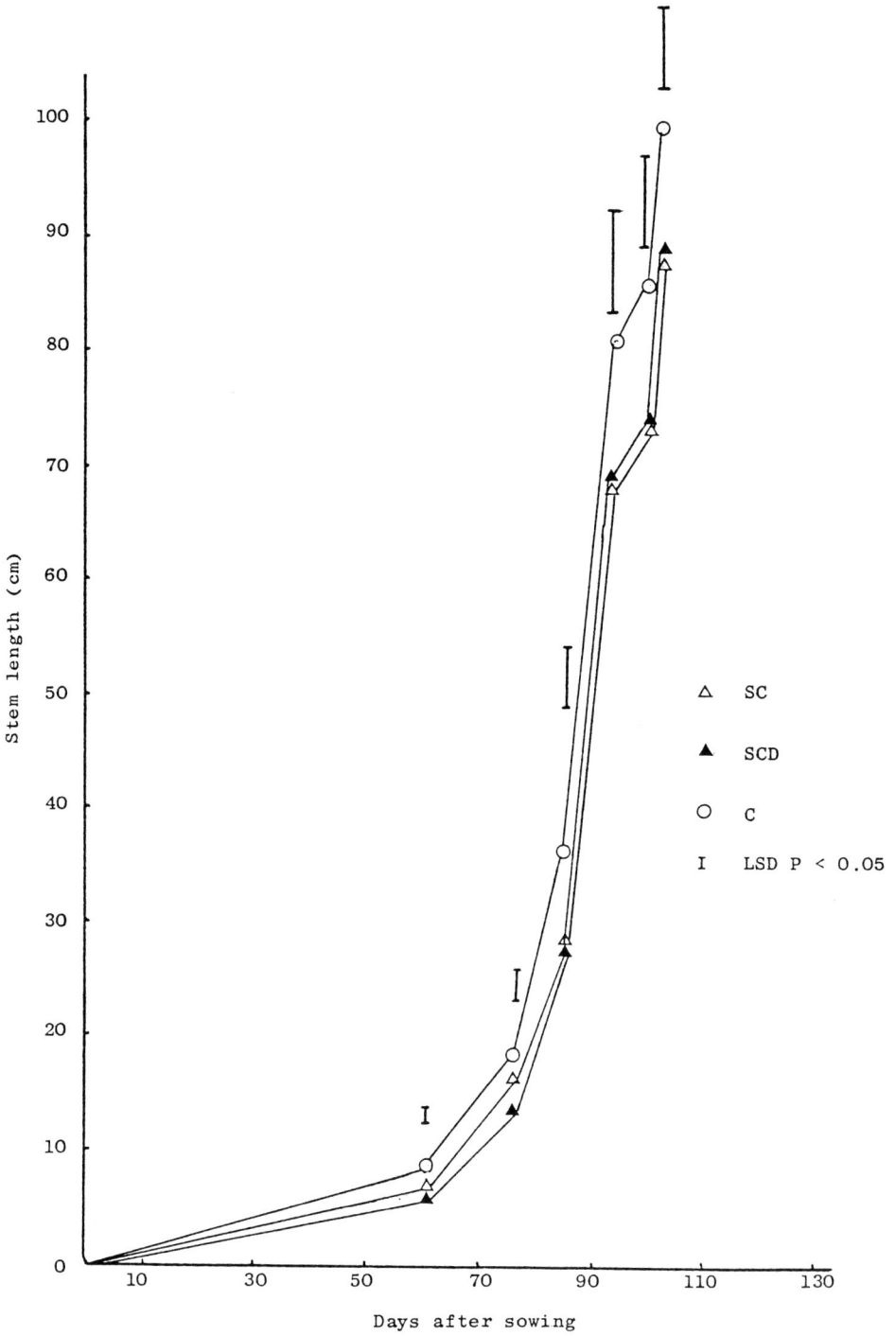

Fig. 2. Stem length as influenced by compaction

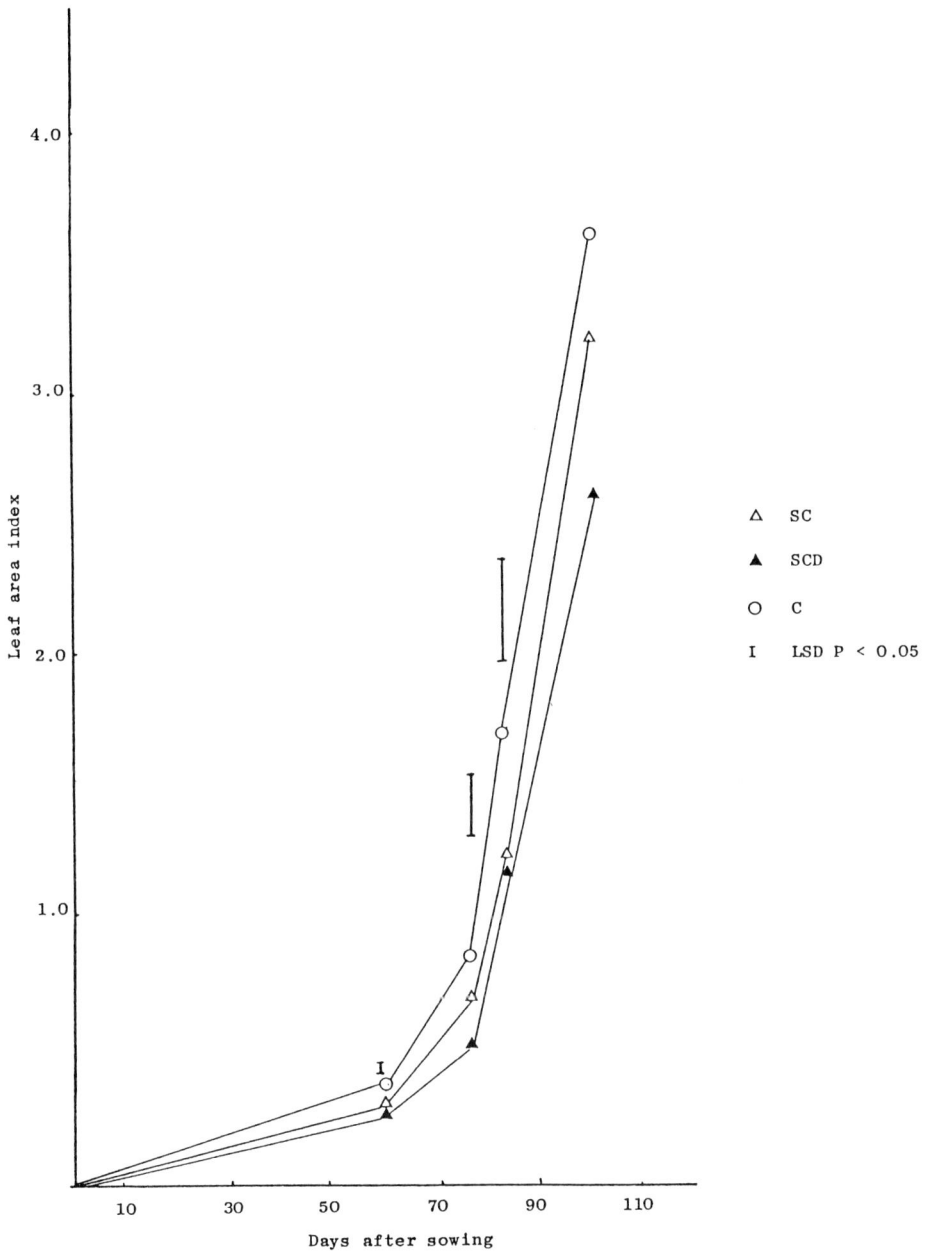

Fig. 3. Leaf area index as influenced by compaction

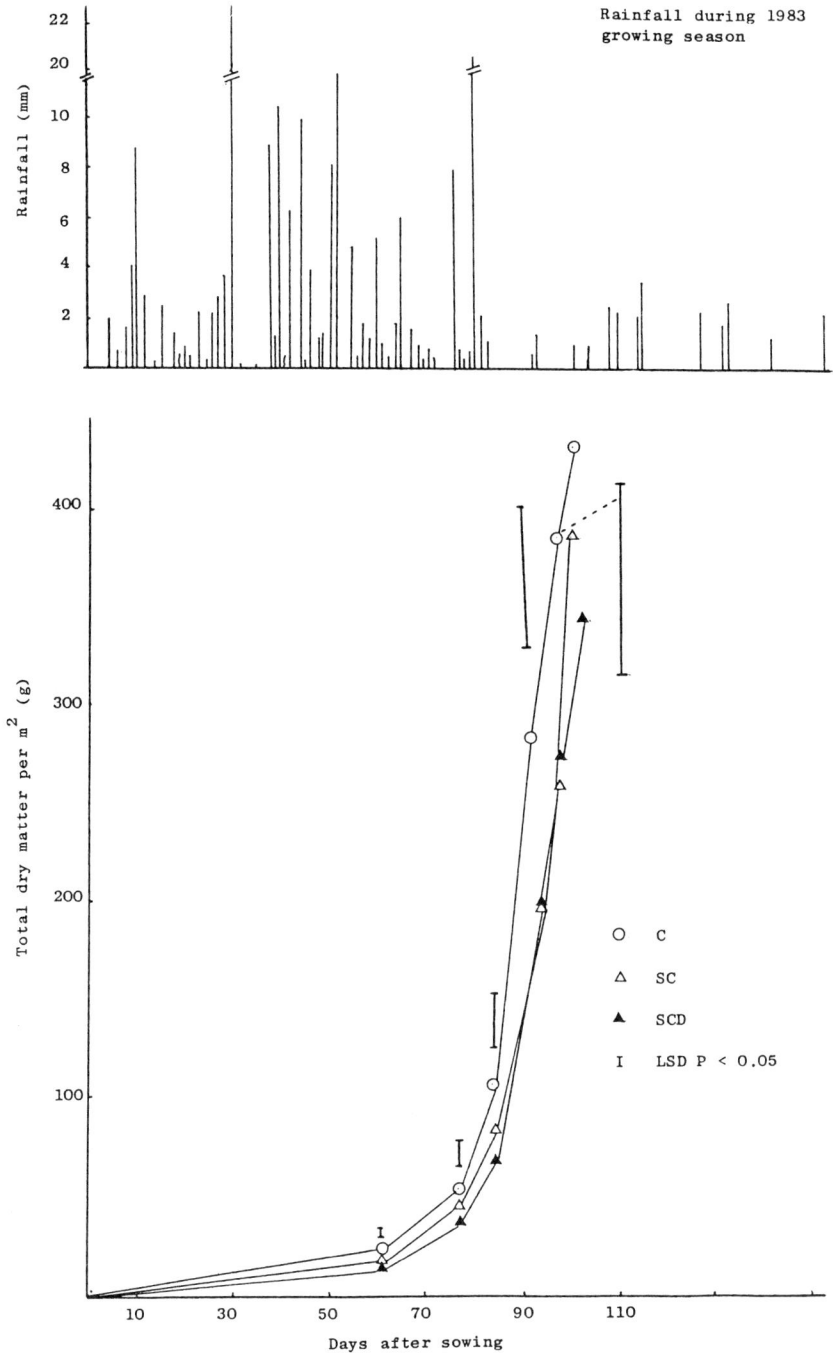

Fig. 4. Dry matter accumulation as influenced by compaction and rainfall

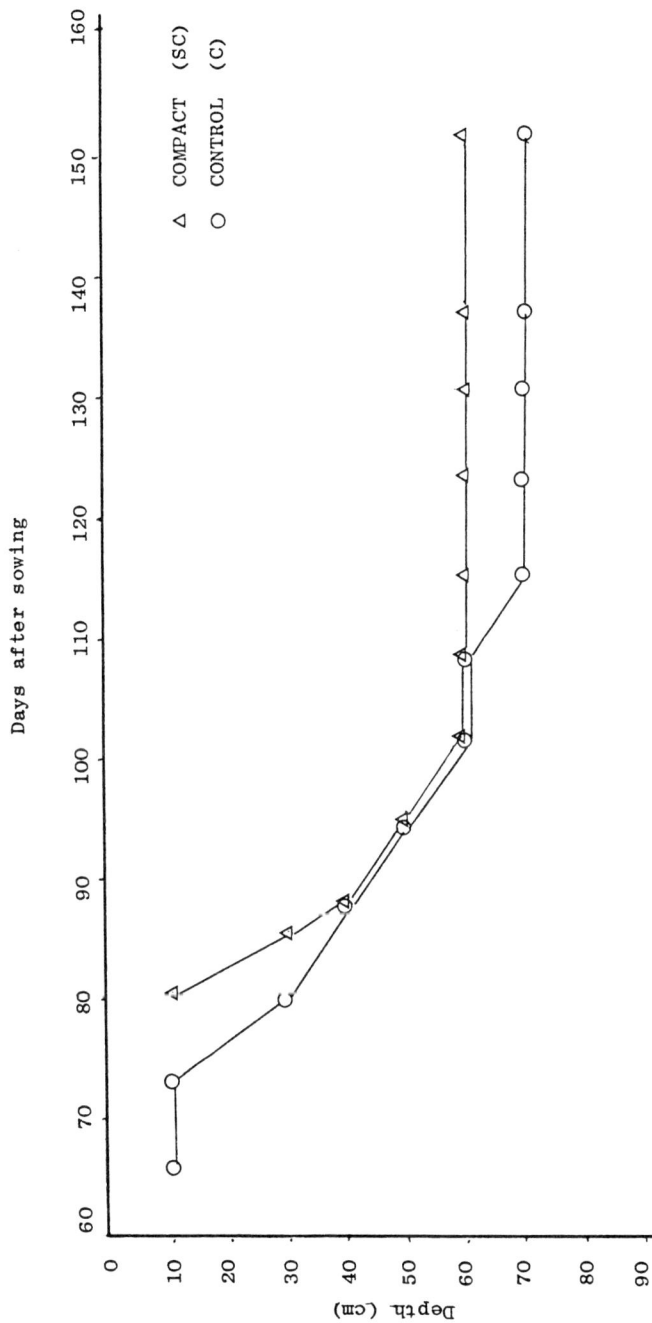

Fig. 5. Changes in effective rooting depth between the SC and C treatments as measured with the neutron probe.

Root distribution (SC)

Root distribution (C)

Fig. 6. Root profiles from profile pits indicating extent of rooting in the SC and C treatments

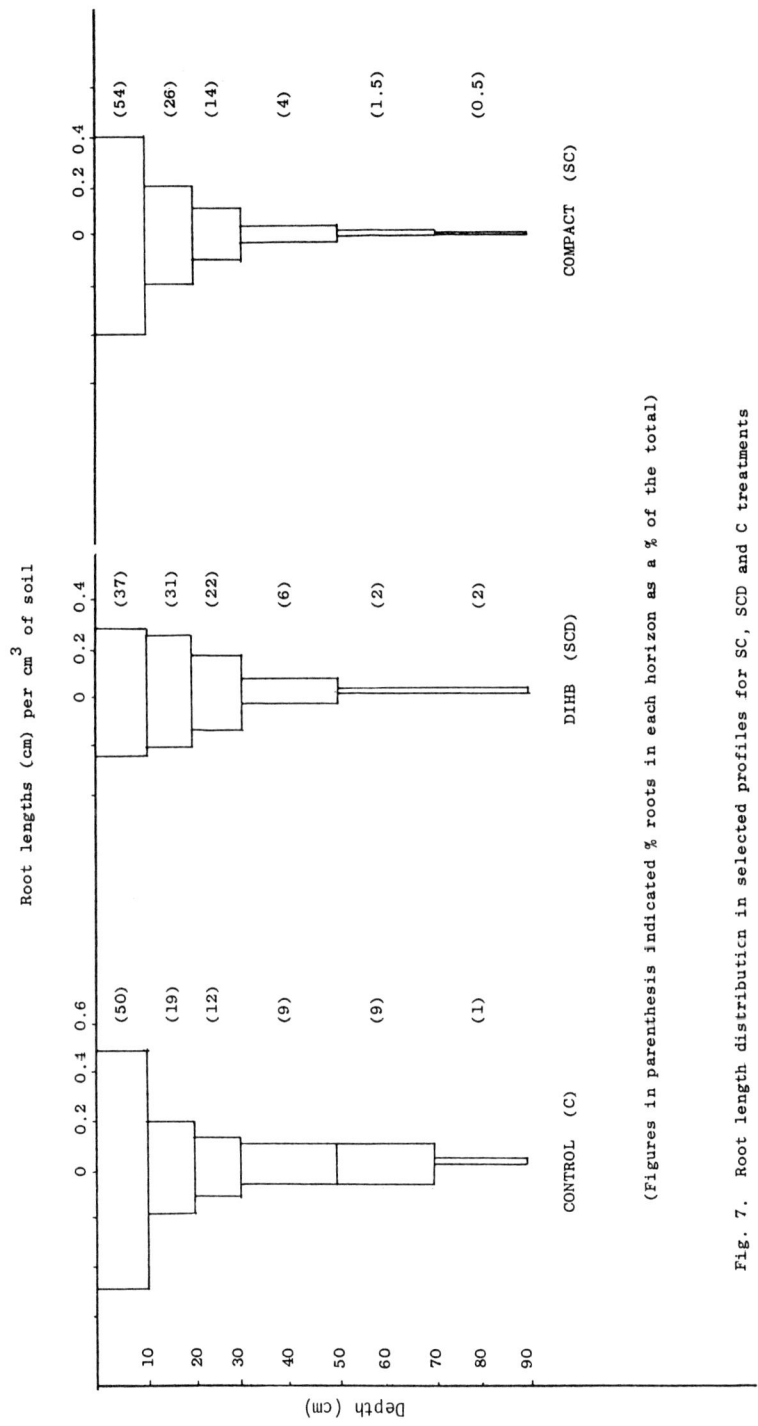

Fig. 7. Root length distribution in selected profiles for SC, SCD and C treatments

(Figures in parenthesis indicated % roots in each horizon as a % of the total)

REFERENCES

Anderson, G., Pidgeon, J.D., Spencer, H.B. and Parks, R. 1980. A new hand-held recording penetrometer for soil studies. J. Soil Sci., 31, 279-296.

Bell, J.P. 1969. A new design principle for neutron soil moisture gauges: the 'Wallingford' neutron probe. Soil Science, 108, 160-164.

Böhm, W. 1976. In situ estimation of root length at natural soil profiles. J. agric. Sci., Camb., 87, 365-360.

Kays, S.J., Nicklow, C.W. and Simons, D.H. 1974. Ethylene in relation to the response of roots to physical impedance. Plant and Soil, 40, 565-571.

McGowan, M. 1974. Depths of water extraction by roots. In "Physics and Irrigation Studies" (International Atomic Agency) Vienna. pp 435-445.

Wilkins, H., Wilkins, S.M. and Wain, R.L. 1976. Can chemicals be used to alleviate the effects of soil compaction on pea seedling growth? Nature. Lond. 259, 392-394.

EFFECT OF SEED SOURCE ON PERFORMANCE OF FABA BEAN VARIETIES

D.A. Bond and M. Pope

Plant Breeding Institute
Maris Lane, Trumpington
Cambridge, U.K.

ABSTRACT

Three seed stocks, obtained from different sources, of each of four winter bean varieties were compared for yield. Significant differences were found between two stocks of the variety Bourdon and between two stocks of Maris Beagle. In neither of these pairs of stocks was the yield difference associated with any difference in plant establishment.

Each stock was also sorted into large and small seeds before sowing. In general large seeds gave plots which were earlier maturing and clearly outyielded those from small and from unsorted seeds.

Present indications are that the yield of a given stock may not always represent the true relative potential of a variety particularly if ungraded stocks vary in seed size or if the seed of some stocks is graded and only the larger seeds sown.

INTRODUCTION

Owing to the possibility of natural outcrossing, much of the seed of faba bean varieties used in trials has to be produced at different locations, and it is sometimes produced in different years. In trials at the Plant Breeding Institute the yields of some varieties in relation to controls have differed significantly from those expected from previous trials, but because of the use of seed from different sources, it has not always been possible to conclude that these changes in relative yield were entirely due to the different responses of the varieties to a new environment.

The trials reported here describe the variation in yield from different sources of seed, (defined here as 'stocks') within varieties. Moreover as different stocks vary in seed-size, the effect of size of seed sown on yielding ability is examined. Very little information is available on the yield of stocks within varieties, and the few experiments which have tested the effects of sowing different seed-sizes within varieties (Salih, 1982) have shown a positive response to selection for large seed in one year's trial but not in another. If a response in yield is to be expected from grading seed then information on the magnitude and frequency of such a response is clearly important.

P.D. Hebblethwaite, T.C.K. Dawkins, M.C. Heath and G. Lockwood (eds.)
Vicia faba: Agronomy, Physiology and Breeding. ISBN 90-247-2964-5.
© 1984, Martinus Nijhoff/Dr W. Junk Publishers. Printed in The Netherlands.

MATERIALS AND METHODS

For each of four winter bean varieties, Bourdon, Banner, Maris
Beagle and Bulldog, all grown at various locations in eastern England,
seed samples of three stocks were obtained. The Bourdon stocks had been
produced in 1975, 1979 and 1980 at certification grades of "breeders",
"prebasic" and "basic" respectively, and for the other three varieties
the harvest year was either 1980 or 1981 and the grades basic or Cl.
Thus the stocks had been through very few generations of multiplication
in which to diverge genetically from the original population produced by
the breeder.

Part of each sample was sorted into (a) large and (b) medium to
small seeds (hereafter referred to as 'small'). The mean seed-weight of
the 'large' and 'small' fractions varied with varieties (see Table 2).
The three seed-size classes, unsorted, large and small of each of the
three stocks of four varieties (= 36 entries) were randomized in a
6 x 6 x 3 lattice design.

The sowing date was 22nd October 1981, with a plot size of 1.5 x
5.3m and seed rate of 25 seeds m^{-2}. Records were taken of the numbers of
plants established (from random 4m samples of row length), flowering and
maturity dates and yield of grain after drying to a constant (15%)
moisture content.

Data were analyzed in two ways: (1) lattice square analysis to give
adjusted entry means, and (2) analysis of variance for main effects and
interactions each tested for effect of the covariate 'number of plants
established'.

RESULTS

In none of the plots were nematodes or seed-borne diseases such as
Ascochyta fabae, or viruses detected in the seedlings; it was assumed
that stocks did not differ in seed health.

Stocks varied in number of plants established but over all plots
there was no relationship between plant establishment and yield, and
covariance analysis allowed no significant adjustment of yield data.
However it is possible that poor plant establishment adversely affected
yield in some plots and stocks, but not in others where tillering, for
example, could have compensated for a low stand. Thus although there was
a significant effect on yield of stocks within varieties (Table 1),

confidence could only be given to significant yield differences between
stocks where differences in plant establishment were very small.

TABLE 1   Analysis of Variance of Yields.

Item	df	ms $(x10^{-3})$	VR $\underline{v}$ Residual	VR $\underline{v}$ interaction
Varieties	3	5487	56.6***	10.8**
Seed size	2	685	7.1**	7.9*
Stocks within varieties	8	508	5.2***	3.6*
Varieties x weights	6	86	0.9	0.6
(Stocks within vars) x sizes	16	142	1.5	1.5
Residual	70	97		

*** P < 0.001, ** P < 0.01, * P < 0.05

Two examples of such pairs of stocks were the stocks MAD 79 and PBI
80 from Bourdon (unsorted seed), and the stocks NSDO 81 and WAL 81 from
Maris Beagle (large seed) (Table 2).

Relationship of mean seed weight as sown and yield

The overall mean yield for large seeds was 7.7 per cent and 5.9 per
cent higher than those for unsorted and small seeds respectively. Also
in five stocks (Bourdon PBI 80, Banner NSDO 80 and NSDO 81, and Bulldog
NSDO 81 and WAL 81), selection for large seeds resulted in significant
increases in yield. Three of the varieties also gave their greatest
yields when grown from large seeds.

The same positive relationship between yield and 100-seed weight
held for all unsorted stocks ($\underline{b}$ = 33.1 ± 12.7, P < 0.05) and for stock
means over all three seed-sizes. However on the means of all three
stocks, the ranking of varieties for yield did not change as between the
three seed-size classes. Yield differences independent of seed weight
also existed; e.g. unsorted seeds of stocks NSDO 80 and WAL 81 within
Bulldog (as well as NSDO 81 and WAL 81 within Beagle) differed
significantly in yield but had almost the same mean seed weights (Table 2).

TABLE 2 100 seed-weight (HSW) as sown, number of plants established and yield (kg ha^{-1}) of winter bean stocks from different sources.

Variety	Stock	Unsorted HSW	Unsorted Plants	Unsorted Yield	Large HSW	Large Plants	Large Yield	Small HSW	Small Plants	Small Yield	MEANS Plants	MEANS Yield
BOURDON	BAL 75	76	120	3641	88	118	3866	67	135	3953	124	3820
	MAD 79	70	159	4265	90	133	4384	66	164	4253	152	4300
	PBI 80	70	161	3743	80	164	4160	56	159	3615	161	3839
	MEAN	72	147	3883	86	138	4137	63	153	3940	146	3987
BANNER	NSDO 80	49	156	3476	70	152	3761	47	157	3040	155	3425
	NSDO 81	54	154	3235	77	157	4040	51	148	3859	161	3711
	DAL 81	58	175	3457	75	153	3662	55	159	3725	160	3614
	MEAN	54	162	3389	74	154	3821	51	160	3541	159	3583
BEAGLE	NSDO 80	64	142	3342	86	140	3383	60	157	3346	146	3357
	NSDO 81	60	163	3517	83	157	3607	56	148	2971	156	3365
	WAL 81	58	156	2965	76	159	3086	53	159	2881	158	2977
	MEAN	61	154	3275	82	152	3359	56	155	3066	153	3233
BULLDOG	NSDO 80	49	164	3029	72	151	3003	45	162	3223	159	3085
	NSDO 81	58	145	2860	76	128	3198	52	143	2958	139	3005
	WAL 81	49	155	2698	69	152	3169	45	153	3075	153	2980
	MEAN	52	155	2865	72	144	3123	47	153	3085	150	3023
Seed Size Means		59	154	3352	78	147	3609	54	155	3408	152	3457

LSD between:	Plants	Yield
any pair of entries in table	19.9	276
seed size means	5.8	96.9

LSD between:	Plants	Yield
stock means	11.5	159
seed sizes within varieties	11.5	159

Flowering and maturity dates

Only differences among varieties were detected in flowering date, but in maturity date there were also differences among seed-size classes ($P < 0.001$) and among stocks within varieties ($P < 0.05$) (Table 3). Large seeds produced plants which were estimated to have matured, on average, 1.4 days earlier than those from unsorted and 1.8 days earlier than those from small seeds. Banner Stock DAL 81 matured 3 days earlier than Stock NSDO 80 when both were grown from unsorted seeds and 2 days earlier from large seeds. In the latter pair, and also between the stocks MAD 79 and PBI 80 of Bourdon, there was an association of high yield and early maturity with almost equal plant establishment. High yield was clearly associated with early maturity in most of the large-seeded entries when compared with plants from small or unsorted seeds of the same stock.

TABLE 3   Estimated maturity dates (day in August) of winter bean stocks from different sources.

Variety	Stock	Seed size as sown			MEAN
		Unsorted	Large	Small	
BOURDON	BAL 75	8.67	8.33	8.00	8.33
	MAD 79	6.33	6.67	8.67	7.22
	PBI 80	9.67	6.33	8.33	8.11
BANNER	NSDO 80	9.33	7.67	11.00	9.33
	NSDO 81	7.67	8.33	9.00	8.33
	DAL 81	6.33	5.67	9.33	7.11
BEAGLE	NSDO 80	7.67	6.67	7.67	7.33
	NSDO 81	7.33	6.00	8.67	7.33
	WAL 81	9.33	8.67	9.00	9.00
BULLDOG	NSDO 80	10.33	7.00	10.00	9.11
	NSDO 81	10.33	6.67	7.67	8.22
	WAL 81	10.33	8.67	10.67	9.89
Seed size	MEAN	8.61	7.22	9.00	8.28

LSD between : any pair of entries in table,   2.58
            : seed size means,   0.74
            : stock means,   1.49

DISCUSSION

The main conclusion about stocks was that yield differences existed within varieties and for two pairs of stocks these were not associated with differences in plant establishment. It was not possible to establish that yield of stocks was related to the mean weight per seed as sown, but selection of large seeds gave a clear improvement in yield for some stocks, for 3 of the 4 varieties and for the mean over all stocks. The latter result confirms that reported by Salih (1982) for one of his two years' trials.

The frequency of differences in yield between stocks where differential establishment was clearly not implicated was two cases out of 12. Other instances of significant differences in yield between stocks may have been associated with differences in plant establishment.

The magnitude of the detected differences between stocks was about 500 kg ha^{-1} or 14% below the better stock. Significant increases in yield as a result of selecting for large seeds varied from 300 to 800 kg ha^{-1} or 11 to 25% over unsorted seed. These results were from only one fairly precise trial but if they were to occur consistently there would be implications for the multiplication and evaluation of faba bean varieties. Although the ranking of varietes on mean yields of all three stocks did not change with seed-size class a given stock may not always represent the potential yield of a variety or selected population and it therefore may be necessary to test a number of stocks (say, three) to obtain a true assessment.

An alternative is to produce the seed of all genotypes to be evaluated, and controls, at one site. Some breeders do this, but it allows some cross pollination between varieties, and even within one field the environment can vary.

The effect of harvest year on seed size is often greater than that of the location of the seed harvest; this was true for yield differences for stocks within Bourdon and Banner but not within Beagle. A possible reduction in differences between stocks might be achieved by using only seed from the same harvest year.

Unless uniform seed-production conditions, without varietal intercrossing, can be provided the conclusion is that varieties which give high and low relative performances in successive trials should be retested with different stocks. Also, it is clear that ungraded seed of one variety cannot be fairly compared with graded seed of another. The results suggest that grading seed might be profitable to the grower, but more information is needed on the size of the large-seeded fraction which can be selected and still give expectation of increased yield.

Further investigations are required to elucidate the causes of differences in yielding ability of seed from different sources, but possibilities to be tested would include the following:

(1) The effect of the seed-producing environment, especially soil fertility, on seed size. Large seeds may produce large plants, which can bear a high yield.

(2) Frequency of natural crossing may be an important cause of variation among seed sources within varieties. A high proportion of crossbred seed would be expected to raise the mean level of yield and autofertility of a population and this might account for the association between large seeds (as sown) and early maturity. Such an effect could occur in a population with considerable genetic variation but not in an inbred line.

(3) Where there is poor pollination sometimes fewer seeds per pod are set but those that do develop are larger on average and could have the same effect as (1) above.

Whatever the cause it is important that seed multiplication and trialling procedures should be such that the genetical potential of new varieties is not obscured by environmental effects on the seed as sown.

We thank Dalgety-Spillers Ltd and the National Seed Development Organisation for supply of seed, and Mrs. J.A. Hall and Mr. C.W. Howes for help with analyses of data.

REFERENCE

Salih, F.A. 1982. Influence of seed size and sowing date on yield and yield components of faba beans. Fabis 4, 38-39.

RECENT DEVELOPMENTS IN FORECASTING APHIS FABAE DAMAGE AND
CONTROL, AND ITS IMPLICATIONS FOR OTHER PESTS OF VICIA FABA

M.E. Cammell & M.J. Way

Imperial College, Silwood Park, Ascot, Berkshire

ABSTRACT

Forecasting the need for and timing of chemical control of the
black bean aphid, Aphis fabae on spring-sown field beans, Vicia faba
is discussed in relation to different forecasting procedures, which
vary from short-term crop monitoring to longer-term monitoring of aphid
populations during dispersal or at source. The merits and limitations
of these different procedures are considered against the background of
inaccuracies involved in establishing economic thresholds and involve
criteria such as the value placed upon the time when the forecast is
available, the level of accuracy required and their cost and conven-
ience. The possible limitations of interpreting area-based dispersal
or source forecasts for individual fields is also considered. It is
concluded that long-term area forecasts based upon numbers of A. fabae
eggs on its overwintering host, Euonymus europaeus provide a sound basis
for advice on chemical control of the aphid.

Changed pesticide usage resulting from the use of forecasting data
may have implications for some other pests of beans, incidentally controlled by chemicals applied for control of A. fabae. Some possibilities of forecasting other pests are also discussed.

INTRODUCTION

It is widely recognised that in order to achieve more effective
pest control and to minimise adverse side effects of pesticide usage,
such as pesticide resistance and environmental hazards, we should aim
to improve forecasting of the need for and the timing of chemical
treatments.

In this paper we will first outline briefly some general guidelines for pest forecasting and monitoring; secondly consider the current status of practical forecasting as developed for the black bean
aphid, Aphis fabae and, thirdly, discuss the possibilities and implications of forecasting for other pests of faba beans.

GENERAL GUIDELINES FOR FORECASTING AND MONITORING

Ideally, a control measure is only applied if there would otherwise
be an economic crop loss. This requires the use of economic threshold
data combined with the forecasting of pest abundance as early as possible before the crop is damaged. In practice, defining economic thresh-

P.D. Hebblethwaite, T.C.K. Dawkins, M.C. Heath and G. Lockwood (eds.)
Vicia faba: Agronomy, Physiology and Breeding. ISBN 90-247-2964-5.
© 1984, Martinus Nijhoff/Dr W. Junk Publishers. Printed in The Netherlands.

holds has proved difficult because of the many biological and economic variables involved. Much data is often required to establish the nature of a particular pest-crop damage interaction under differing weather and cropping conditions. Errors may also arise because of varying sale price and yield of the crop and varying costs of pesticide treatment. However, despite such complexities, useful thresholds have been developed, often based initially on empirical evidence and then subsequently refined as the nature of particular pest-crop relationships become better understood.

Forecasting procedures for airborne pests and diseases may be conveniently considered as follows:- (a) after arrival on the crop; (b) during dispersal to the crop or during previous migrations; and (c) at the source of the pest before departing to attack the crop.

Irrespective of the system chosen, the forecasting data will need to be applied at the individual field level. The most accurate data should be available from monitoring pest numbers within individual crops. Crop monitoring techniques can often be usefully developed from relatively limited biological data. However, counts are often laborious and also an almost immediate pesticide application is required if numbers reach the economic injury threshold. In contrast, longer-term forecasts allow a longer interval between the forecast and the pesticide application and have obvious advantages to pesticide suppliers and advisors as well as growers. Such forecasts made before the pest enters the crop, particularly those based on source monitoring, require a more complete understanding of the pest ecology and therefore often need a longer period for research and development before their feasibility and value can be adequately assessed. However, such an approach provides opportunities for understanding the nature of the overall population dynamics of the pest and thus identifying factors which influence pest incidence, the sources of infestation and assessing the importance of short and long range migration, all of which can provide a sound base for developing optimal control strategies. A potentially major limitation of source monitoring is the accuracy with which such area, regional or national-based forecasts relate to the individual field.

FORECASTING AND MONITORING OF A. FABAE POPULATIONS ON SPRING BEANS

The most important pest of spring-sown beans in Britain and temp-

erate Europe is the black bean aphid, _Aphis fabae_ which may cause severe
crop losses, primarily by direct feeding damage. In the absence of
forecasting data, growers tend to use a routine preventive chemical
treatment. During the past 30 years there has been considerable research on the ecology and control of the aphid and such research has
formed the basis for developing rational control strategies based on
forecasting data.

In Britain and temperate Europe, _A. fabae_ overwinters as án egg
almost exclusively on the spindle tree, _Euonymus europaeus_ and migrants
disperse to beans and other summer hosts over a well-defined limited
period between mid May – Mid June. Several generations occur on a wide
range of herbaceous secondary hosts until late Summer – early Autumn
when migrants return to spindle.

Forecasting and Monitoring Methods and Results, 1970 - 1982
(a) Short-term crop monitoring

Damage assessment studies of _A. fabae_ on spring-sown field beans
have shown that the principal damage to the crop is caused by populations which develop on young plants before flowering and are initially colonised by primary migrants from spindle during late May –
mid June. Aphid populations initiated after flowering cause little
damage. Plot experiments provided the basis for establishing an economic injury threshold of 5% of bean stems infested by primary migrants
on the S.W. headland of fields in early-mid June (Way and Cammell, 1973).
Since 1970, the economic threshold of 5% has been used as the basis for
advice although taking into account yield variation, change in control
costs and variation in sale price the threshold calculated retrospect-
ively for the period 1970 - 76 varied from about 6% - 13%. Since
variables such as crop yield can seldom be pre-determined, such data
illustrates the extent of inevitable inaccuracies.

(b) Short and long term monitoring during aphid dispersal

Samples of _A. fabae_ migrants were collected daily by 'Rothamsted'
suction traps operating at a height of 12.2m ('40' ft traps) at various
sites in Britain (Taylor, 1979). The total numbers trapped have been
analysed for two main migration periods, namely from mid Sept - early
Nov (Autumn migration), representing numbers of gynoparae returning to
the winter host and from May - mid June (Spring migration), correspond-

ing to numbers of alatae migrating from spindle to beans and other early summer hosts. Total numbers trapped during these periods were compared with subsequent mean % bean stems initially infested on local sample bean crops. The results from 1970 - 82 show that useful forecasts can be made on the basis of thresholds of 15 and 5 total alatae trapped during the Autumn and Spring migrations respectively.

Autumn trap catches are particularly useful as very early forecasts of likely very large or very small populations on·beans about 7-8 months later although they lack precision when populations are of intermediate size. Trap sampling of the spring migration gives a more precise forecast and also provides an indication of the correct timing of chemical treatment, if necessary, but the forecast is available only shortly before treatment may be required (Way et al, 1981).

(c) Long-term source monitoring

Since 1970, counts of eggs in winter on selected spindle bushes have been used to make separate forecasts in each of 16 areas in England, South of the Humber, (Way et al, 1977). After correction of the data in areas of high spindle abundance, forecasts for spring-sown field beans fall into one of three categories: (i) economic damage unlikely, chemical treatment unnecessary; (ii) damage possible, treatment necessary on some fields, and (iii) damage probable, chemical treatment necessary on nearly all fields. A further sampling of the aphid population on spindle about mid May also provides information on when to apply a chemical, if necessary, for a single insecticide treatment applied to coincide with the end of the primary migration can give good control for the whole season.

Since 1970 the accuracy of area forecasts have been checked against the size of initial infestations on sample bean fields in corresponding forecasting areas. Initial infestations on beans have varied considerably both from year-to-year and also between different areas. In some years, such as 1973 and 1974 there were widespread, potentially damaging infestations in all areas, whereas in years such as 1975 and 1980 all infestations were below the 5% economic threshold. In other years such as 1970 and 1981 potentially damaging infestations were confined to certain areas which clearly demonstrated the need for area-based forecasts. In general,   area initial infestations on beans closely reflected the forecasts based on numbers of overwintering eggs (Fig. 1).

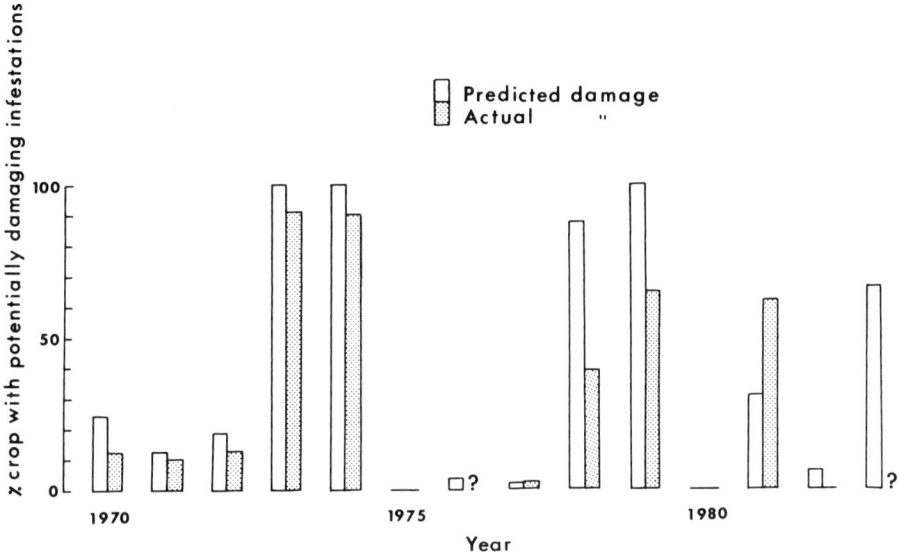

Fig. 1. Comparison of the predicted and actual percentage of the spring-sown field bean crop with potentially damaging infestation (ie. 5% or more of bean stems infested on the S.W. headland in early- mid June), 1970-83.

At the individual field level from 1970-75, <1%, 43% and 86% of cropped areas would have required treatment following forecasts of unlikely, possible and probable damage respectively. In general, the area forecasts adequately predict relatively heavy attacks or absence of attack on individual fields but there is need for improvement in relating borderline area forecasts (ie. possible forecasts) to individual fields. During the period 1970-82 possible forecasts were given for approximately 10% of the total cropped area and, at present, following such a forecast either requires timely crop monitoring to assess whether a particular field exceeds the economic threshold or, in the absence of crop monitoring, must be taken as advice to treat. Current work is aimed at defining those characteristics which make an individual crop more, or less at risk from aphid attack and damage than the average.

These include crop density and sowing date, local topography, wind-
breaks and field shape and size. Thus sparse, late-sown crops grown in
small, sheltered fields are probably at greatest risk. Such criteria
still need to be validated but they indicate a potential for greatly
improving the applicability of area forecasts to individual fields.
Furthermore, such studies have important implications for other pests.

In conclusion, the source monitoring system has proved convenient
and sufficiently accurate and cost effective when compared with other
control strategies (Cammell and Way, 1977) and is the current basis for
Ministry of Agriculture, Fisheries and Food advice to growers on chemical control of A. fabae on spring-sown field beans. Furthermore, in
areas where 'Rothamsted' suction traps are sited these provide valuable complementary advice.

POSSIBILITIES AND IMPLICATIONS OF FORECASTING FOR SOME OTHER PESTS OF BEANS

Aphids

In Britain, two other aphid species occasionally attack field
beans, namely the pea aphid, Acyrthosiphon pisum and the vetch aphid,
Megoura viciae. Neither rarely reaches sufficiently large numbers to
cause direct economic damage but A. pisum, in particular, is an important vector of several virus diseases of beans (Heathcote & Gibbs,
1962). Both species overwinter on various leguminous plants and mi-
grate to beans during May - June. Therefore, routine preventive treatments applied in June to control A. fabae incidentally control potentially damaging populations of pea and vetch aphids. However, if
aphicides are only applied when A. fabae numbers exceed the economic
threshold there may be circumstances where the chemical treatment is
not applied but nevertheless may have usefully controlled populations
of these other aphid species. Therefore, more data is required on the
population dynamics of these species in relation to virus transmission
and whether changed pesticide usage will alter pest status.

Weevils (Curculionidae)

The bean flower weevil, Apion vorax is the main vector of broad
bean stain and broad bean true mosaic viruses, which in recent years
have been two of the most common viruses in spring-sown beans in S.E.
England (Cockbain, 1971). In Britain, the adults overwinter on various

trees and herbaceous plants in woodland and marginal land and adults
fly to beans during April-May where they feed on the foliage, although
such damage is of little significance compared with the adults' importance as a virus vector.

Numbers on beans may vary considerably from year to year. Adults
are most common in woodland and hedgerows bordering fields where beans
were previously grown thus indicating that dispersal from beans the
previous summer is very localised (Cockbain et al, 1982). Both virus
diseases are seed-borne and, in the absence of vectors and spread between plants infection is of little significance. Insecticides to control the vectors have met with varying success and developing economic
thresholds is complicated by the virus-vector relationship. More data
is obviously required on the ecology of A. vorax especially on adult
dispersal in spring and the timing of control measures. More information is also required on factors determining overall weevil abundance
in the crop such as the importance of natural enemies and of the use of
insecticides applied to control A. fabae (Cockbain et al, 1982).

In Britain, the pea and bean weevil, Sitona lineatus is by far the
most common weevil on beans. The adult overwinters among leaf debris
or on various leguminous crops such as clover and lucerne and in early
spring they fly or crawl into pea and bean crops. Adults feed on leaves
and may cause damage to very young plants. However, larval damage to
root nodules is of greater importance and chemical control of larvae
has been shown to increase yields, on average, by about 7% (Bardner
et al, 1979).

Bardner, et al (1979) suggested that severe attacks of S. lineatus
might be predicted from counts of adults dispersing in late summer, of
overwintering adults or those dispersing to crops in spring, using
numbers caught by suction and light traps and, possibly by pheromone
traps. 'Rothamsted' suction trap catches of adults are greatest in
areas where peas and beans are grown intensively and in spring are
earliest in warm, sunny seasons. Unfortunately, there is no correlation
between late summer catches and numbers trapped the following spring
and also adults appear in the crop in spring before detected by suction
traps (Hamon et al, 1982). Furthermore, it has not yet been possible
to correlate intensity of adult attack with subsequent losses caused
by larval feeding because mortality of eggs and larvae is often high but

142

variable. Obviously more work is required on S. lineatus particularly
in relation to damage assessment, larval and egg mortality and adult
dispersal before prospects for forecasting can be adequately assessed.

In conclusion, work on forecasting A. fabae illustrates the possibilities for developing forecasting procedures for other pests given
the relevant biological background. The change in pesticide usage associated with A. fabae forecasting has implications for control on other
pests. Ultimately the control of all pests and pathogens of beans must
be considered within an overall control strategy.

REFERENCES
Bardner, R., Fletcher, K.E. and Griffiths, D.C. 1979 . Problems in the
        control of the pea and bean weevil (Sitona lineatus). Proc. Brit.
        Crop. Prot. Conf. - Pests and Diseases 1, 223-229.
Cammell, M.E. and Way, M.J. 1977 . Economics of forecasting for
        chemical control of the black bean aphid, Aphis fabae on the field
        bean, Vicia faba. Ann. appl. Biol. 85, 333-343.
Cockbain, A.J. 1971 . Epidemiology and Control of weevil-transmitted
        viruses in field beans. Proc. 6th Brit. Insecticide & Fungicide
        Conf. 1, 302-306.
Cockbain, A.J., Bowen, R. and Bartlett, P.W. 1982 . Observations on
        the biology and ecology of Apion vorax (Coleoptera:Apionidae), a
        vector of broad bean stain and broad bean true mosaic viruses.
        Ann. appl. Biol. 101, 449-457.
Hamon, N., Bardner, R. and Fletcher, K.E. 1982 . Monitoring and Fore-
        casting Sitona. Ann. Rep. Rothamsted Expt. Sta. for 1981,
        (Part 1), 101.
Heathcote, G.D. and Gibbs, A.J. 1962 . Virus diseases in British Crops
        of field beans (Vicia faba L). Pl. Path. 11, 69-73.
Taylor, L.R. 1979 . The Rothamsted Insect Survey; an approach to the
        theory and practice of pest forecasting in agriculture. In
        "Movements of Highly Mobile Insects; Concepts and Methodology in
        Research"pp. 148-85. (Ed. G.G. Kennedy & R.L. Rabb)North Carolina
        State University : Raleigh.
Way, M.J. and Cammell, M.E. 1973 . The problem of pest and disease
        forecasting - possibilities and limitations as exemplified by work
        on the bean aphid, Aphis fabae. Proc. 7th Insecticide & Fungicide
        Conf. 3, 933-954.
Way, M.J., Cammell, M.E., Alford, D.V., Gould, H.J., Graham, C.W.,
        Lane, A., Light, W.I. St. G., Rayner, J.M., Heathcote, G.D.,
        Fletcher, K.E. and Seal, K. 1977 . Use of forecasting in chemical
        control of the black bean aphid, Aphis fabae Scop., on spring-sown
        field beans, Vicia faba. Pl. Path. 26, 1-7.
Way, M.J., Cammell, M.E., Taylor, L.R. and Woiwod, I.P. 1981 . The
        use of egg counts and suction trap samples to forecast the infest-
        ation of spring-sown field beans, Vicia faba, by the black bean
        aphid, Aphis fabae. Ann. appl. Biol. 98, 21-34.

THE POTENTIAL OF WHOLE-CROP VICIA FABA FOR ENSILAGE

J. S. Faulkner* and R. W. J. Steen**

Department of Agriculture for Northern Ireland
*Plant Breeding Station, Loughgall, Co. Armagh
**Agricultural Research Institute, Hillsborough, Co. Down

ABSTRACT

The potential of spring-sown faba beans (Vicia faba) for whole-crop ensilage has been investigated in Northern Ireland. Using seed rates of 240 kg ha^{-1}, tic bean cultivars (var. minor) have given DM yields of about 10 t ha^{-1} at the end of July or in early August. Well preserved silages with pH 3.8-4.1 and an average of 15% crude protein in the DM were obtained using a formic acid additive. In a feeding trial with Friesian steers, intake of two faba bean silages (7.7 and 7.8 kg d^{-1}) and of an aftermath grass silage (7.9) were higher than that of a cereal with pea silage (7.3). Carcase gains were greatest on grass silage (0.81 kg d^{-1}), followed by the later cut bean silage (0.66), the earlier cut bean silage (0.59) and the cereal with pea silage (0.57). Establishment of undersown grass in faba beans was normally good, but was favoured by lower bean seeding rates and earlier cutting. Whole-crop faba beans appear to be an attractive alternative to whole-crop cereals as an annual silage crop.

INTRODUCTION

Conventional crops of faba beans for livestock feeding are extravagant. Under British conditions they occupy the ground for nearly two months after they have stopped growing, and when they are eventually harvested more than half the crop - the haulm or straw - is worthless or even a liability. Growing beans need not entail such extravagance. Where the crop is grown for feeding ruminants, it could be harvested immature as a forage and made into silage. This would reduce the time for which the crop occupies the ground and ensure utilization of all above ground parts of the crop.

The idea of growing faba beans for ensilage seemed particularly appropriate in Northern Ireland, where cattle farming is the major agricultural enterprise and many farmers are experienced at ensiling grass and whole-crop cereals. A series of field plot trials was begun in 1980 at Loughgall (54° 25' N, 6° 36' W), and a feeding trial was carried out in 1982 at Hillsborough, 40 km to the east of Loughgall. Mean monthly figures for rainfall and temperature at Loughgall are given in Figure 1.

Sowing normally took place between mid-March and mid-April into ground which had been ploughed and harrowed. Neither herbicide nor N fertiliser

P.D. Hebblethwaite, T.C.K. Dawkins, M.C. Heath and G. Lockwood (eds.)
Vicia faba: Agronomy, Physiology and Breeding. ISBN 90-247-2964-5.

were used, but P and K were applied to the seedbeds. In some trials the
faba beans were grown mixed with oats or barley, or with undersown grass.

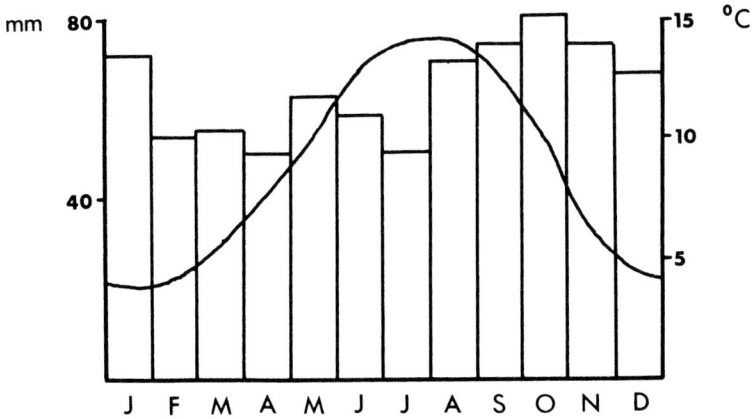

Fig. 1   Mean monthly rainfall totals (mm - columns) and screen air
temperatures (°C - curve) at Loughgall, Northern Ireland.

The series of trials is still continuing, and the account presented
here should be regarded as a progress report.

CROP GROWTH AND YIELDS

No problems were encountered in establishing a vigorous crop of faba
beans, except on one occasion when there was severe bird damage to a trial
sown in February. Isolated arable crops in a mainly grassland area are
very attractive to rooks (Corvus frugilegus), especially if they are sown
early, so thorough protective measures are vital. Disease levels were
consistently low, and Aphis fabae was never sufficiently numerous to
warrant insecticide spraying.

During June and early July, the crop has a very low dry matter
content. During the latter part of July and August, dry matter content
increases steadily as the crop matures. The progressive drying out of two
crops of V. faba cv. Blaze is illustrated in Figure 2: both of these
crops were grown in warmer and drier than average summers, so they may have
dried out more rapidly than normal.

The crop was normally cut in late July or early August. Crops
remaining in the ground until August gave greater dry matter yields -
usually over 10 t ha^{-1} - than crops harvested in July. Dry matter yields
at Loughgall appeared to be significantly higher than those reported in

regions with a more continental climate, such as East England (Toynbee-Clarke, 1973) or France (Plancquaert, 1971).

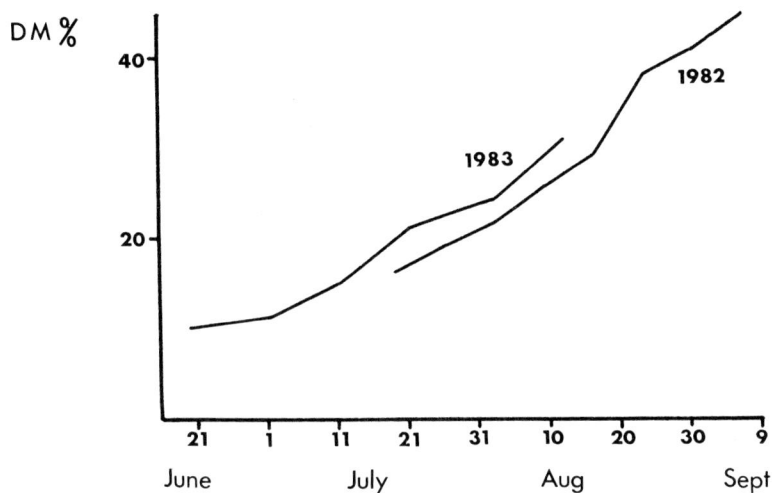

Fig. 2   Percentage dry matter content in the above-ground parts of two crops of Vicia faba cv. Blaze, sown at Loughgall on 30 March 1982 and 15 March 1983.

Seeding rate affected yields. The greatest yields were obtained after sowing at 240 kg ha^{-1}, a seeding rate commonly used for grain crops. Lower seeding rates gave somewhat lighter crops, though the yield was not reduced pro rata (Table 1). Replacing some of the faba bean seed by either barley or oats reduced yield as compared with sowing 240 kg ha^{-1} of beans alone, although the dry matter percentage of the whole crop was increased (Faulkner, unpublished).

In a comparison of cultivars, all sown at 240 kg ha^{-1}, cultivars of V. faba var. minor, being smaller seeded, produced a denser stand of plants and tended to yield more than var. equina or var. major. Although this trial was sown in spring, some winter cultivars were included but they did not perform particularly well (Table 2).

The yield disadvantage of the larger seeded cultivars could probably be overcome by sowing a fixed number of seeds rather than a fixed weight. However this would mean sowing the larger seeded cultivars at a heavier rate, thus encountering the deterrent of additional seed costs.

TABLE 1   Effect of seeding rate on dry matter yield of whole-crop
faba beans cv. Blaze (4 separate comparisons)

Date of cut	Seed rate kg ha^{-1}	DM yield t ha^{-1} mean	SE of mean
31.7.1980	240*	9.57	0.47
	160	8.00	
31.7.1981	240	10.04	0.29
	150	8.00	
19.7.1983	240	8.44	
	180	7.69	0.29
	120	7.04	
29.7.1983	240	9.56	
	180	9.19	0.20
	120	7.94	

*   All crops except this one were undersown with grass

TABLE 2   Whole-crop DM yield, and DM content of 4 classes of V. faba
cultivar sown on 4 April 1980 at 240 kg seed ha^{-1} and harvested on
31 July 1980

Class	No. of cultivars	No. plants m^{-2}	DM yield t ha^{-1}	DM %
var. minor	9	60.9	9.90	15.7
var. equina (spring)	1	50.5	8.60	15.8
var. equina (winter)	3	46.9	7.84	15.2
var. major	5	23.6	7.85	15.2

No differences in whole-crop yield were found among the cultivars of
var. minor currently recommended for grain production in England, but two
older cultivars, Deiniol and Albyn Tick, and the short early Finnish
cultivar Mikko gave significantly lower yields in one trial.

ENSILING QUALITY
    Bulked plot samples of whole-crop faba bean were ensiled in batches
of 0.6 t fresh weight in small scale silos. The forage was chopped,
treated with 4 litres of formic acid per tonne, and compacted by treading.

Satisfactory fermentations took place, producing a blackish-brown silage with a sweet or somewhat acid smell.

TABLE 3   Analyses of five silage fermentations of faba bean cv. Blaze: pH; percentage dry matter (DM %); percentage crude protein in dry matter (CP %); percentage modified acid detergent fibre in dry matter (MADF %); and litres of effluent per silo

	Date sown	Date cut	pH	DM %	CP %	MADF %	Effluent
1	2.4.1981	31.7.1981	3.87	19.3	15.3	38.9	unrecorded
2	30.3.1982	19.7.1982	3.78	19.3	14.5	38.6	48
3	30.3.1982	19.7.1982	3.89	19.0	14.9	39.1	46
4	30.3.1982	12.8.1982	4.06	23.2	15.5	35.7	0
5	30.3.1982	12.8.1982	4.12	26.6	15.0	36.6	< 1

The pH levels, ranging from 3.78 to 4.12 (Table 3), were regarded as indicating well-preserved silages. On exposure to air, however, deterioration appeared to be rather rapid. Dry matter contents were low in silages 1-3 as a result of cutting the crops rather early, and a substantial amount of effluent was produced. The crude protein levels, averaging about 15%, were good by the standards of grass silages and far above the level for whole-crop cereals. Modified acid detergent fibre contents of silages 1-3, at about 39%, were intermediate between the figures expected for grass and whole-crop cereals, but the lower figures of about 36% in silages 4 and 5 would be typical of an average grass silage.

Ensiling a mixture of faba beans and cereal would have produced a silage of lower quality. Dried mixed forages from plots comprised of about 50% faba bean and 50% oats in 1981 were 3.5% lower in crude protein than pure faba bean forage. Pepsin-cellulase digestibility was also slightly lower, but dry matter content was 3% higher (Faulkner, unpublished).

FEEDING TRIAL, 1982

The feeding value to beef cattle of two faba bean silages was compared with that of silages made from grass or from whole-crop cereal

with peas. The faba bean silages were produced from crops of cv. Blaze
sown on 6 April and cut on 20 July (early) or 12 August (late). The
grass silage was made from a sward of Lolium perenne with Trifolium repens
cut on 1 September after a regrowth period of 7 weeks, and the cereal/pea
silage from a crop sown with 30% oats, 46% barley and 24% peas on 6 April
and cut on 20 July. The grass silage was regarded as representative of a
late-season regrowth crop, and the cereal/pea silage as representative of
a type of nurse crop frequently used for reseeding grassland in Northern
Ireland. No N fertiliser was applied to the faba beans, but 60 kg ha^{-1}
was given to the cereal/pea crop. All crops were cut by precision-chop
forage harvester, and had 3 litres formic acid added per tonne of fresh
weight.

The 4 silages were offered ad libitum with 3.5 kg barley per head
daily to 36 Friesian steers, 9 for each silage. The steers weighed an
average of 510 kg at the start, and were slaughtered at the end of the 85
day trial period.

The early faba bean crop contained only 12% dry matter at cutting so
it was left in the field for two days to wilt by which time the dry matter
had reached 21%. However, there was a considerable loss of material
because of leaf shatter when this crop was lifted and it is believed that
this may partly explain why a much lower dry matter yield (5.1 t ha^{-1}) was
obtained than from the late faba beans (10.5 t ha^{-1}). The cereal/pea crop
yielded 9.5 t ha^{-1}.

Intake of the bean silages was almost as high as of the grass silage,
but intake of the cereal/pea silage was significantly less (Table 4).
Animal performance as measured by carcase gain was 42% better on grass
silage than on cereal/pea silage. The bean silages were intermediate,
although the early one was not significantly better than the cereal/pea
mixture.

In Manitoba (Canada), Ingalls et al. (1979) found faba bean silage to
be superior to maize, barley or grass/legume silages for wintering beef
calves or finishing beef steers. Intake of bean silage was high, leading
to greater liveweight gains than with the other silages. For dairy
heifers and lactating cows, faba bean silage was of similar value to
grass/legume silage, intake and milk production being about equal.

McKnight and Macleod (1972) fed faba bean and grass/legume silages
to lactating cows in Ontario (Canada). The faba bean silage had 4% more
crude protein, but this advantage was offset by feeding a concentrate

containing extra protein to the cows eating grass/legume silage. Cows fed on faba bean silage ingested more silage and increased in body weight more than cows fed on grass/legume silage. Milk yields, protein content and total solids content were similar, but the fat content in milk from the faba bean silage was slightly higher.

TABLE 4   Mean intake of silage and weight gains (kg d^{-1}) by Friesian steers fed on four types of silage

| | Silage type | | | | |
	Early beans	Late beans	Cereal + peas	Grass	SE mean
Silage intake	7.7	7.8	7.3	7.9	0.16
Liveweight gain	1.02	1.18	1.10	1.30	0.038
Carcase gain	0.59	0.66	0.57	0.81	0.026

UNDERSOWING WITH GRASS

In regions like Northern Ireland where livestock farming on grass leys is important, whole-crop faba beans may be of value as a nurse crop for reseeding grassland. With this possibility in mind, most of our experimental crops of faba beans were undersown with a grass mixture (Lolium perenne with Trifolium repens). Sowing of the grass mixture took place immediately after sowing the beans, except in one experiment in 1983 when it was delayed by four weeks.

On most occasions, a good uniform establishment of the grass was obtained and regrowth yields of up to 2.0 t dry matter per hectare were recorded in October. Regrowth forage was comprised almost entirely of L. perenne and T. repens, with only a few scattered faba bean shoots. In 1982, following a very dry period in mid-summer, establishment under beans cut on 19 July was moderate, but under beans cut on 12 August most of the grass died out. In a 1981 trial, faba beans sown at 150 kg ha^{-1} were followed by a significantly better regrowth of grass than faba beans sown at 240 kg ha^{-1}.

CONCLUSIONS

Under Northern Ireland conditions, faba beans give a high yield of forage suitable for ensilage, without any need for N fertiliser. Cultivars of V. faba var. minor appear to be the most suitable.

Although protein levels are good, the feeding value of faba bean silage is probably lower than that of good grass silages. Whole-crop faba beans are not likely to be an attractive alternative to perennial grass leys for routine silage production unless the relative cost of N fertiliser increases. However whole-crop faba beans may be of value either as an arable break crop to be followed by an autumn sown cereal, or as a nurse crop for a grassland reseed. As compared with cereal silage as a nurse crop, faba beans require no N fertiliser and produce a better quality silage.

Undersown grass normally establishes well under faba beans but the ideal managements for the grass and the beans are different. Managing an undersown crop of faba beans involves balancing the benefits of a high seeding rate and late harvest on the yield and dry matter of the silage against the benefits of a low seeding rate and early harvest on the establishing grass sward. The inclusion of some cereal with the faba beans might permit earlier harvesting by increasing the dry matter content, but at the expense of silage quality.

REFERENCES

Ingalls, J.R., Sharma, H.R., Devlin, T.J., Bareeba, F.B. and Clarke, K.W. 1979. Evaluation of whole plant faba bean forage in ruminant rations. Can. J. Anim. Sci., 59, 291-301.

Plancquaert, P. 1971. Études comparative de plantes fourragères annuelles, 1: semis de printemps et début d'été. Institut Technique des Cereales et des Fourrages, Paris.

McKnight, D.R. and Macleod, G.K. 1977. Value of whole plant faba bean silage as the sole forage for lactating cows. Can. J. Anim. Sci., 57, 601-603.

Toynbee-Clarke, G. 1973. Whole-crop Vicia faba for conservation. A comparison between winter-, horse-, tick-, and broad bean cultivars. J. Br. Grassld Soc., 28, 69-72.

SOME RESULTS OBTAINED IN BELGIUM
ON GROWTH, YIELD AND QUALITY OF VICIA FABA
HARVESTED IN GRAIN OR FOR WHOLE CROP SILAGE

A. Falisse[1], F. Cors[1], R. Biston[2] N. Bartiaux-Thill[2]

(1) Faculté des Sciences agronomiques - B-5800 - Gembloux
(2) Centre de Recherches agronomiques - B-5800 - Gembloux

ABSTRACT
    Since 1978, trials on field beans have been set up in two sites in
Belgium. Grain yield comparison between varieties shows Maxime and
Exelle to be superior for both yield and yield stability. When harvested
green, Maxime, Exelle and Maris Bead have given yields between 8 and 14.5
tons of D.M./ha with crude protein contents of about 16%. Nutritive
values before ensiling were about 725 VEM/kg D.M. and the silage
assessments were good. Chemical composition of the silage is given with
and without addition of formic acid whose use is recommended. In that
case, the average nutritive values after ensiling determined using sheep
were about 700 VEM/kg D.M. and 100 g DCP/kg D.M. Digestibility for dry
matter and organic matter were respectively 58% and 61%; dry matter
intake averages 65 g D.M./kg W0.75.

INTRODUCTION
    Trials on field beans have been conducted since 1978 at two sites in
Belgium. The main research topics were the study of the whole plant
productivity and nutritive value after ensiling. This article deals with
yield comparisons and stability of varieties harvested both for grain and
as a whole crop (part 1).
    For one variety the main assessments on the whole crop including the
nutritional values are given before, during and after ensiling (part 2).

1. RESULTS AT HARVEST

1.1. Experimental conditions
    Plot trials were located at two sites:
- at Gembloux, in the loamy region of Central Belgium, altitude 150 m,
  cereal and sugar beet crops;
- at Libramont, in the Southern part of Belgium, called "the Ardennes",
  altitude above 500 m, cold winter and spring, grassland region.
    Separate experiments have determined the main agronomic aspects of
drilling and crop establishment which in the reported trials were as
follows:

P.D. Hebblethwaite, T.C.K. Dawkins, M.C. Heath and G. Lockwood (eds.)
Vicia faba: Agronomy, Physiology and Breeding. ISBN 90-247-2964-5.
© 1984, Martinus Nijhoff/Dr W. Junk Publishers. Printed in The Netherlands.

- sowing dates between end of February and early April for Gembloux;
  between beginning and mid-May for Libramont.
- seed rates between 40 and 50 seeds/m^2, both for grains and whole
  plant harvest.
- inter-rows width : about 30 cm.
- all crops have been protected against aphids by a spray of Pyrimor at
  the beginning of flowering; an average increase of 10 to 15% for
  grains was attributable to this treatment.

1.2 Grains harvest

Average yields obtained for several varieties and their inter-annual
variations are given in Table 1. Compared with cv. Maxime as a standard
all varieties have given lower average yields and higher variation, apart
from Danas and Exelle. Average yields have reached levels higher than 5
tons/ha indicating interest for some farming situations.

The main problem remains the important inter-annual variation
particularly when the results of 1980, included only in Table 3, are
taken into account.

1.3. Whole crop harvest

Average yields of varieties are given in Table 2. In Gembloux, all
varieties in trials have given yields higher than 10 tonnes of dry
matter/ha with inter-annual variations of about 20%, mainly due to the
poor 1980 results. In Libramont, yields were close to 9 tonnes D.M./ha.
Harvests were made between August 9th and September 20th at 18,4 to 20,5%
D.M., earlier or at higher D.M. contents in Gembloux

The protein contents varied from 15.6 to 16.8 and were higher for
Exelle. Between 1.70 and 2.26 tons/ha of crude protein were produced in
Gembloux, 1.42 tons/ha in Libramont.

Maxime, Exelle and Maris Bead were amonst the best varieties both
for dry matter and crude protein yields and stability. Other varieties,
Danas, Dacre, Blaze, ..., which were tested gave lower yields and their
results are not presented.

Table 1. Grain yields of some field beans varieties (1978 to 1982).

Varieties and years	Grain yields at 13 % H_2O	
	Average yields $t. ha^{-1}$	Interannual Coeft. of variation %
1978-79-81-82		
Maxime	5.95	11
Dacre	5.38	18
1978-79-81		
Maxime	6.32	8
Blaze	5.90	8
Maris Bead	5.83	10
Danas	5.80	4
1981-82		
Maxime	6.00	15
Exelle	5.70	7

Table 2. Whole plant harvest of field beans varieties (1978 to 1982).

Varieties, years and location	Dry matter		Crude Protein	
	Average yields $t. ha^{-1}$	Interannual c.v %	Content % DM	Yields $t. ha^{-1}$
Maxime (1) 79-80-81-82	10.9	22	15.6	1.70
Maxime (2) 78-79-80-81-82	8.9	8	16.0	1.42
Maris Bead (1) 79-80-81	11.0	32	16.8	1.85
Exelle (1) 81-82	12.5	5	18.1	2.26
Wierboon (1) 78-82	10.8	24	16.0	1.73

(1) Trials located at Gembloux, in the central area of Belgium called "loamy region" (alt 150 m).
(2) Trials at Libramont, in the southern area of Belgium (alt 500 m).

Table 3. Yields for grains and whole crop harvests in 5 years trials
(means of all varieties).

Years	Grain Yields t. ha^{-1} Gembloux	Whole crop D.M. yields t. ha^{-1} Gembloux	Libramont
1978	5.94	13.3	8.2
1979	5.62	10.1	9.5
1980	2.63	7.8	7.9
1981	6.23	14.5	9.7
1982	4.67	10.0	9.1
Means (t.ha^{-1})	5.02	11.1	8.9
c.v. (%)	26	22	8

Data for individual years are given in Table 3; they confirm the
yield levels and show that the inter-annual variation is lower for
material harvested green than for grain harvest.

2. WHOLE CROP SILAGE

2.1. Materials and methods

Faba bean forage was cut when the leaves started to turn black (dry
matter 18-20%) and ensiled in a vertical silo with and without formic
acid (5%).

After 4 months of storage, feeding trials were conducted using sheep
to evaluate the nutritive value, the digestibility and the intake of the
whole crop silage. The feeding level chosen for the measurements was
near to maintenance level.

Feed samples were taken before and after conservation for the
determination of the chemical composition. Dry matter crude protein,
ether extract, crude fibre, ash, soluble sugar were determined according
to standardized analytical methods. In addition, samples of fresh silage
were taken for determination of pH, volatile fatty acids and soluble
nitrogen.

2.2. Results

(i) Whole crop composition and silage ability assessments (Table 4).
At the considered time for harvesting, the crude protein content was in

Table 4. Whole crop chemical composition, nutritive value and silage ability assessments.

	1978	1979	1980	1981	1982
Dry matter at harvest (%)	18.8	20.5	19.1	18.4	20.5
Chemical composition (% DM)					
- Crude protein	16.3	14.8	15.1	15.8	17.8
- Crude fibre	31.5	34.1	31.2	41.2	25.7
- Ash	6.2	8.0	7.0	7.3	9.5
- Soluble sugar	10.8	9.6	10.5	8.6	-
- N-free extract	46.0	43.1	46.7	35.7	47.0
Nutritive value					
- VEM /kg DM (*)	743	717	738	694	734
- DCP g/kg DM	117	107	109	114	128
Silage ability					
- Buffer power	36	33	39	33	-
- Sugar /CP ratio	0.66	0.64	0.69	0.54	-
- % soluble N	33.4	33.7	33.0	35.0	-

Table 5. Whole crop silage : conservation quality assessments.

	1978		1979		1980	1981	
	(1)	(2)	(1)	(2)	(2)	(1)	(2)
pH	3.7	3.8	4.5	3.9	3.7	3.9	4.1
NNH$_3$ / N Tot. ratio	9.5	6.4	13.4	10.1	5.7	9.0	6.3
% soluble N	37.2	39.9	39.1	45.0	39.5	38.7	40.3
Fatty acids (% F.M.)							
- formic	0.02	0.48	0.02	0.42	0.32	0.01	0.30
- acetic	0.34	0.16	0.41	0.25	0.10	0.29	0.11
- propionic	0.07	0.06	0.06	0.07	0.07	0.02	0.05
- butyric	0	0	0.30	0.03	0.02	0	0
- lactic	2.08	0.86	1.27	1.20	0.20	2.16	0.32

(1) Without formic acid

(2) With 5 ‰ formic acid

(*) VEM = Voeder eenheid melk (milk feed unit), dutch system for energy
    content of feed.

the range 15 to 17% of dry matter; the fibre content varied more widely (27 to 41%), and is dependant on the climatic conditions of the year. Dry matter varied according to the season and was in the range 18 to 20%. The soluble sugar content was stable, on average at 10% of the dry matter. The energetic value and the digestible crude proteins did not vary according to the harvest. They were on average, 725 VEM/kg dry matter and 115 g/kg dry matter respectively. The weak buffer power and the high sugar protein ratio showed that this material was suitable for silage.

(ii) Whole crop silage : conservation quality assessments (Table 5).

The fermentation features; pH, ammoniacal and soluble nitrogen and the organic acids content, indicated that the conserved material was of good quality. Without formic acid, the quality was good except in 1979 when poor fermentation occurred. This may have resulted from insufficient compression or a contamination by butyric spores. The amount of lactic acid produced confirms estimations based on the sugar protein ratio. The silages prepared with formic acid were excellent. This additive allows a reduction of the ammoniacal nitrogen.

In both cases, the soluble nitrogen amount remained largely inferior to the recommended value of 50%. The total losses varied from 11 to 14%. The addition of formic acid largely reduced the losses by fermentation.

(iii) The chemical composition of the silage (Table 6) did not vary much in comparison with the green forage.

The energetic value of the silage was, on average 625 VEM/kg D.M. It was weaker than one given for the green faba bean and varies according to the quality of conservation. The digestibility of the organic matter and the intake measured on adult sheep were 0.6 (0.53 to 0.63) and 65 g/kg W.$^{0.75}$ respectively. The dispersed data result from the fibre content variation, which is itself depending on the morphological composition and the climatic conditions during the growing season.

The addition of formic acid induced a positive effect on the intake.

3. CONCLUSIONS

The faba bean with its high yields, appears to be an interesting alternative to other forage crops. The optimum harvesting time for silage allows, in the same year, the seeding of a winter cereal. The conservation balance is positive especially with the addition of formic acid.

The digestibility and intake values, measured in our trials are lower than the results reported in the literature.

Table 6. Whole plant silage : chemical composition, digestibility, intake and nutritive value.

	1978 (1)	1978 (2)	1979 (1)	1979 (2)	1980 (2)	1981 (1)	1981 (2)
Dry matter (corrected) %	18.9	20.0	21.7	22.9	20.9	20.2	20.8
Chemical composition (% DM)							
- crude protein	15.8	16.4	15.3	14.9	14.2	14.3	14.3
- crude fibre (WEENDE)	32.1	30.4	35.3	34.3	33.6	38.7	37.5
- Ash	7.7	6.8	9.5	9.5	15.4	7.6	7.8
- Soluble sugar	1.4	4.6	1.4	4.2	3.4	1.5	3.3
- N-free extract	44.4	46.4	39.9	41.3	36.8	39.4	40.4
- NDF	43.8	44.0	54.5	51.6	47.3	54.4	53.8
- ADF	33.9	33.7	44.6	43.4	40.9	46.9	45.6
- Lignin (H_2SO_4)	7.5	7.7	8.8	8.5	10.1	10.3	10.1
Digestibility (%)							
- Dry matter	57.5	61.0	51.8	56.2	55.6	60.5	54.9
- organic matter	59.9	63.2	52.9	61.2	60.6	62.7	57.9
Intake ($gDM/kgW^{0,75}$)							
- Dry matter	68.9	76.3	57.4	60.8	72.6	51.0	–
- organic matter	64.3	71.4	52.3	55.5	61.9	53.3	–
Nutritive value							
- VEM/kg DM (*)	688	747	584	687	637	722	652
- DCP g/kg DM	103	107	99	97	92	93	93

(1) Without formic acid

(2) With 5 0/00 formic acid

These two criteria depend on the fibre content, which is itself dependant on the climatic conditions and on the morphological evolution of the plants.

158

For the years when the pod's formation is slow, earlier crop followed by a wilting, would allow an improvement of these coefficients.

(*) VEM = Voeder eenheid melk (milk feed unit), dutch system for energy content of feed.

REFERENCES

Cors, F. and Falisse, A. 1983. La fèvérole récoltée en graines ou plant entière : techniques culturales et rendements. In: "La févérole". Faculté des Sciences Agronomiques Gembloux (Belgium). pp. 142.

Biston, R. and Bartiaux-Thill, N. 1983. La févérole plante entière: production, valeur alimentaire avant et apres ensilage. In: "La févérole". Faculté des Sciences Agronomiques Gembloux (Belgium) pp. 142.

INFLUENCE OF DENSITY ON THE GROWTH AND DEVELOPMENT

OF WINTER FIELD BEAN (Vicia Faba)

D. POULAIN

Ecole Nationale Supérieure Agronomique
Chaire de Phytotechnie
65, rue de Saint-Brieuc
35000 RENNES
FRANCE

ABSTRACT

Plots were sown with a variety of field bean (S 45) at different plant densities (8, 15, 25 and 45 plants/m^2). Measurements of different parameters of growth and development were made including number of stems per plant, height, leaf area index, dry weight of leaves, stems, flowers, pods and seeds.

The highest yield was obtained with 25 plants/m^2, which corresponds to the density normally used.

The effects of an increased plant density were accompanied by an increase in plant height, a decrease (or suppression) of ramification, an increase in leaf area index (values up to 5), as well as an absence of a plateau for the maximum leaf area. Individually the plants produced much lower quantities of dry matter but the efficiency of this is more important. Finally, the dry matter losses appeared lower during ripening.

INTRODUCTION

Like most cultivated plants, the field bean modifies its morphological structure according to the space available to each individual plant. The aim of this study was to determine the parameters of growth and development which respond to crop density modification and the importance of their fluctuations in field conditions.

MATERIAL AND METHODS

The almost complete success in obtaining homogeneity of the plants analysed, led us to eliminate the classic varieties because of their synthetic genetic origin, and to utilise a pure line, thereby conferring upon each plant the same genetic potential.

The trial, which was set up on the grounds belonging to the Ecole Nationale Supérieure Agronomique of Rennes, consisted of 3 replications, each having 4 theoretical densities : D1 = 8   D2 = 15   D3 = 30   D4 = 60 plants per square metre. Each plot comprised 4 rows 25 metres long and 50 cm apart from one another. The size of these plots made it possible to take samples

P.D. Hebblethwaite, T.C.K. Dawkins, M.C. Heath and G. Lockwood (eds.)
Vicia faba: Agronomy, Physiology and Breeding. ISBN 90-247-2964-5.
© 1984, Martinus Nijhoff/Dr W. Junk Publishers. Printed in The Netherlands.

of one square metre solely from the 2 middle rows, by spacing them in such a manner as not to affect the surrounding plants by the "gap" which resulted from samples taken.

The trial was set up on 24 november 1981, on a field where corn for silage had been recently harvested; sowing was carried out by means of a pneumatic drill. The highest plant population was obtained by passing twice over, using the drill set at 30 plants/m^2.

Flowering began 5 May and ended 6 June. The final harvest took place on 12 August.

Each sampling brought all the plants of the same plot together and focused on all of the following observations :

1°) number of plants

2°) total number of stems

3°) average height of plants

4°) total leaf surface

5°) dry weight of leaves

6°) dry weight of stems

7°) dry weight of flowers

8°) dry weight of pods

9°) dry weight of seeds

Eight samplings were made starting in spring (at the time of new plant growth) until harvest time : 6 April, 11 May, 27 May, 9 June, 24 June, 7 July, 27 July, 12 August.

The surface remaining after the last sampling was harvested by means of a combine in order to determine the yield.

RESULTS

Plant population

If the small number of plants obtained (D1 = 8   D2 = 15 plants/m^2) correspond in number to the seeds sown, such is not the case for higher densities where the "losses" were greater : thus, for the desired plant populations of 30 and 60 plants/m^2, only 25 and 45 respectively were attained.

Ramification

The average coefficient of ramification for one sampling was obtained by dividing the total number of stems by the number of plants. At the end of winter (6 April) the effect of density did not have a noticeable effect on ramification (average value : 2.63 ± 0.20). Afterwards, ramification continued in sparse plant populations, whereas one could observe the disappearance of certain stems in dense plant populations. (Figure 1)

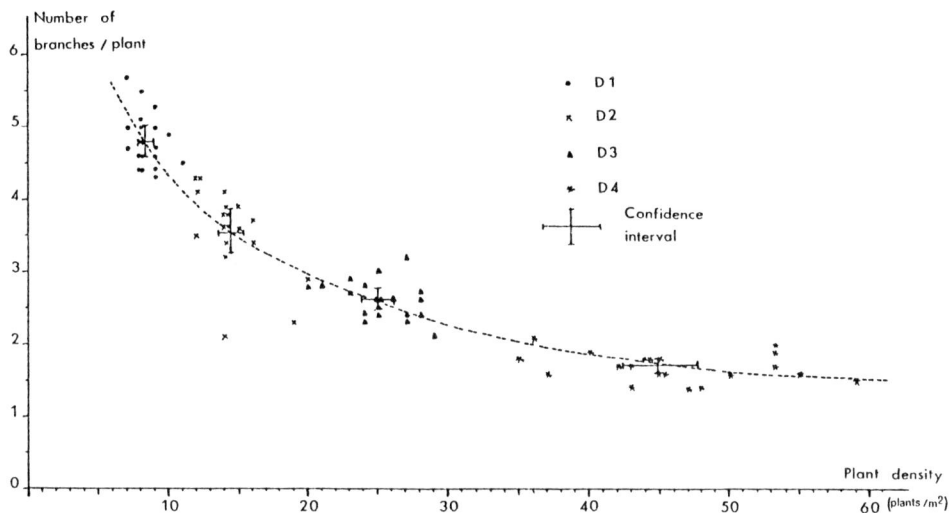

Figure 1. : Influence of plant density on the branching of winter field bean. (D1 = 8 plants/m^2 ; D2 = 15 pl./m^2 ; D3 = 25 pl./m^2 ; D4 = 45 pl/m^2).

Height

The highest plant density was characterised very early by its greater size. (Figure 2). During the month of April, as stem growth increased so, too, did the density. However, at the end of the month of May, all the plants were characterised by an identical growth rate, regardless of the plant population. Growth was approximately 2.2 cm/day ; the maximum height, reached at the beginning of the month of June, was therefore dependent on density. On the other hand, the number of internodes, was not subject to the influence of plant density.

Leaf surface

The leaf area index (figure 3) developed very rapidly : from the start of plant growth in spring, as well as during the intense growth period, the leaf area indices depended on density. The maximum figures were attained towards the end of the month of may, although not all of them were significantly different from one another. The greatest values (up to 5, which is very high to obtain an optimal photosynthetic activity) were obtained among the highest plant populations. On the other hand, the lowest density only reach a value of approximately 3.5, a figure which is much more satisfactory for photosynthetic activity.

It was possible to notice a better persistance of the maximum values of the leaf area index among the sparser plant populations.

The individual leaf surface (obtained by dividing the total leaf sur- face of a sampling, by the number of plants) varies considerably according to the density (figure 3b). Values which were significantly different were obtained from the active growth phase (and the beginning of May) until the end of the month of June, when leaves fell in large quantities.

As it is the case for the leaf area index, the persistance of a maximum leaf surface is greater in low densities.

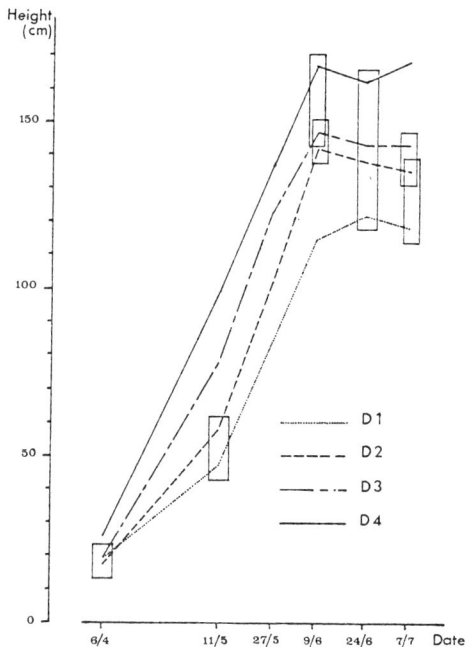

Figure 2. : Influence of plant density on the height of winter field bean (D1 = 8 Plants/m^2 ; D2 = 15 pl/m^2 ; D3 = 25 pl/m^2) ; D4 = 40 pl/m^2) ; points located in a given rectangle are not statistically different.

Growth of the various organs

_ The total production of dry matter per unit of surface increased sli- ghtly with density. Whatever the density of the plants, the maximum was attained about 90 days after the return of spring vegetation. (In our trial, this was shortly before the end of June). It is possible to note that the maximum was reached exponentially for sparse plant populations, and in a more linear manner as density increased. (Figure 4).

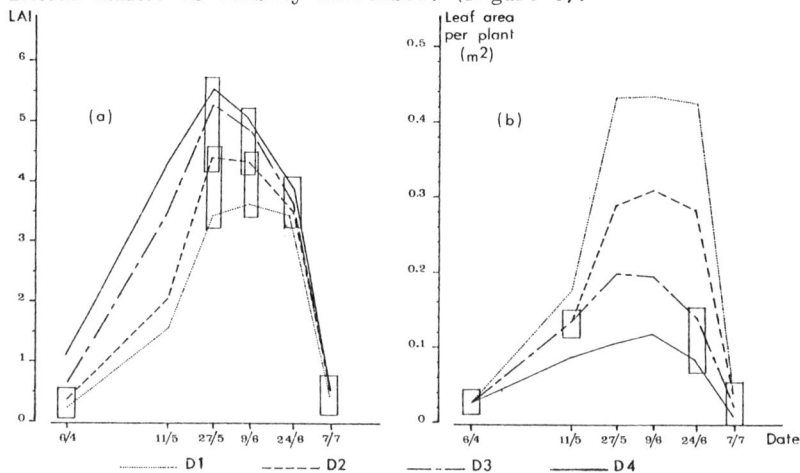

Figure 3. : Influence of plant density on (a) leaf area index (LAI) and (b) leaf area perplant (D1 = 8 pl/m^2 ; D2 = 15 pl/m^2 ; D3 = 25 pl/m^2 ; D4 = 45 pl/m^2). Points located in a given rectangle not statistically different.

- On an individual scale, plant morphology varied, however, according to the space available. For example, the lowest plant density produced plants whose dry matter production was approximately 5 times greater than that of the highest plant density.

- The higher the plant population, the more rapidly the combined weight of stems + leaves ceases to increase : the maximum was reached at the beginning of June with a high density compared to a low density by the end of June.

- The falling of the leaves explains the decrease in the dry weight of these organs. Parallel to this, it is possible to see that the stems show the same tendency, although not always significantly. However, the corresponding dry matter was not transferred to the seeds, because the combined dry weight of stems + pods + seeds decreased significantly in the course of ripening. (Figure 5).

- Moreover, for 3 of the densities studied, the total pods + seeds showed a significant drop in dry matter in the course of ripening. One can assume that the green pods gave up dry matter to the seeds because a slight

164

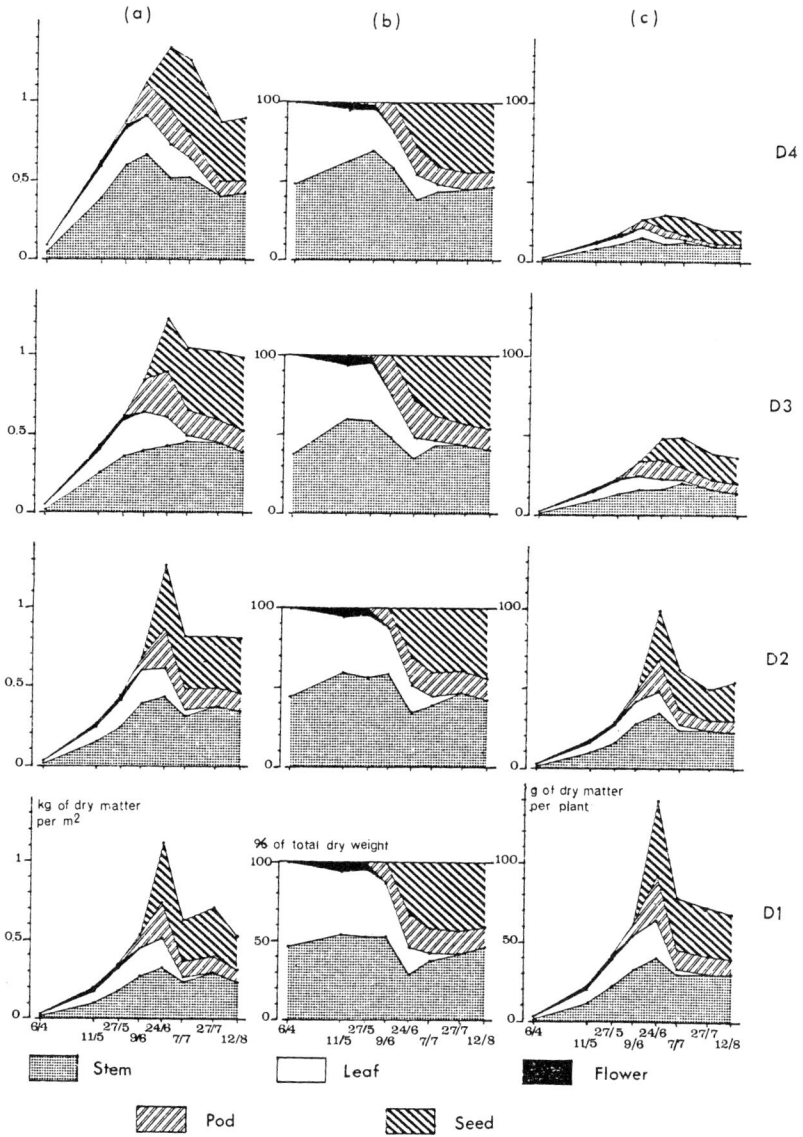

figure 4. : Influence of plant density on the evolution of dry matter distribution. (a) : per unit of surface (m^2) ; (b) : percentage of the total dry matter ; (c) : per plant.

(D1 = 8 plants/m^2 ; D2 = 15 pl/m^2 ; D3 = 25 pl/m^2 ; D4 = 40 pl/m^2).

decrease in the dry weight of the pods can be noted parallel to an increase in the weight of the seeds.

- At an individual level, high densities did not provoke losses of dry matter as great as those which are found in low densities. The "efficiency" of the dry matter formed (= dry weight of harvested seeds/maximum dry weight of the whole plant formed in the vegetation process) varied according to densities : 20 % for the sparsest plant densities and as much as 30 % in the highest densities.

- The final dry weight of seeds can pratically be considered as attained by the end of June.

- The respective contributions of the various organs to the production of total dry matter (expressed in percentages) were not strongly modified by the increase in density, with the exception, perhaps, of the part of the stems which were slightly greater in higher plant concentrations. The analysis of the relationship dry matter of stems/dry matter leaves shows that its value rose with time and that the highest values were obtained (though not always significantly) among higher plant populations.

- Analysed in a different way the criteria "leaf density" (mg/cm^2) and "stem density" (mg/cm) did not reveal any significant differences among the 4 plant groups. For each plant population we can establish a similar evolution :

  * leaf density was at its lowest level (4 to 4.5 mg/cm^2) towards the end of May, at which time the pods began to form. The maximum values were attained (8 to 9 mg/cm^2) shortly before the leaves fell.

  * "stem density" increased at the start of stem growth. The values obtained fluctuated noticeably, and did not make it possible to define clearly the role of plant density. We may at least suppose that the greatest density results quite often in the lowest values.

Yields

Combine harvest at the time of the last sampling showed that yields from the lowest to the highest densities ranged respectively as follows : 25.9 q/ha ; 33.8 q/ha ; 44.4 q/ha ; and 38.7 q/ha of dry seeds. D2 and D3 were not significantly different, as well as D3 and D4.

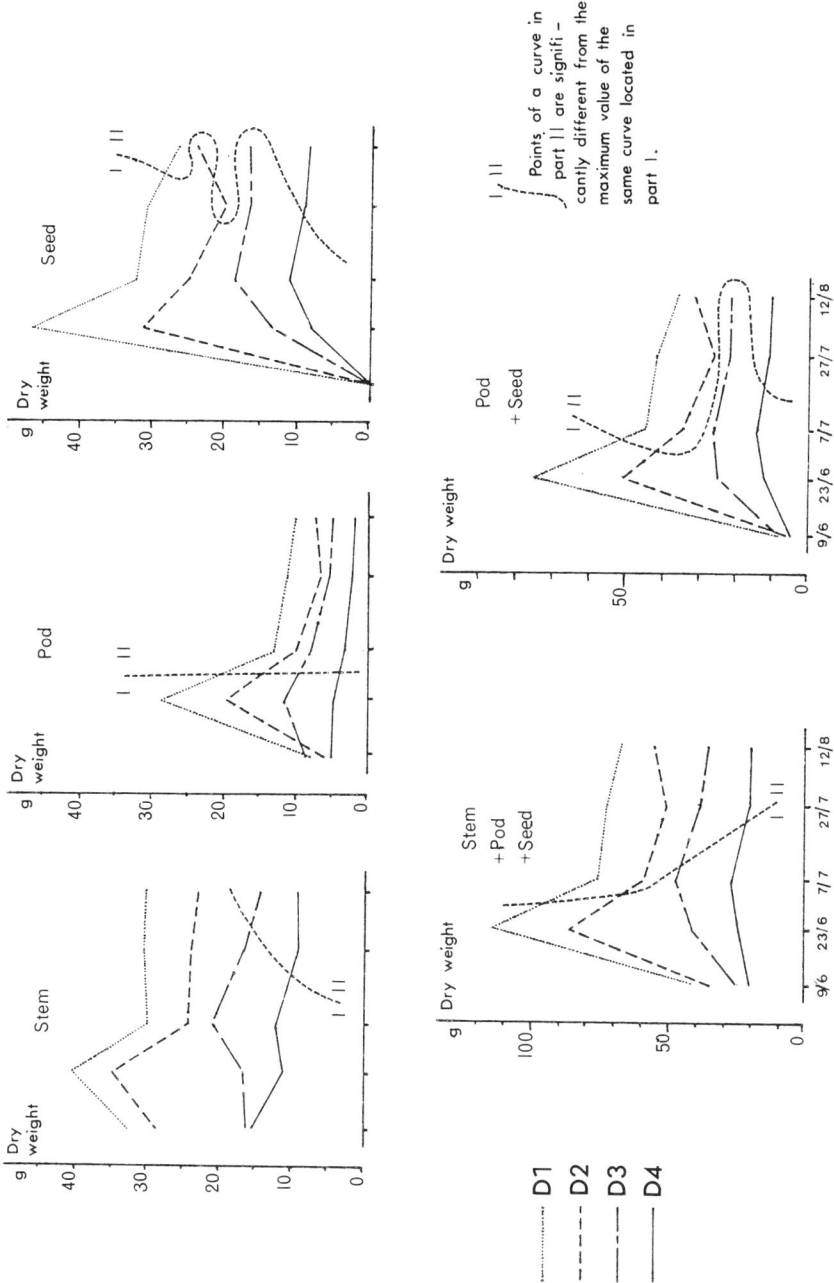

Figure 5. : Influence of plant density on the dry weight evolution of different parts of a plant.

CONCLUSIONS

The variability of the yields obtained (26–44 q/ha) was lower than that of the plant populations under study (8–45 plants/m^2). The plant is therefore likely to adapt its morphology and the factors of its yield according to plant density.

Ramification was perhaps the factor most sensitive to variations in densities. The other parameters seem to reflect a modification of plant morphology to a greater extent. In view of this, the ratio (dry matter seeds/dry matter whole plant) was nearly identical for all densities (0.40 – 0.45).

One notes however, that the best yield was obtained with the D3 density despite the fact that high densities present a certain number of disadvantages :

- plants which branch out very little (the capacity of the winter field bean to ramify is not used here)

- plants of great size (increase in sensitivity to beating down, which did not occur in this trial)

- high leaf area index (losses through respiration, micro climate favoring diseases)

The advantages of this density, however, seemed to outweigh the disadvantages :

- the great amount of total dry matter (since the percentage of grains fluctuates little, the yield is augmented)

- lower losses in dry matter in the course of ripening (which could facilitate the migration and recovery of mineral elements).

RESULTS OF THE JOINT FABA BEAN AND PEA TRIALS

OF THE YEARS 1980-1982

E. Ebmeyer

Institut für Pflanzenbau und Pflanzenzüchtung,
Universität Göttingen, Von-Siebold-Strasse 8
D-3400 Göttingen, Federal Republic of Germany

ABSTRACT

During the years 1980-1982 a second series of the EEC Joint Faba Bean Test was carried out. Eight bean varieties, representing different plant types, and three pea varieties were tested at nine locations in Western Europe. In the average of all environments involved, the highest yielding bean variety was 'Minica' which reaches its yield by a higher harvest index as compared to other varieties. The latter indicates a more efficient distribution of dry matter between generative and vegetative parts of the plant. Comparing yielding capacity of faba beans and peas, the results show that some locations seem to be more suitable for growing peas than for growing beans.

INTRODUCTION

In order to estimate yield capacity and yield stability under West-European conditions a first series of the EEC Joint Faba Bean Test was carried out during the years 1977-1979. The results, published by Dantuma et al. (1983) showed that in general large seeded varieties of faba beans yielded more and were more responsive to higher yielding environments than small seeded ones Furthermore it was concluded from a limited number of data that a semidwarf plant type with reduced vegetative growth should gain special attention in faba bean breeding.

With special emphasis on plant type and harvest index a second series of the Joint Faba Bean Test was carried out during the years 1980-1982. In this series 3 pea varieties were included in order to compare yielding capacity of peas and beans under different climatic conditions.

MATERIALS AND METHODS

The trials were conducted at 8 locations in Great Britain, The Nether-lands, France, Austria and Germany (Table 1). The varieties (Table 1) re-presented different plant types. 'Minica' and 'Montica' are semidwarf types, cultivated mainly for canning industry. Together with 'Wierboon' they belong to the large seeded group. All other varieties are small seeded field beans. The pea varieties are dried pea types with seed weights

P.D. Hebblethwaite, T.C.K. Dawkins, M.C. Heath and G. Lockwood (eds.)
Vicia faba: Agronomy, Physiology and Breeding. ISBN 90-247-2964-5.
© 1984, Martinus Nijhoff/Dr W. Junk Publishers. Printed in The Netherlands.

between 250 and 300 mg.

TABLE 1   Locations and varieties of the Joint Faba Bean Test 1980-1982

Locations	Varieties			
Cambridge (UK)	Faba beans:	Kristall (D)	Peas:	Finale (NL)
Dundee (UK)		Herz Freya (D)		Amino (F)
Wageningen (NL)		Strubes (F)		Columba (D)
N.O.Polder (NL)		Maris Bead (UK)		
Dijon (F)		Deiniol (UK)		
Fuchsenbigl (A) irrigated/not irrigated		Minica (NL)		
		Wierboon (NL)		
Göttingen (D)		Montica (NL)		
Hohenheim (D)				

The experimental layout of the Joint Faba Bean Test was a randomized block design with at least 3 replications at each location. Plot sizes ranged from 6.75 to 13.0 m². The combined analysis of variance in which "Fuchsenbigl-irrigated" and "Fuchsenbigl-not irrigated" were regarded as two different locations, was carried out according to Cochran and Cox (1957, chapter 14). Results of an analysis of yield stability will be reported by Bond (1984). Because of a lot of missing values in the data set the number of environments (year-location combination) is different for the characters. Missing plot values were estimated only for seed yield in order to obtain balancy for the calculation of mean squares due to effects of locations and years separately.

RESULTS

The means of 14 measured or calculated characters over varying numbers of environments are given for each faba bean variety in Table 2. The highest mean yield (50.6 dt/ha) was produced by 'Minica', the lowest (33.4 dt/ha) by 'Montica', both varieties of ssp. major. Within the small seeded group 'Strubes', which is more an equina type, gave the best seed yield. Regarding the other characters the data show that the large seeded varieties have a different structure of yield as compared to the small seeded ones. In particular 'Minica' and 'Wierboon' have an earlier beginning of flowering, and same (Minica) or later (Wierboon) maturity date which consequently leads to a longer generative phase. Furthermore the large seeded

TABLE 2 Means of 14 measured or calculated characters of eight faba bean varieties

Varieties:	Beg.of Flowering (days from sowing)	Duration of Flow. (days)	Generative phase (days)	Maturity (days from sowing)	Plant Length (cm)	No.of pod bearing stems per plant	No.of pod bearing nodes per plant	No.of pods per plant	Daily product. of seed yield (kg/ha)	1000 seed weight (g)	Crude protein cont. of seed (%)	Seed yield (dt/ha at 86% d.m.)	Harvest index (%)	Total yield* (dt/ha harvested dry matter)
No. of environments	25	25	23	23	23	23	25	25	23	25	25	27	25	25
Kristall	72	26	75	147	143.8	1.06	9.1	14.3	59.5	413	30.0	42.1	46	82.9
Herz Freya	72	24	75	146	136.0	1.08	8.1	15.4	60.6	385	28.7	41.9	47	81.1
Strubes	68	26	80	147	131.2	1.09	7.7	10.5	60.2	646	31.0	45.6	50	83.4
Maris Bead	74	27	78	151	134.0	1.21	8.0	14.0	53.9	393	30.8	39.1	46	77.4
Deiniol	73	27	80	152	128.9	1.21	8.8	14.1	55.6	412	31.7	41.3	47	80.4
Minica	65	25	81	146	92.5	1.34	5.1	6.4	65.5	893	29.6	50.6	56	81.7
Wierboon	66	28	87	153	129.1	1.29	6.1	7.6	56.8	1182	28.4	46.5	46	90.2
Montica	65	25	77	142	66.5	1.62	6.0	6.9	45.9	869	28.9	33.4	54	54.6
Mean	69	26	79	148	120.1	1.24	7.4	11.2	57.2	649	29.9	42.6	49	78.9
LSD$_{5\%}$	0.8	1.4	2.8	2.6	5.1	0.09	0.5	1.0	4.2	31	0.6	3.9	1.8	5.4

* calculated from seed yield and harvest index

TABLE 3   Mean performance data of 'Kristall' (K), 'Minica' (M) and 'Wierboon' (W) at eight locations

Character		Cambridge	Wageningen	N.O.Polder	Dijon	Dundee	Fuchsenbigl NI*	Fuchsenbigl I*	Göttingen	Hohenheim
Seed yield	K	36.7	52.8	56.8	51.0	46.0	27.7	31.3	**40.4**	**37.9**
(dt/ha)	M	50.9	64.6	73.2	50.5	49.1	28.2	39.5	57.1	49.0
	W	48.2	62.4	66.9	49.7	52.0	28.3	32.3	43.5	39.8
Harvest	K	46	50	45	53	46	49	37	43	-
index	M	57	63	58	62	55	49	49	59	-
(%)	W	49	50	47	53	50	41	39	45	-
Total yield	K	69.0	91.4	107.8	83.1	86.6	48.4	72.6	80.2	-
(dt/ha)	M	77.1	87.7	108.2	69.7	76.9	49.2	69.8	83.1	-
	W	85.3	108.0	123.2	80.6	90.0	59.1	71.5	83.7	-
Duration of	K	80	81	89	70	82	51	54	84	-
generative	M	91	92	97	73	87	54	58	88	-
phase (days)	W	93	96	105	82	91	60	64	98	-

* Mean of two years only

varieties have more pod bearing stems but less pod bearing nodes and far less pods per plant. The latter, however, is compensated by a higher thousand grain weight.

A comparison of the two highest yielding varieties 'Minica' and 'Wierboon' shows remarcable differences in harvest index and total plant yield. 'Minica' has an average total plant yield not differing from that of the small seeded varieties, but a high harvest index which causes the high grain yield. 'Wierboon' on the other hand has a low harvest index, which in combination with a very high total plant yield leads to a good grain yield too. 'Montica' again is a variety with a high harvest index, but its total dry matter production of only 54.6 dt/ha is too low to reach an adequate grain yield.

For each location Table 3 contains means of 4 characters of 'Kristall', 'Minica' and 'Wierboon', which represent varieties of different yield structure. First of all the data show large differences between locations in the average of 3 years. N.O.Polder was the place with the highest and "Fuchsenbigl-not irrigated" with the lowest productivity for grain yield as well as for total plant yield. In contrast Dijon reached the highest values for harvest index, indicating that this location favours generative growth. In nearly all environments 'Minica' yielded far more than 'Kristall', although total dry matter production was equal or even less. The late ripening variety 'Wierboon' was superior to 'Kristall' particulary at locations which allow a long generative phase and which favour total plant growth.

TABLE 4 Mean seed yields (dt/ha) of bean and pea varieties tested together at eight locations in three years

Bean varieties		Pea varieties	
Kristall	43.4	Finale	51.1
Herz Freya	43.5	Amino	49.1
Strubes	47.1	Columba	43.7
Maris Bead	40.4		
Deiniol	43.0		
Minica	52.5		
Wierboon	48.4		
Montica	33.9		
Mean	44.0		48.0

$LSD_{5\%}$ for comparison between and within beans and peas = 5.0

As already mentioned also 3 pea varieties were included in the Joint
Field Bean Test at all locations, except "Fuchsenbigl-irrigated". Table 4
shows their mean yields in comparison to those of the bean varieties.
'Finale' and 'Amino' produced grain yields of about 50 dt/ha which were
equal or in most cases even higher than those of the beans. A comparison of
the best bean variety and the best pea variety at each location, as shown
in Table 5 gives a slightly different picture. At Cambridge, Wageningen,
Dijon, Dundee and Hohenheim no differences were found between yields of the
best bean and pea variety respectively. N.O.Polder and Göttingen are locations which appear to be better suited for bean production whereas Fuchsen-
bigl is a location more suitable for growing peas. Here the best pea
variety yielded nearly twice as much as the best bean variety.

TABLE 5   Mean seed yield (dt/ha) of the best bean variety in comparison
to the best pea variety at eight locations in three years

Location	Mean of location		Best variety at each location				Difference best bean – best pea variety
	beans	peas	beans		peas		
Cambridge	38.9	45.5	50.9	Minica	50.7	Amino	0.2
Wageningen	55.1	58.2	64.5	Minica	63.4	Finale	1.1
N.O.Polder	58.6	53.9	73.2	Minica	61.6	Finale	11.6**
Dijon	47.4	47.3	53.1	Herz Freya	52.3	Amino	0.8
Dundee	46.3	48.0	53.2	Strubes	51.5	Amino	1.7
Fuchsenbigl NI	25.1	46.3	26.5	Deiniol	48.1	Finale	−21.6***
Göttingen	41.1	40.7	57.1	Minica	48.7	Finale	8.4
Hohenheim	39.7	43.8	49.0	Minica	48.2	Finale	0.8
Mean	44.0	48.0					

, * significance at 5% and 1% level respectively

The analysis of variance in Table 6, calculated for beans and peas
together and seperately, gives a survey of the influence of the different
factors. For beans only as well as beans and peas together the main effect
of years and their interactions with varieties was not significant, whereas the influence of locations and their interactions with varieties and
years was highly significant. The strong influence of locations was already
mentioned in Table 3. The significant interaction "locations x years x
varieties" indicates that at some locations large effects of years on seed
yield do exist. In the case of peas no evidence was found for the main

175

TABLE 6 Analysis of variance for seed yields of beans and peas

Source	Beans and peas together			Beans			Peas		
	DF	MS	F-test	DF	MS	F-test	DF	MS	F-test
Locations (L)	7	7513.65	3.25**	8	8215.91	5.57***	7	859.47	0.57
Years (Y)	2	1174.11	0.59	2	681.19	0.53	2	662.05	0.43
L x Y	14	2107.42	11.30***	16	1342.46	12.95***	14	1461.89	19.82***
Varieties (V)	10	2027.49	6.06***	7	2177.75	10.46***	2	1043.08	3.77*
V x L	70	247.23	1.78***	56	144.26	2.30***	14	113.40	2.87***
V x Y	20	109.91	0.79	14	69.70	1.11	4	174.18	4.32***
V x L x Y	137	138.29	8.41***	108	62.64	5.14***	27	40.28	1.78**

Slight differences in MS compared to Dr. Bond are due to different estimations of missing plot values. For the F-tests the "random model" (EISENHART, 1947) was assumed.

effect of locations., indicating that the pea varieties seem to be better adapted to a wider range of environments than beans. Due to this, peas showed a higher mean performance than beans (Table 4).

CONCLUSIONS

The comparison of varieties with different plant type revealed that high seed yields are realized in two ways: (1) by a high total dry matter production combined with low harvest index and (2) by an average total dry matter production combined with a high harvest index. A high total production, however, seems to be combined with a long duration of the growing period which leads to late harvesting. Therefore, an improvement in seed yield should preferably be realized by an increase in harvest index. This means that the plant type has to be changed in such a way that the total dry matter produced, is more efficiently distributed between generative and vegetative parts. This is particulary important under unfavourable conditions (locations).

An alternative crop for seed legume production may be dried peas which exhibited similar yielding capacity as faba beans. The results indicate that some locations are more suited for growing peas than for beans. Comparing the two crops it has to be considered, however, that up to now peas do not meet the requirements of modern farming with respect to difficulties during harvesting, which may cause losses of already produced grain yield. On the other hand peas reach their grain yields in a much shorter period than beans, providing them with a very favourable position as a preceeding crop in the rotation.

REFERENCES

Bond, D.A. 1984. Yield stability of faba beans and peas in EEC joint trials 1980-82. EEC Seminar, Nottingham 14-16 September (in this book).

Cochran, W.G. and Cox, G.M. 1957. Experimental designs, 2nd ed. John Wiley & Sons, Inc. New York - London - Sidney.

Dantuma, G., von Kittlitz, E., Frauen, M. and Bond, D.A. 1983. Yield, yield stability and measurements of morphological and phenological characters of faba bean (Vicia faba L.) varieties grown in a wide range of environments in Western Europe. Z.Pflanzenzüchtg., 90, 85-105

Eisenhart, C. 1947. The assumptions underlying the analysis of variance. Biometrics 3, 1-21

YIELD STABILITY OF FABA BEANS AND PEAS IN EEC JOINT TRIALS 1980-82

D.A. Bond

Plant Breeding Institute
Maris Lane, Trumpington
Cambridge, U.K.

ABSTRACT

Eight faba bean varieties (G) in 27 environments (E) (comprising 9 locations in 3 years) showed highly significant G x E interaction, some of which was attributable to heterogeneity of regression of varieties on environmental means. As in the 1977-79 series of trials (Dantuma et al., 1983) the larger-seeded (major) varieties gave high mean yields, due to greater responsiveness to high-yielding environments, as compared with the more stable but lower yields of the minor types. The variety Strubes, an equina type not tested in 1977-79, responded in the same way as the major types. Minica had a high mean yield but made a large contribution to the environment x variety interaction.

In a combined analysis of the eight faba bean and three pea varieties over 24 environments, regressions did not account for a significant proportion of the G x E interaction but the pea variety Finale had a greater mean yield and a lower coefficient of variation than the major-type bean, Wierboon. In both mean yield and yield stability there were greater differences among bean or pea varieties than between the means of beans and of peas. Regressions of pea varieties on to environmental means of peas accounted for a much higher proportion of the variance than on to environmental means composed mainly of beans; i.e. one crop was not a satisfactory index of the environment suited to the other.

INTRODUCTION

The means and analyses of variance of yield and other characters recorded in the 1980-82 series of EEC faba bean and pea trials have been reported by Ebmeyer (1984). Another objective of these trials was to assess the relative yield stability of varieties over a range of environments. The principles considered were similar to those outlined by Dantuma et al. (1983) for the joint trials conducted in 1977-79, though in addition there was the intention of comparing beans and peas. The contrasting locations were again considered to approximate to a sample of environments expected over a number of years at a given location, and thereby to allow an estimate of the likely stability of varieties in future years as well as their geographical range of suitability.

P.D. Hebblethwaite, T.C.K. Dawkins, M.C. Heath and G. Lockwood (eds.)
Vicia faba: Agronomy, Physiology and Breeding. ISBN 90-247-2964-5.
© 1984, Martinus Nijhoff/Dr W. Junk Publishers. Printed in The Netherlands.

MATERIALS AND METHODS

The analyses reported by Ebmeyer (1984) showed, in addition to main effects of varieties, locations and years, a highly significant interaction of varieties x years x locations, with relatively little evidence of interactions of varieties x locations or varieties x years. The analyses of stability therefore utilised environments defined as each location in each year, a total of 3 x 9 = 27 environments for beans and 24 for beans plus peas. No attempt was made to estimate stability for locations only or for years only.

As with analyses reported by Ebmeyer (1984), the data were analysed for (a) beans and peas together (b) beans only and (c) peas only. Varieties of beans were also grouped into two classes: (i) minor types, (ii) major types.

For the purpose of this exercise, Strubes, which is an equina type, was classed along with the four minor varieties Maris Bead, Deiniol, Herz Freya and Kristall. The three varieties classed as major types (Montica, Minica and Wierboon) were large-seeded or vegetable types originating in the Netherlands.

The stability parameters estimated were those described by Dantuma et al. (1983), namely:

1. Standard error (SE) and coefficient of variation (CV) about the variety mean.
2. Contribution of the variety to genotype x environment (G x E) interaction (= 'ecovalence' of Wricke, 1962).
3. Regression (b), and SE of b, of each variety on the environmental means.

The computer program used was written by the Statistics Department of the Plant Breeding Institute, Cambridge and was based on the method of Finlay and Wilkinson (1963).

RESULTS

For all three analyses of variance the main effects of environments (E), varieties (G) and their interaction (G x E) were highly significant (Table 1). For the analysis of 'beans only' a significant proportion of the G x E interaction was attributable to heterogeneity of regression on environmental means, but for beans and peas together this was just short of significance and for 'peas only' no indication of differences between regressions was discerned.

TABLE 1   Analyses of variance of yields of beans and peas in EEC
joint trials 1980-82.   *** P < 0.001, * P < 0.05.

	Beans + peas		Beans only		Peas only	
	DF	MS	DF	MS	DF	MS
Environments	23	364.3***	26	347.0***	23	124.4***
Residual	48	4.9	54	4.2	48	3.5
Varieties	10	209.8***	7	242.6***	2	102.9***
Env x Variety	227	16.7***	178	8.3***	45	7.3***
Regression	10	29.1	7	20.0*	2	6.1
Deviation	217	16.1***	171	7.8***	43	7.4***
Residual	474	1.6	370	1.2	94	2.3

Small differences in mean squares, compared to those reported by
Ebmeyer (1984), are due to different estimations of missing-plot
values.

Beans and peas together

As in the trials reported by Dantuma et al. (1983), minor-type beans
had lower SE and CVs than major types. The three pea varieties had lower
CVs than the major beans and were about equal to the minor beans (Table
2). In general a high SE was associated with a high mean yield (e.g.
Minica), and expressing the SE in relation to the mean yield (i.e. as
CV), only diluted this association, it did not remove it.

TABLE 2   Mean yield, some measures of stability and response
to environment in the EEC joint faba bean and pea test 1980-82.
Beans and peas analysed together. (Location Fuchsenbigl-irrigated,
excluded)

Variety (B = Bean) (P = Pea)	Mean Yield kg/ha @ 86% DM	SE	CV	Ecovalence	% variance due to regression	Regression coefficient (b) * P < 0.05 b ≠ 1
Maris Bead (B)	4039	131	3.24	141	82	0.947 ± 0.091
Deiniol (B)	4296	139	3.23	115	87	1.014 ± 0.083
Herz Freya (B)	4353	146	3.35	268	73	0.997 ± 0.126
Kristall (B)	4344	142	3.27	126	87	1.055 ± 0.086
Strubes (B)	4709	161	3.42	200	86	1.198 ± 0.101
Montica (B)	3358	147	4.37	310	67	0.965 ± 0.141
Minica (B)	5247	189	3.60	537	71	1.286 ± 0.168
Wierboon (B)	4885	172	3.53	285	86	1.245 ± 0.105*
Finale (P)	5113	166	3.25	652	47	0.901 ± 0.196
Amino (P)	4833	148	3.06	719	40	0.707 ± 0.175
Columba (P)	4365	143	3.28	609	38	0.686 ± 0.178

The contributions of peas as well as of major beans to the G x E sum of squares (ecovalence) were greater than those for minor beans. Thus the peas were as stable as the most stable beans in terms of CV, but not in terms of contribution to the G x E interaction. The latter effect was associated with a wide scatter about the pea regression lines, shown by the high SE of b, and no pea regression accounted for as much as 50 per cent of the variance.

The above results from individual varieties were confirmed by analyses of all the data grouped into the three classes (Table 3). There were interactions of classes x environments but regressions did not account for a significant proportion, probably due to the high variances of the pea regressions.

TABLE 3   Yield and stability of yield of variety groups (analysis of beans and peas together).

	Mean Yield kg/ha	SE	CV	Ecovalence	% variance due to regression	Regression coefficient (b)
Minor beans	4348	65	1.50	488	89	1.033 ± 0.076
Major beans	4496	115	2.58	500	97	1.160 ± 0.093
Peas	4769	92	1.92	1562	47	0.785 ± 0.170

Beans only

Values for the individual varieties and the means of the minor and major classes (Table 4) generally confirmed the results for the 1977-79 trials (Dantuma et al., 1983) showing lower SEs and CVs for minor than major varieties. High yielding varieties mainly had high SEs and CVs, except for Montica, a major type with low yield and high CV. The SE of the highest yielding minor type, Strubes, was high, but its CV was intermediate.

In all varieties except Montica, regressions on environments accounted for over 85% of the variance, and heterogeneity of regression mainly involved the contrast of Minica, Wierboon and Strubes with the remaining four minor varieties. Maris Bead, for example, with a slope significantly below 1, was different from Minica with b > 1.

TABLE 4   Mean yields, some measures of stability and response to environment in the EEC joint faba bean test 1980-82. (Beans only)

Variety mi = minor ma = major	Mean yield kg/ha 86% DM	SE	CV	Ecovalence	% variance due to regression	Regression coefficient (b) * P < 0.05 b ≠ 1
Maris Bead (mi)	3910	124	3.17	126	89	0.856 ± 0.060*
Deiniol (mi)	4132	134	3.24	79	93	0.942 ± 0.052
Herz Freya (mi)	4189	141	3.37	177	85	0.958 ± 0.079
Kristall (mi)	4207	134	3.19	60	95	0.958 ± 0.045
Strubes (mi)	4560	151	3.31	130	91	1.067 ± 0.067
Montica (ma)	3243	141	4.35	376	69	0.852 ± 0.111
Minica (ma)	5063	180	3.55	349	85	1.222 ± 0.102
Wierboon (ma)	4688	164	3.50	181	90	1.144 ± 0.075
Minor mean	4199	62	1.47	150	97	0.965 ± 0.032
Major mean	4348	110	2.52	261	94	1.059 ± 0.054

The regression coefficients plotted against mean yield reveal a strong positive association between responsiveness to good environments and mean yield (Fig. 1). As a consequence there was no group of varieties with similar mean yields within which selection could be made for a low b value. Wierboon had the lowest SE of b among the major types

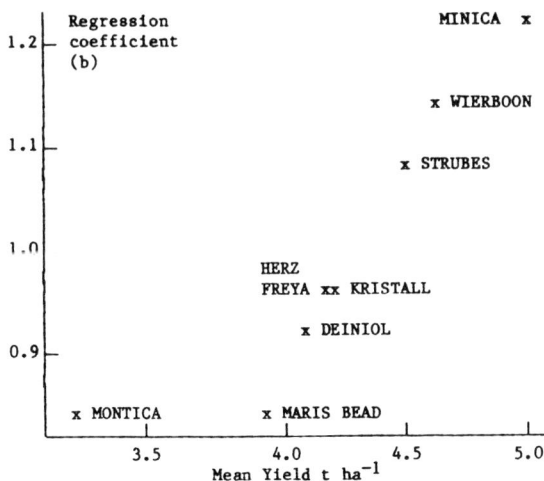

Fig. 1   Relationship of mean yield and regression on environmental means for 8 faba bean varieties in EEC trials.

and Kristall among the minor types, but taking into account the data for 1977-79 (Dantuma et al., 1983), their mean values for SE of $\underline{b}$, for the six years were no better than average for all varieties in their respective groups.

Peas only

The CV of Finale was no higher than that of Columba despite the high SE of the former, but as in the beans, the highest yielding variety contributed most to the G x E interaction and had the steepest regression (Table 5).

It is important to note that regressions on to environmental means measured only by pea yields accounted for a much higher proportion of the variances (over 87%) than when the environmental index was measured largely by bean yields (then under 50%).

TABLE 5   Mean yield, some measures of stability, and response to environment in the EEC joint pea test 1980-82.   (Peas only)

Variety	Mean Yield kg/ha 86% DM	SE	CV		% variance due to regression	Regression coefficient (b)
Finale	5113	166	3.25	164	87	1.076 ± 0.087
Amino	4833	148	3.06	68	93	1.007 ± 0.057
Columba	4365	143	3.28	97	89	0.917 ± 0.066

DISCUSSION

In the joint analysis of 'peas and beans' together, peas made large contributions to the G x E interaction, and had low $\underline{b}$ values (Table 2). But it is clear from the contrast of variances due to regression in the analysis of 'beans and peas together' compared with those in 'peas only' (Table 3) that environmental mean yields of beans + peas (of which 8/11 were beans) were not as good an environmental index for peas as for beans.   The $\underline{b}$ values for peas were, therefore, subject to bias, and regression analyses of 'peas and beans together' difficult to interpret. Probably the environment of each species can only be measured satisfactorily by its own mean yields.

Thus the only stability parameters valid for comparisons between

beans and peas were the SEs and CVs of the variety and class means, these
being independent of each other. For a combination of high yield and low
CV the pea varieties Finale and Amino compared favourably with the best
beans. Also, although only a limited range of pea and bean varieties was
tested, the class means (Table 3) confirmed this conclusion. However,
the difference in yield stability between peas and beans was probably too
small to be a major factor influencing choice of crop. Other factors,
such as agronomic considerations (e.g. straw strength and harvest date)
and demands of the market, are likely to be more important to a farmer.
Moreover the difference between the environmental indices confirms that
peas and beans are best suited to different environments, and therefore
at many locations the two crops are not likely to be in competition.

The main conclusion from the bean yields is similar to that from the
1977-79 series of trials, namely that major types were rather less
stable, higher yielding, more responsive, but less predictably so (higher
SE of b), than minor types. However Montica was low yielding, relatively
unstable and unresponsive. Strubes, which is an equina type, was similar
to Minica and Wierboon in being responsive with a moderately high yield,
but its values of CV and SE suggest it would be more reliable than them.
As in 1977-79, Kristall again had the best combination of mean yield and
stability among the small-seeded varieties.

The results offer only limited scope for selection for stability at
any one level of yield and the choice for a grower is therefore mainly
between high but rather unpredictable yield and lower yield with greater
reliability. At a favourable location it is clearly wise to choose
varieties such as Minica which can respond to the good conditions even
though this might result in higher variances of yields over years at that
location. At a poor location a grower of Minica would not, on average,
suffer any greater yield loss than with a lower yielding more stable
variety such as Kristall (i.e. the regression lines of these two
varieties did not cross over), but, because of the greater instability of
Minica, his investment would be at greater risk.

This factor is compounded, in the case of the varieties tested in
this series of trials, by the large seed-size of the responsive,
high-yielding varieties, (Minica, Wierboon and Strubes). Their extra
seed-costs, including in some cases requirements for specialist machines,
means that more capital is put at risk, and such varieties may be of
interest to growers only at locations where the high yields of these

varieties are frequently expressed.

However a policy of minimum outlay on seed at the poor and average locations means that the full potential production of faba beans will not be realised at those places in favourable years. Ebmeyer (1984) has shown the large effect of years on yield at some locations, and the environments utilized in the stability analyses represented the effects of years as well as of locations. There is therefore a case for encouraging investment in the use of the responsive varieties at the average, or even poor locations as well as at the good ones. Reductions in factors limiting yield (e.g. control of disease, drought, etc.) would also be more likely to be worthwhile where responsive varieties are grown, but at present many growers do not want to risk the outlay.

The dilemma between "high yield and high risk" and "low cost with safe but low return" might be partly resolved if breeders could produce small-seeded responsive varieties, and eventually stable, high-yielding varieties.

I thank Dr. E. Ebmeyer and all the participants in the EEC faba bean and pea trials for supplying the data, Mr. C. Howes for writing the computer program, Mrs. J.A. Hall for running the data on this program and Dr. A.J. Wright for advice on stability parameters.

REFERENCES

Dantuma, G., von Kittlitz, E., Frauen, M. and Bond, D.A. 1983. Yield, yield stability and measurements of morphological and phenological characters of faba bean (Vicia faba L.) varieties grown in a wide range of environments in Western Europe. Z. Pflanzenzüchtg. 90, 85-105.
Ebmeyer, E. 1984. Results of the joint faba bean and pea trials of the years 1980-82. EEC Seminar, Nottingham 14-16 September 1983.
Finlay, D.W. and Wilkinson, G.N. 1963. The analysis of adaptation in a plant breeding programme. Austral. J. Agric. Sci. 14, 742-754.
Wricke, G. 1962. Über eine Methode zur Erfassung der ökologischen Streubreite in Feldversuchen. Z. Pflanzenzüchtg. 53, 266-343.

FABA BEANS IN CHINA : SOME FACTS AND
PERSONAL OBSERVATIONS, PARTICULARLY
CONCERNING VIRUS DISEASES

A.J. Cockbain

Rothamsted Experimental Station
Harpenden, Hertfordshire, England

ABSTRACT

The area of faba bean in China represents about 60% of the total
world area of the crop. The main regions of cultivation are in the
southern provinces of Yunnan, Sichuan, Hubei,. Zhejiang and Jiangsu
(southern part), and in the northern provinces of Qinghai, Gansu,
Ningxia and Inner Mongolia (western part). In the south the crop is sown
in the autumn, and in the north in spring. In Nanjing, Jiangsu Province,
the crop is mainly restricted to small fields, gardens and waste land, or
is interplanted with other crops. The mean yield reported for faba beans
in south China is about 3000 kg/ha, and in the north 2250-3000 kg/ha.
Yields of up to 8000 kg/ha have been reported for the Qinghai - Tibetan
plateau.

Up to 1982 fourteen viruses or virus-like diseases had been reported
in faba beans in different parts of China. In surveys of autumn-sown
crops in the Nanjing area in March and April 1983, five viruses were
detected, namely broad bean wilt, bean yellow mosaic, broad bean luteo-
type, broad bean vein mottle and broad bean spherical (the last three
names are provisional only). Of these, broad bean wilt virus was the
most common and widespread. All except broad bean spherical virus were
readily transmitted by Aphis craccivora. The seed- and pollen-borne
vicia cryptic virus was detected in glasshouse-grown seedlings of three
Chinese cultivars.

INTRODUCTION

This year, at the invitation of Professor Fang Chong-tah, I spent

four months working in the Plant Protection Department at Nanjing

Agricultural College, Jiangsu Province. The main purpose of my visit was

to initiate work on virus diseases of faba bean (Vicia faba) in the

Plant Protection Department and to identify or characterize viruses

isolated from faba bean crops in the Nanjing area. The visit was

supported in part by a grant from the British Council under the Academic

Links with China Scheme and in part by funds provided by Nanjing

Agricultural College. In this paper I summarize the initial results of

the virus study, I report briefly on other observations I made on faba

beans in the Nanjing area, and I quote some facts concerning faba bean

P.D. Hebblethwaite, T.C.K. Dawkins, M.C. Heath and G. Lockwood (eds.)
Vicia faba: Agronomy, Physiology and Breeding. ISBN 90-247-2964-5.
© 1984, Martinus Nijhoff/Dr W. Junk Publishers. Printed in The Netherlands.

cultivation in China.

General accounts of faba bean production and cropping systems in China are given by Zhou (1981) and Jing (1982).

DISTRIBUTION AND CULTIVATION OF FABA BEANS IN CHINA

According to FAO figures the estimated area of faba beans in China in 1981 was 2.2 million ha (Anon., 1982); this represents about 60% of the total world area of the crop. The main regions of cultivation are in the southern provinces of Yunnan, Sichuan, Hubei, Zhejiang and Jiangsu (southern part), and in the northern provinces of Qinghai, Gansu, Ningxia and Inner Mongolia (western part). In the south, where the crop usually follows rice or is interplanted with cotton or maize, it is sown in autumn (usually in October) and harvested about May; in the north, where it is usually grown in rotation with wheat, it is sown in spring (usually in March) and harvested in July or early August.

CULTIVATION OF FABA BEANS IN THE NANJING AREA

Faba beans are widely grown in southern Jiangsu but in the Nanjing area the crop is mainly restricted to small fields, gardens, ditch banks and other waste land. All crops I saw were autumn-sown and all were of green-seeded cultivars of the multi-branch type (seed weights, 750-1050 g/1000). The largest pure stands I saw were about 1-2 ha, and similar sized crops were seen where the beans had been interplanted with peas or in peach orchards. (In Hangzhou, Zhejiang Province, I saw faba beans interplanted in mulberry plantations). Larger crops were seen where the beans were interplanted with autumn-sown wheat.

Most faba bean crops I saw, including all pure stands, were grown as a vegetable for human consumption, but many of the crops interplanted with wheat were grown as a green manure for cotton (Table 1). For the latter purpose the beans are dug in when in flower in April and cotton seedlings are transplanted in May. I was informed by Mr Xu Shing-ghir (Director, August 1st Commune) that the cost of faba bean as a green manure for cotton is about 45 yuan (£15)/ha whereas the cost of artificial fertilizer for equivalent benefit is 270 yuan/ha.

TABLE 1   Some features of three communes near Nanjing visited in 1983

| | Commune | | |
	August 1st	East Bank	October 1st
Total arable land (ha)	2030	860	820
Main crops	Wheat, rice and cotton	Vegetables	Wheat, rice and vegetables
Faba bean area (ha)	167	67	25
Main use	Green manure	Vegetable	Vegetable
Mean seed yield (kg/ha)	2300	–	3000

YIELDS OF FABA BEANS IN CHINA

FAO data indicate that the mean yield of faba beans in China in the 3-year period 1979–81 was 1190 kg/ha/year (Anon., 1982); this compares with a mean yield of 1350 kg/ha in thirteen European countries during the same period (range in the different countries, 540–3670 kg/ha). However, Zhou (1981) reports that the mean yield of faba beans in south China is about 3000 kg/ha, and in the north about 2250–3000 kg/ha. Further, he reports yields of up to 4500 kg/ha for the south, and 7500 kg/ha for the north, and Jing (1982) reports yields of up to 8000 kg/ha for the Qinghai – Tibetan plateau.

Mean yields of faba beans quoted to me on visits to communes in the Nanjing area ranged from 2300–3000 kg/ha (Table 1). However, much higher yields have been reported in Qidong County in south-east Jiangsu (near Nantong) where about 27000 ha are sown to faba beans each year. Thus in a field trial in Qidong County in 1978/79 the mean yield of six faba bean cultivars ranged from 4020–4575 kg/ha (Table 2) (Chen and Lo, 1979).

VIRUSES REPORTED IN FABA BEANS IN CHINA

Yu (1979) described ten viruses or virus-like diseases that have been recorded in faba beans in China, namely bean leaf roll, bean yellow mosaic, broad bean acropetal necrosis, broad bean mild mosaic, broad bean red blotch, broad bean yellow rosette, dolichos ringspot mottle, pea enation mosaic, tomato spotted wilt and tooth tumour-like swelling vein. He also reported a virus-induced wilt of faba bean in Yunnan Province, but was uncertain if the causal virus was broad bean wilt.

TABLE 2   Yields of six faba bean cultivars in an experiment at the
Agricultural Institute of Qidong County, Jiangsu, in
1978/79 (from Chen & Lo, 1979)

Cultivar	Mean number seeds/shoot	Mean number shoots/ha (x 1000)	Mean seed yield (kg/ha)
Duo jia xing	9.8	449	4328
Duo li xing	13.3	426	4343
Qidong no. 1	9.6	438	4080
Qidong no. 2	13.6	438	4575
75-27	10.3	436	4020
75-37	11.8	462	4140

(The above experiment was sown in October 1978 at a seed rate of
98 kg/ha = 2 seeds per hole at a spacing of 20 and 83 cm.   Seed
weights for the different cultivars at harvest ranged from 765-
865 g/1000)

Since 1979 another three viruses have been reported in faba beans
in China, namely broad bean fireblight in Yunnan Province (Kang & Zhou,
1981) and broad bean mottle and broad bean true mosaic in Zhejiang
Province (Ford et al., 1981).   The report of the latter two viruses is
based on symptoms only (R.E. Ford, pers. comm.) and needs to be confirmed
by serology or other means.

VIRUSES DETECTED IN FABA BEANS IN 1983

In surveys in March and April, five viruses were detected in autumn-
sown crops at six sites in the Nanjing area (Table 3).   Virus incidence
was low (fewer than 1% of plants with symptoms) in crops examined in mid-
March but up to 5% of plants showed symptoms in crops examined when in
flower in mid-April.

TABLE 3   Viruses isolated from faba bean crops in the Nanjing area
in 1983

Virus	Particles	Site
Broad bean wilt	Spheres (25 nm)	1, 2, 3, 4, 5, 6
Broad bean luteo-type*	None seen	2, 3, 4, 5
Broad bean vein mottle*	Rods (860 nm)	4, 5, 6
Bean yellow mosaic	Rods (760 nm)	2, 4
Broad bean spherical*	Spheres (35 nm)	4

Sites:

1) August 1st Commune (30 km north-east of Nanjing)

2) East Bank Commune (western suburbs of Nanjing)

3) Jiangsu Institute of Agriculture (eastern suburbs of Nanjing)

4) Nanjing Agricultural College (eastern suburbs of Nanjing)

5) North Bridge Brigade (north of Yangtze Bridge)

6) October 1st Commune (28 km east of Nanjing)

* Provisional name only

The viruses are:

(i) Broad bean wilt virus. This was the most common virus and was
detected in crops at each site. Infected plants showed a range of symptoms
including mosaic, leaflet distortion, leaf and stem necrosis and stunting.
The virus was identified by double diffusion serology and was readily
transmitted by sap inoculation and by aphids (A. craccivora) in the non-
persistent manner.

(ii) Broad bean luteo-type virus (provisional name only). This virus
was detected in plants showing yellowing, stunting and a rolling and
thickening of the leaves in crops at four sites. The virus was transmitted in the persistent manner by A. craccivora but not by sap inoculation.
No virus particles were seen in extracts of crude sap and none were
detected in extracts of dried leaves by immunosorbent electron microscopy
using antisera to some luteoviruses including bean leaf roll and barley

yellow dwarf. In preliminary host range studies the virus caused systemic infection in cowpea and soybean.

(iii) Broad bean vein mottle virus (provisional name only). This virus, with filamentous particles about 860 nm long, was detected at three sites in plants showing a mild yellowish mottle and yellow vein-banding. The virus was readily transmitted by sap inoculation and by aphids (A. craccivora) in the non-persistent manner. Preliminary host range studies suggest that plants with a yellowish mottle and reddish ringspots may have been infected with the same virus.

(iv) Bean yellow mosaic virus. This virus was detected in plants showing mild mosaic symptoms at two sites and was identified by micro-precipitation serology tests. It was readily transmitted by sap inoculation and by aphids (A. craccivora) in the non-persistent manner.

(v) Broad bean spherical virus (provisional name only). This virus, with spherical particles about 35 nm in diameter, was detected by electron microscopical examination of sap from a plant that was also infected with broad bean vein mottle virus in a crop at Nanjing Agricultural College. Its manner of transmission is not yet known.

In addition to the above, the seed- and pollen-borne vicia cryptic virus (Kenten, Cockbain & Woods, 1978; Cockbain, 1980) was detected by immunosorbent electron microscopy in seedlings of three Chinese faba bean cultivars, namely Mao Zi Qing, San Bai and You Xi Number 50 (all from Jiangsu Province). The virus was not detected in seedlings of cultivar Mi Hou Ben (from Fujian Province) and Qinghai Number 5 (Qinghai Province).

PESTS AND DISEASES OTHER THAN VIRUS DISEASES

Of the many pests and diseases of faba beans in China, aphids and bean beetle (Bruchus rufimanus) are the main insect pests (Guan Zhi-he, pers. comm.) and chocolate spot (Botrytis fabae) and root rot and wilt induced by Fusarium spp. the main fungal diseases (Yu, 1979; Zhou, 1981). Sitonia weevils seem not to be important and there are no reports of stem eelworm (Ditylenchus dipsaci).

A. craccivora is the most common and damaging aphid species but Acyrthosiphon pisum is also sometimes common and Megoura japonica is a pest on faba beans in Yunnan Province. Aphis fabae seems not to occur in China (Guan Zhi-he, pers. comm.).

The only aphid species I observed on faba beans in the Nanjing area in 1983 was A. craccivora (identification confirmed by V.F. Eastop). The aphid was common and was causing obvious damage in some crops examined in April but I was informed that beans are rarely sprayed to check aphid infestation.

Studies on the viruses found in faba bean crops in the Nanjing area in 1983 are being continued at Nanjing Agricultural College and Rothamsted Experimental Station. More detailed results of these studies will be published later.

ACKNOWLEDGEMENTS

I thank Professor Fang Chong-tah for inviting me to Nanjing Agricultural College, Professor Chen Yong-xuan and Mr. Xu Zhi-gong for help with the virus surveys, Mr. Sun Chi-Ching for arranging visits to communes, Professor Zhon Xie and Mr. Chiang Chialiang for translating Chinese papers, and Mrs. Sheila E.L. Roberts and Mr. R.D. Woods for electron microscopy. I also thank the British Council and Nanjing Agricultural College for financial support during my visit to China.

REFERENCES

Anonymous 1982. FAO Production Yearbook, 1981. (Food and Agriculture Organization of the United Nations, Rome). Vol. 35.
Chen, J.H. and Lo, Z.Z. 1979. The cultivation test for the bumper harvest of broad bean. In "Collection of Reports by the Agricultural Institute of Qidong County, 1979". (In Chinese)
Cockbain, A.J. 1980. Viruses of spring-sown field beans (Vicia faba) in Great Britain. In "Vicia faba: Feeding Value, Processing and Viruses" (Ed. D.A. Bond). (Martinus Nijhoff, The Hague). pp. 297-308.
Ford, R.E., Bissonnette, H.L., Horsfall, J.G., Millar, R.L., Schlegel, D., Tweedy, B.G. and Weathers, L.G. 1981. Plant Pathology in China, 1980. Plant Dis., 65, 706-714.
Jing, H.X. 1982. A survey of the cropping systems of faba bean (Vicia faba) in China. Fabis, No. 4, 9-10.
Kang, L.Y. and Zhou, C.C. 1981. Notes on the viral agent of fireblight disease of broad bean in Yunnan, China. Microbiol., 8, 153-155. (In Chinese).
Kenten, R.H., Cockbain, A.J. and Woods, R.D. 1978. Vicia cryptic virus. In "Rothamsted Experimental Station Report for 1977". p. 222.
Yu, T.F. 1979. Vicia faba Diseases. (Scientific Press, Peking). 168 pp. (In Chinese).
Zhou, X.T. 1981. Faba bean production and research in China. Fabis, No.3, 24-25.

SELECTION METHODS FOR YIELD IMPROVEMENT IN FABA BEANS.

L.M. Monti and L. Frusciante

Istituto di Agronomia, Miglioramento Genetico,
Università di Napoli

ABSTRACT

On the basis of the results obtained in a long-run experiment on polli-nation in <u>Vicia faba</u> carried out in Southern Italy, a scheme of modified selection was outlined: this is based on a four-year cycle, two of self- and two of open-pollinations, with selection applied in the 2nd and 4th year, respectively, for self-fertility and for combining ability. By using bees for intercrossing after some selfed pods have already been obtained, the cy-cle was reduced to three years. The progress obtained by using these proce-dures is reported.

INTRODUCTION

Several ecotypes of broad and field beans are grown in Italy, many of them with low uniformity of agronomic traits and with poor yield (Monti,1977; Scarascia-Mugnozza <u>et al</u>.,1979). Selection programmes are in progress at Portici with the main aim of improving the yield and the yield stability of some of these types. Modified selection schemes were developed based on the results of our previous research into the genetic structure of the local populations and the influence of presence or absence of pollinating insects on yield.

This paper reports data concerning a long-run experiment on pollination in <u>Vicia faba</u>, the methods followed and the results obtained so far in sele-ction programmes. All the experiments were carried out in Southern Italy in the province of Salerno.

INFLUENCE OF SELF- AND OPEN-POLLINATION ON YIELD

Investigation of some ecotypes of <u>V</u>. <u>faba</u> (Porceddu <u>et al</u>.,1980) revea-led that the percentage of outcrossing ranged between 15 and 46%, with an in-crease in the crossing rate as plant densities increased. Analysis of plants derived from seeds segregating within pods indicated that when pollinating insects visit flowers, both cross-pollination and self-pollination occur in nearly equal proportions. It was concluded that <u>V</u>. <u>faba</u> is to be considered as a species having a rate of cross-pollination between 0 and 50% depending

P.D. Hebblethwaite, T.C.K. Dawkins, M.C. Heath and G. Lockwood (eds.)
Vicia faba: Agronomy, Physiology and Breeding. ISBN 90-247-2964-5.
© 1984, Martinus Nijhoff/Dr W. Junk Publishers. Printed in The Netherlands.

on the number and activity of pollinating insects.

In 1977 a long-term experiment was started and is still continuing.
Plants of the cv. Scuro Torre Lama were grown without bees in net cages and
with bees in open fields; in the following three years, the seed derived
from the selfed plants was divided and tested under both self- and open-pollination and this was repeated with the seed of the open-pollinated plants.
In the last three years, alternate generations of self- and cross-pollina-
tions were no longer followed. Split-plot designs were used each year with
the main plot based on the system of pollination during the experiment and
the sub-plot based on the origin of the material; each sub-plot was 5 m^2 in
area.

In the complete absence of pollinating insects, seed yield was always
greatly reduced due to lack of tripping and consequent drop of flowers. The
research clearly showed that the bee-dependence of V. faba not only affects
the present crop but also the following generation, owing to the combined
effects of inbreeding depression and the lower ability of inbred plants to
self-fertilize. Inbred offspring are obtained at higher rates from plants
grown without bees in comparison with plants grown in open fields (Frusciante et al.,1980). The average values of yield reduction were estimated as 36%,
22% and 19%, due respectively to lack of tripping, to inbreeding depression
and to the reduced self-fertility of inbreds (Monti et al.,1982).

The effect of one self-pollination applied in different generations was
studied by comparing homogenous populations of common origin; Table 1 shows
that yield was reduced by 40% when selfing was applied in the last generation, and by 15% when it was applied in the last but-one generation, but no
effect was obtained when selfing was applied at the last but-two generation.

The effect on yield of one open-pollination was examined in a similar
way (Table 1). Increased seed yield was obtained when open-pollination was
applied in the last generation (89%) and in the last but-one generation
(22%), while no effect was observed when open-pollination was applied two
generations earlier.

TABLE 1   Influence of self- and open-pollination on the yield of dry
seed (Kg).

Period Population	1977–1979	1978–1980	1979–1981	1980–1982	1981–1983	Mean yield %
PPP	2.68	3.36	3.65	2.12	2.72	100
PPS	1.80	1.98	2.39	1.62	0.98	60
PSP	2.22	2.97	___	2.06	2.60	85
SPP	2.40	3.60	___	2.07	2.83	94
SSS	1.01	1.46	1.74	1.19	1.25	100
SSP	1.99	2.70	3.17	1.92	2.80	189
SPS	1.37	2.62	___	1.30	1.18	122
PSS	0.98	1.63	___	1.51	0.94	95

P open-pollination   S self-pollination

The above results show that the yield of SP population (an open-pollinated population coming from a selfed population in the previous year) is
still affected by the negative effect of self-pollination and that in a PS
population (a selfed population coming from an open-pollinated population in
the previous year) the yield is still positively influenced by the open-pollination.

When bees were provided in the net cages, yields were higher than those
in open fields (Frusciante et al.,1980), which showed the possibility of u-
sing bees in order to bring about random cross-fertilization among the
plants in breeding programmes.

SELECTION PROCEDURES

The dependence of our ecotypes on bee visitation to ensure a high yield
and the occurence of inbreeding depression and heterotic effects determined
that the main aim of our selection programme is to obtain improved populations based on inbred lines with good self-fertility and with high combining
ability. A different selection technique is followed, as already proposed by
Rachie et al. (1975), but with a modified procedure, aiming at progressively
increasing the frequency of alleles favourable to both self-fertility and to

combining ability.

Four-year cycle

The results previously reported show that selection for self-fertility is most effective when performed on plants grown in the absence of pollinating insects and derived from plants which were self-pollinated in the previous year, that is in a SS population. In the same way, it is better to select for combining ability by inter-crossing plants obtained from open-pollinated plants, that is in a PP population.

In the scheme which we are following each cycle is therefore based on four years, two of selfing and two of inter-crossing (SSPP), with selection performed in the 2nd and in the 4th year (Fig. 1).

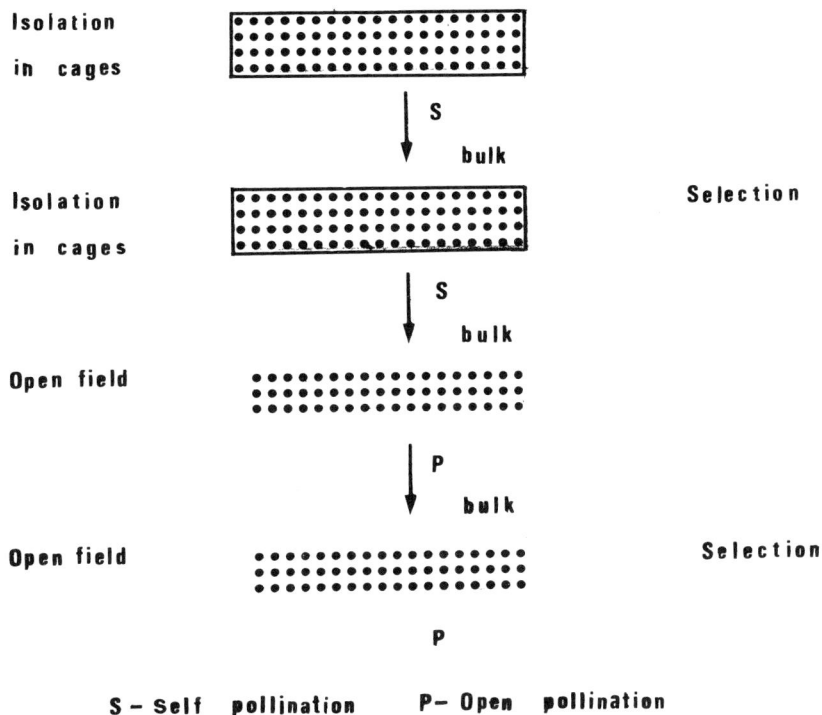

S - Self pollination   P- Open pollination

Fig. 1   Scheme of a modified selection programme in V. faba in four-year cycles.

The results reported in Table 2 support use of this method. Four populations of V. faba were obtained by different sequences of open- and self-pollinated generations, selection was applied in the last self-generation and followed by two generations of open-pollination. The four populations were tested in a randomised block design. There were significant differences in dry seed yield, the highest yield being obtained from the SPPSSPP population. This high yielding population was characterized by two consecutive years of P, which gave good intercrossing among the plants, and by two consecutive years of S in the second of which it was possible to apply effective selection for self-fertility.

TABLE 2 Yields of four populations of V. faba of different origin (see text). Selection was performed on the last S generation (plot size = 5 m^2).

Populations	Dry seed yield (Kg)	Dry plant yield (Kg)	Harvest index (%)
SPPSSPP	2.53	8.53	29.66
SSSSSPP	2.31	8.33	27.73
SPPPSPP	2.32	8.60	26.98
SPSPSPP	2.22	8.00	27.75
Mean	2.35	8.36	28.07
LSD (P = 0.05)	0.14	NS	NS

  P open-pollination   S self-pollination

Three-year cycle

A modified procedure (Fig. 2 and Table 3) of the previously described method was followed in order to reduce each cycle from four to three years. This selection procedure was applied on one cv. (Chiaro Torre Lama) and on three populations, identified as 3, 24 and 27, derived from North African accessions.

In the first year, 600 plants were grown without bees in net cages and only the poorest plants were eliminated. The remaining plants were bulked

198

and grown under net cages in the second year. After pod-setting had occurred
at the first three flowering nodes, bees were introduced in order to ensure
that the plants would inter-pollinate. At harvest, plants were selected on
the basis of the numbers of selfed pods at the first three nodes and cross-
pollinated pods at the higher nodes.

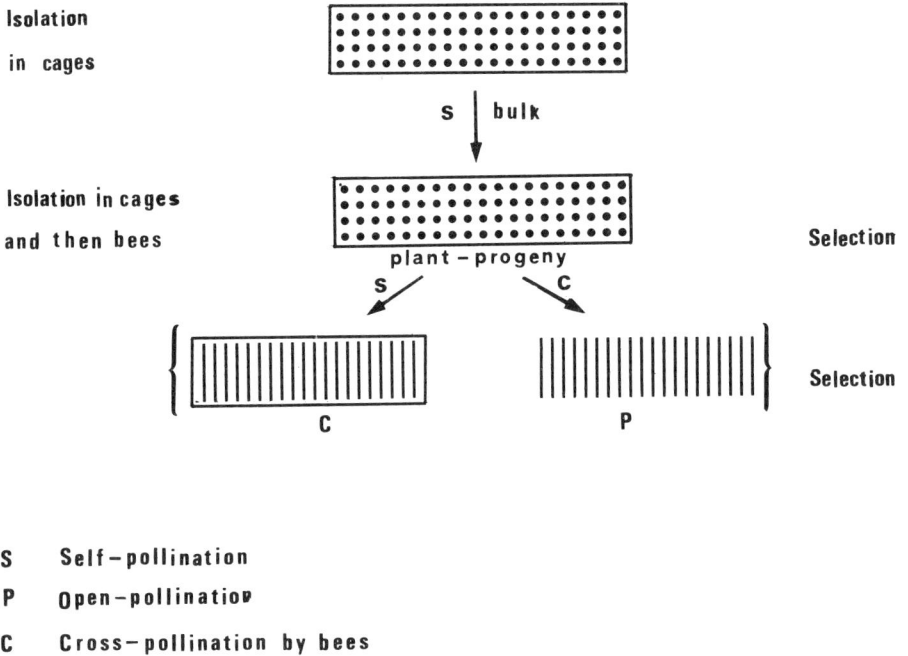

Isolation

in cages

S | bulk

Isolation in cages

and then bees   Selection

plant - progeny

S   C

C   P   Selection

S   Self-pollination
P   Open-pollination
C   Cross-pollination by bees

Fig. 2   Scheme of a modified selection programme in V. faba in three
year cycles.

On 40 plants, pods were taken separately from the first three nodes and
from the higher ones. In the following year, the selfed-seed was again grown
in net cages, and the cross-pollinated seed in open fields, giving rise to
two sister progenies for each plant. At the beginning of flowering, the progenies in the net cages were selected for morphological traits, taking into
account the performance of the sister material grown outside.

Selection for the same characters was then repeated within the 20 remaining progenies before bees were introduced and the plants were allowed to inter-pollinate.

TABLE 3   Number of selected progenies and plants and selection criteria used in a modified selection programme with V. faba.

Year	Progenies analysed (no.)	Plants analysed (no.)	Progenies selected (no.)	Plants selected (no.)	Selection criteria
1	—	600	—	480	Elimination of poor plants
2	—	600	—	40	Selection based on plant fertility
3	40	800	20	350	Selection based on morphological traits of the progenies both in isolation and in open-pollination

TABLE 4   Yield of four selected populations of V. faba in comparison with unselected controls.

		Chiaro T.L.	3	24	27	Mean	LSD Treatment
Dry seed yield (Kg)	Control	2.98	2.60	2.89	2.86	2.83	
	Selected	3.82	3.25	3.02	3.13	3.31	0.25
	Mean	3.40	2.93	2.96	3.00		
LSD Lines NS							
LSD Interaction NS							
Dry plant yield (Kg)	Control	9.07	7.33	8.20	8.67	8.19	
	Selected	10.20	7.93	7.17	7.47	8.32	NS
	Mean	9.63	7.63	7.68	8.07		
LSD Lines 0.680							
LSD Interaction 0.622							
Harvest index (%)	Control	32.8	35.4	35.9	32.9	34.3	
	Selected	37.40	40.9	42.1	41.9	40.8	2.9
	Mean	36.0	38.2	38.7	37.3		
LSD Lines NS							
LSD Interaction NS							

Least significance differences (LSD) at p = 0.05, NS not significant.

The four selected populations were compared with the unselected base populations in a split-plot design with three replications and plots 5 m^2 in area.

Data on dry seed yield, dry plant yield and harvest index are given in Table 4. There were significant differences between the selected populations and each control in seed yield and in harvest index, yield having increased by an average of about 12% and harvest index by about 13%.

This progress seems to confirm the possibility of utilizing a three-year cycle, achieving concurrent selection for self-fertility and for combining ability in the same year by introducing bees to the net cages only after some pods have set by self-pollination.

In addition to saving one year, the proposed scheme offers the possibility of a further enhancement of the genetic gain by screening the selfed material in the 3rd year for other characteristics, such as disease or pest resistance, which remain primary objectives in all faba bean breeding programmes (Hawtin,1982).

REFERENCES

Frusciante, L. and Monti, L.M. (1980). Direct and indirect effects of insect pollination on the yield of field beans. Z. Pflanzenzuchtg, 84, 323-328.

Hawtin, G.C. (1982). The genetic improvement of faba beans. In: Proceedings of the Faba Bean Conference, Cairo, 1981. ICARDA, 15-32.

Monti, L.M. (1977). Le leguminose da granella per l'alimentazione animale: il miglioramento genetico della Vicia faba L. e sua importanza per l'Italia. Riv. di Agronomia, 4, 229-235.

Monti, L.M. and Frusciante, L. (1982). Pollination studies on faba beans. In: Proceedings of the Faba Bean Conference, Cairo, 1981. ICARDA, 33-39.

Porceddu, E., Monti, L.M., Frusciante, L., Volpe, N. (1980). Analysis of cross-pollination in Vicia faba L. Z. Pflanzenzuchtg, 84, 313-322.

Rachie, K.O. and Gardner, C.O. (1975). Increasing efficiency in breeding partially outcrossing grain legumes. In: International Workshop on Grain Legumes. ICRISAT, 285-297.

Scarascia-Mugnozza, G.T., Porceddu, E., Monti, L.M. (1979). Stato attuale del miglioramento genetico delle leguminose da granella. Riv. di Agronomia, 1, 24-40.

BARRIERS TO INTERSPECIFIC HYBRIDISATION BETWEEN
VICIA FABA AND OTHER SPECIES OF SECTION FABA

G. Ramsay, B. Pickersgill, J.K. Jones,
L. Hammond and M.H. Stewart

Department of Agricultural Botany,
Plant Science Laboratories, University of Reading,
Whiteknights, Reading RG6 2AS, U.K.

ABSTRACT

The barriers to hybridisation between V. faba and other species of section Faba were studied. Pollen tubes penetrated stigmas in all combinations. When V. faba was pollinated by V. bithynica, or used as a pollen parent with V. galilaea or V. johannis, the pollen tubes ceased growth in the style. Pollen tubes reached the ovary in reciprocal crosses of V. faba and V. narbonensis but fertilisations were uncommon. Three combinations of parents, V. faba pollinated by V. galilaea and V. johannis, or used as a pollen parent with V. bithynica, resulted in many ovaries with one or more fertilised ovules. All of the embryos produced by these fertilisations aborted at an early stage. A decrease in crossability with increase in seed size of the V. faba parent may reflect greater divergence of the larger seeded varieties from the wild species.

INTRODUCTION

Faba bean breeding would be likely to benefit from the use of hybrids between V. faba L. and its wild relatives. Wild Vicia species are either known or suspected to harbour resistance to Orobanche, aphids, Botrytis and Ascochyta and also show adaptation to poor soils and drought (Cubero 1982). Hybrids could also be used to interpret chromosome homologies between V. faba and related species.

Grain legumes in general appear to be difficult to hybridise (Smartt 1979). Apart from one early and somewhat doubtful Pisum sativum x V. faba hybrid (Sobolev et al., quoted by McComb 1974), no interspecific hybrid plants with V. faba as one parent have yet been produced (Cubero 1982).

Interspecific crosses may fail for various reasons. Pollen may not germinate, pollen tubes may cease growth before reaching the ovary or may not be guided to micropyles and embryo sacs after reaching the ovary. The only data on pre-fertilisation barriers in Vicia section Faba are Van Cruchten's observations

P.D. Hebblethwaite, T.C.K. Dawkins, M.C. Heath and G. Lockwood (eds.)
Vicia faba: Agronomy, Physiology and Breeding. ISBN 90-247-2964-5.
© 1984, Martinus Nijhoff/Dr W. Junk Publishers. Printed in The Netherlands.

that pollen tubes did enter ovules in some interspecific crosses
involving V. faba. After fertilisation, the hybrid embryo may
develop slowly or irregularly, accompanied or caused by abnormal-
ities in the endosperm. Nothing is known about interspecific
hybrids in Vicia in this respect.

A first step to overcoming barriers to interspecific hybridisation is to determine which barriers are operating. They are
likely to vary with different combinations of parental species
and, within each species, some genotypes may prove more readily
crossable than others. It should be possible to select and use
particular combinations of parents together with techniques
developed to overcome or bypass the barriers involved. These
techniques include the use of plant growth substances, to
alleviate problems with pollen tube growth or prolong embryo
development, and the excision and culture of young embryos
(Pickersgill et al. 1983).

Species used in this study are those conventionally included
with V. faba in section Faba. V. faba is, however, so distict
that it should perhaps form a section of its own (Ladizinsky 1975
Cubero 1982). V. bithynica (L.)L. may also not belong in section
Faba. The remaining species, V. narbonensis L., V. galilaea Plit
et Zoh., V. hyaeniscyamus Mouterde and V. johannis Tamamschian,
form a natural group. All four varieties of V. faba (vars.
paucijuga, minor, equina and major) and all species from section
Faba except V. hyaeniscyamus were represented in our crossing
programme. Previous workers have not used either the putatively
primitive V. faba var. paucijuga or V. bithynica.

MATERIALS AND METHODS

V. faba (127 accessions) and wild species (37 accessions)
were grown in a greenhouse at Reading.

Flowers were fixed 24 hours after pollination to observe
pollen tubes (methyl blue and UV illumination) and 7 days after
pollination to observe embryos and endosperm (sectioned in
paraffin wax and stained by the Feulgen method).

RESULTS

Pollen grains were seen to germinate and penetrate the

stigma in all combinations of parents, including reciprocals.
Pollen tubes could not be observed in the whole length of the
style because of its thickness and fluorescence but they could
be distinguished clearly in the ovary.

The proportion of ovaries with one or more pollen tubes
around the ovules and the proportion of ovaries with one or more
fertilised ovules was recorded (Table 1).

V. faba as seed parent

V. galilaea and V. johannis pollen tubes were present in
about half the V. faba ovaries and about half the ovaries from
these crosses contained fertilised ovules. This indicates that
there was no effective barrier to fertilisation and the formation
of hybrid embryos and endosperm.

V. narbonensis crosses in this direction also had pollen
tubes in about half of the ovaries but the percentage of ovaries
with fertilised ovules was significantly lower (p < 0.001),
indicating the presence of a barrier to fertilisation.

V. bithynica pollen tubes, though able to penetrate V. faba
stigmas, reached the ovary in about 10% of the crosses made and
achieved a similarly low percentage of ovaries with fertilised
ovules. Stylar barriers were evidently more severe in this
cross than in the other combinations so far discussed.

V. faba as pollen parent

V. faba pollen tubes never reached the ovaries of either
V. galilaea or V. johannis and hence no fertilisations were
observed in these crosses. Stylar barriers were more severe in
these crosses than in their reciprocals.

V. narbonensis pistils supported the growth of V. faba pollen
tubes but the pollen tubes reached the ovary less frequently than
in the reciprocal cross. The frequency of ovaries with fertilised
ovules was again significantly lower (p < 0.01) than the frequency
of ovaries with pollen tubes which indicates that a similar
barrier operates as in the reciprocal cross.

Pollination of V. bithynica by V. faba gave good pollen tube
growth and a high level of fertilisation, in marked contrast to
the reciprocal cross of this combination.

TABLE 1   Results of interspecific pollinations in *Vicia* section *Faba*.

		Nos. of ovaries with pollen tubes and with fertilised ovules							
		V. galilaea		V. johannis		V. narbonensis		V. bithynica	
		Nos.	%	Nos.	%	Nos.	%	Nos.	%
V. faba as seed parent	with pollen tubes around ovules	42/85	49	17/40	42	64/131	49	5/45	10
	with fertilised ovules	18/45	40	26/76	34	25/255	10	3/43	7
	X^2	0.640 NS		0.424 NS		89.2 ***		0.004 NS	
V. faba as pollen parent	with pollen tubes around ovules	0/9	0	0/24	0	13/50	26	11/20	55
	with fertilised ovules	0/26	0	0/18	0	3/52	6	10/16	62
	X^2	-		-		6.74 **		0.01 NS	

Note: 2x2 tables were used to obtain a X^2 value to test for differences between nos. of ovaries with pollen tubes and nos. with fertilisations within each cross combination. Weighting the results for different numbers of crosses with different varieties of *V. faba* does not alter the conclusions.

Differences within V. faba

The results of all crosses with V. faba as seed parent were
analysed for intraspecific differences in V. faba. The V. faba
parents were divided into six groups on the basis of seed length
(major, equina, and large, medium and small minor) and plant
morphology (paucijuga). V. faba var. paucijuga and small minor
were of similar seed size. The results were also similar and
therefore were pooled. The percentage of pistils with a) germinating pollen grains, b) pollen tubes in the ovary and c) one
or more fertilised ovules were plotted against seed size (fig. 1).

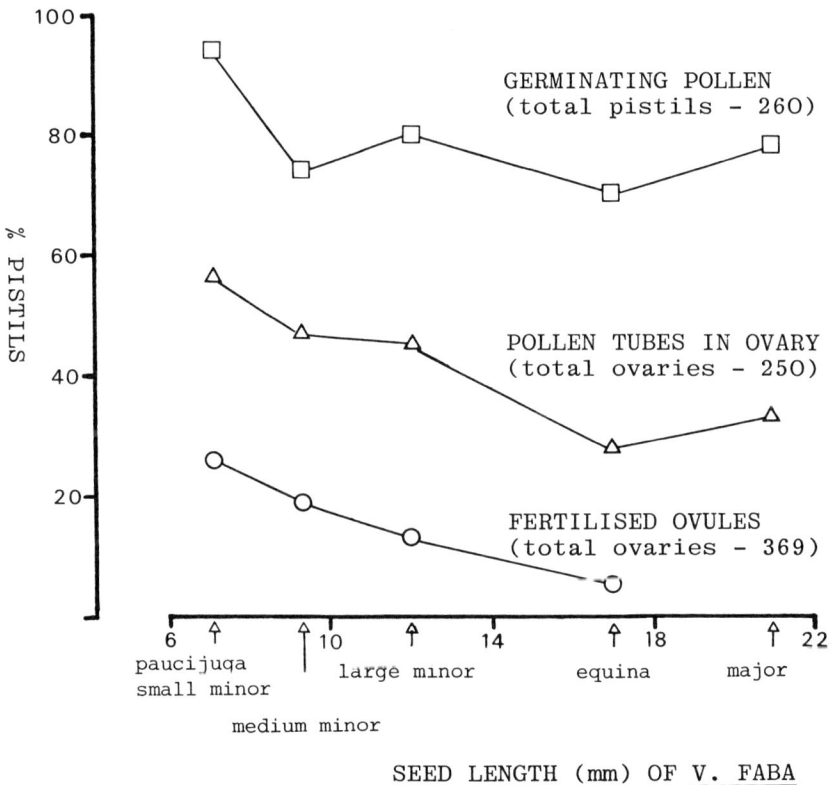

Fig. 1 Success of different varieties of V. faba in
interspecific crosses.

Each parameter decreases as the seed size of the V. faba
parent increases. This trend is more marked for pollen tube

growth and fertilisation than for pollen grain germination. The same trends were found when section Faba species were considered individually (e.g. V. narbonensis pollen fertilised 27% of the pauci juga and small minor ovaries, 19% of the medium minor, 7% of the large minor and none of the equina).

Reciprocal crosses, with V. faba as pollen parent, showed a similar trend.

Subsequent development of fertilised ovules

A limited number of those crosses in which fertilisation occurred most frequently were left longer than 7 days on the plant. Embryos reached their maximum size (about 40 cells) by 14 days after pollination and subsequently degenerated.

DISCUSSION

There were several different barriers to crossing V. faba with related species (Table 2). Pollen tubes were able to germinate and penetrate the stigma in all crosses, but in crosses with V. bithynica as pollen parent and with V. galilaea and V. johannis as seed parents, pollen tubes became arrested in the style which is the normal site of self-incompatibility in the Leguminosae (Heslop-Harrison and Heslop-Harrison 1982).

TABLE 2   Barriers to interspecific hybridisation in Vicia section Faba.

Barrier	Cross	
	Female	Male
Pollen grains fail to germinate	-	-
Pollen tubes arrested in style	faba johannis galilaea	bithynica faba faba
Pollen tubes fail to enter micropyle	faba narbonensis	narbonensis faba
Embryos abort	faba faba bithynica	johannis galilaea faba

When V. narbonensis was used as seed or pollen parent the frequencies of ovaries with fertilised ovules were much less than the frequencies of ovaries containing pollen tubes. This suggests some failure of the mechanism that governs fertilisation, possibly because the chemical basis of attraction and/or recognition between pollen tubes and ovules is different in V. narbonensis and V. faba.

V. galilaea and V. johannis as pollen parents and V. bithynica as seed parent had about half of all ovaries with pollen tubes and also about half of all ovaries with fertilised ovules. These numbers were comparable to those reported for field-grown V. faba (Rowland et al. 1983) which suggests that the barriers to pollen tube growth and fertilisation were not greater in these interspecific hybrids. The hybrid embryos, however, aborted at an early stage.

The results presented here show a trend of increasing crossability with decreasing seed size of the V. faba parent. The smallest seeded variety, paucijuga, is thought to be more closely related than the other varieties of V. faba to wild species of Vicia (Cubero and Suso 1981). It is apparently easier to achieve fertilisation when the putatively primitive variety of V. faba is used in these interspecific crosses.

Interspecific crosses with V. faba which frequently gave hybrid embryos have now been identified - notably V. faba var. paucijuga as a seed parent with V. galilaea and V. johannis and as a pollen parent with V. bithynica. All embryos from these crosses aborted after about two weeks, before they were large enough to culture. Work is now progressing on prolonging the growth of these embryos with chemical treatments and developing culturing techniques able to cope with very small embryos.

ACKNOWLEDGEMENTS
We are very grateful to the Overseas Development Administration for financial support and to all those who supplied seed.

REFERENCES
Cubero, J.I. 1982. Interspecific hybridisation in Vicia. In "Faba Bean Improvement." Proceedings of the Faba Bean

Conference held in Cairo, Egypt, March 7-11, 1981. (Eds. G. Hawtin and C. Webb) (Martinus Nijhoff, The Hague).

Cubero, J.I., and Suso, M.-J. 1981. Primitive and modern forms of *Vicia faba*. Kulturpflanze 29, 137-145.

Heslop-Harrison, J., and Heslop-Harrison, Y. 1982. Pollen - stigma interactions in the Leguminosae: Constituents of the stylar fluid and stigma secretion of *Trifolium pratense* L. Ann. Bot. 49, 729-735.

Ladizinsky, G. 1975. On the origin of broad bean, *Vicia faba* L Israel J. Bot. 24, 80-88.

Mc Comb, J.A. 1975 Is intergeneric hybridisation in the Leguminosae possible? Euphytica 24, 497-502.

Pickersgill, B., Jones, J.K., Ramsay, G., and Stewart, H. 1983. Problems and prospects of wide crossing in the genus *Vicia* for the improvement of faba bean. Proceedings of the International Workshop on Faba Beans, Kabuli Chickpeas and Lentils in the 1980's, 16-20 May 1983, ICARDA, Aleppo, Syria.

Rowland, G.G., Bond, D.A., and Parker, M.L. 1983. Estimates of the frequency of fertilisation in field beans (*Vicia faba* L.). J. Agric. Sci. 100, 25-35.

Smartt, J. 1979. Interspecific hybridisation in the grain legumes - a review. Econ. Bot. 33, 329-337.

Van Cruchten, C. 1974. Étude de la croissance des tubes polliniques dans les croisements intra et inter spécifiques de *Vicia faba* et *Vicia narbonensis*. Internal Report. Station D'Amélioration des Plantes, Dijon - Cedex.

BREEDING FOR SELF-FERTILITY

J.I. Cubero*, M.T. Moreno**

*Dpto de Genetica, E.T.S.I.A., Apartado 3048, Cordoba
**Unidad de Leguminosas, CRIDA-10, INIA, Cordoba

ABSTRACT

Advantages and disadvantages of self-fertile cultivars are discussed, and a method to avoid the complete loss of heterosis is suggested. Nine components of self-fertility related with the processes of pod and seed production were analysed from a genetic point of view. The parameters obtained suggest that transmission of self-fertility is not difficult The genetic control of these characters is also briefly presented and discussed, together with the disadvantages detected when studying it by means of F_1 diallel crosses and F_2 segregations. From a practical point of view, pods/node and seeds/pod (in the absence of insects) seem to be the most convenient indexes to be used to select for self-fertility, according to their correlations with the other parameters studied.

INTRODUCTION

Self-fertility (or autofertility) is the inherent ability of some allogamous (total or partially) plant species to produce seeds when self-pollinated. Total absence of self-fertility, that is, strict self-sterility is only present when strong genes for self-incompatibility are acting (as in the case of many Trifolium species for example). Self-fertility is a condition which may have evolved from weak genetic systems for self-incompatibility, this weakness showing different degrees even within the same species. Both natural and artificial selection can act on these systems awakening or reinforcing them according to environmental conditions or human selection.

Obviously the degree of self-fertility (not only in intensity but also in range) can vary from one crop species to another, depending on the origin and the history of these crops: alfalfa shows a low degree of self-fertility; Lupinus, in contrast shows a low rate of self-sterility. In faba beans self-sterility is moderate but can range as much as from nil to total. Total self-fertility here means only functional and not obligate autogamy that is, the plant can fertilize up to 100% of its flowers in the absence of the pollinating agent, in this case, insects. Only a perfect expression of the "closed flower" mutant could produce an obligate autogamous cultivar. It is worthy of note that from a practical point of view, the rate of 5% allogamy (the rate of allogamy is the

P.D. Hebblethwaite, T.C.K. Dawkins, M.C. Heath and G. Lockwood (eds.)
Vicia faba: Agronomy, Physiology and Breeding. ISBN 90-247-2964-5.
© 1984, Martinus Nijhoff/Dr W. Junk Publishers. Printed in The Netherlands.

210

proportion of zygotes produced by alien pollen) is usually taken as the
barrier separating autogamous from allogamous crops from the viewpoint
of breeding methodology.

A crop with more or less weak genes for self-incompatibility, hence
allowing self-fertilization, can be handled by breeders in two ways; to
select for self-sterility or to select for self-fertility. The former
could be advisable to ensure cross-pollination, for example, to obtain
synthetic varieties. The latter selection method will be used to obtain
a variety yielding in the absence of pollinating insects, that is, a
self-fertile cultivar. This one is not an autogamous cultivar: in the
absence of insects it will yield more than a cultivar not selected for
that character, but if pollinators are present the rate of outcrossing
can be high depending on a number of factors such as the floral biology
of the cultivar, ecological conditions favouring the synchrony of
flowering and insect activity,type of pollinators and climate.

ADVANTAGES AND DISADVANTAGES OF SELF-FERTILE CULTIVARS

It is well known that one of the major problems of Vicia faba as a
crop is lack of yield stability and a solution to this problem is
essential for ensuring a future for the crop. Self-fertile cultivars may
help in the sense of producing a certain degree of independence from
pollinators but total self-fertility is difficult to obtain in commercial
varieties. This represents a base line for yield, that is, a predictable
minimum yield. If conditions favour the presence of active pollinators,
the yield will be higher, but at least one of the many factors causing
instability will have been controlled. This kind of cultivar will be
particularly useful in regions of erratic climate such as the
Mediterranean and, in general, the dry areas. It is not surprising to find
a much higher degree of self-fertility in land-races from these areas
than in those coming from humid regions. These races could also be useful
in regions where the crop is not yet grown, where there are few
pollinators (Lim Eng Siong and Knight, 1979).

It must be stressed, however, that the relative importance of self-
fertility does not determine in its own right the method of breeding. A
high degree of self-sterility can be used to obtain a new standard
cultivar, which is not necessarily a synthetic. Conversely, a high degree
of self-fertility is compatible with a synthetic: mixing different
genotypes showing good general combining ability (g.c.a) and very similar

agronomical traits will allow for a certain minimum yield if there are few insects and for a greater yield if they are numerous. The decision on the breeding method must consider self-fertility as one more variable, perhaps an important one in conjunction with the ecological conditions, but not decisive.

Let us now consider the main disadvantage, the lack of heterotic effects. One third of the characteristics of faba beans show heterosis (Cubero and Martin, 1981), and in that figure the characters related to yield are well represented (yield, seeds and pods per plant, for example). A cultivar showing a low rate of outcrossing will counter, to some extent, the increases due to the heterotic effects (the use of the "closed flower" will counter them absolutely). But the low rate of outcrossing is a fact that can be attributed more to ecological effects than to the self-fertile nature of the cultivar. In any event it must be recognized that a self-fertile cultivar will produce more selfings than outcrossings, even in the presence of pollinators. This is because if the first flower(s) are not visited, a self-fertile cultivar will self pollinate them, thus, if the next flower is outcrossed by a pollinator, the pod produced will be disadvantaged with regard to those selfed earlier, because of intra-plant and intra-node competition. In contrast, the flowers of a self-sterile plant will await a pollinator or, if selfed, the pods produced will be disadvantaged compared to those derived from earlier out-crosses.

METHODOLOGY; SOURCES OF GENES FOR SELF-FERTILITY

Thus, if the method of self-fertile varieties is chosen it will be because the breeder has thought that it is preferable to ensure a minimum production than to be concerned with the uncertain increase in yield resulting from outcrossing. This strategy does not preclude the possibility of using heterotic effects in favourable conditions. As stated previously there are possibilities of obtaining self-fertile synthetic varieties. To obtain measures of g.c.a. in a crop like faba beans is time consuming if the material is very self-fertile. It is possible to perform a double selection in consecutive years in two different conditions: namely, under insect free conditions using cages and in the open field. The former will allow selection for self-fertility. Two consecutive years are advisable in this phase, because the most fertile plants in the first year under insect free conditions could be hybrids from the previous season. Thus, the selection would be performed in the second year using

a cage. The third year, under open field conditions, will be used to
select those plants showing the best yield indices. In our current work
we select for the simultaneous presence of a good pod-set and a good
number of pods/node (Cubero and Martin, 1981; also because of the results
of the present work). Thus plants which allow outcrossing will be
selected. This system is in fact a recurrent selection, perhaps slower in
grouping favourable genes than the strictest method, but much easier to
manage. The method can be applied to a single original population or to a
mixture of several taking care not to mix material which is too
heterogeneous in morphology.

A different way of breeding for self-fertility is to introduce "self-
fertile" genes from convenient material. Mediterranean land-races are
good candidates for this. The reason for this is the historical handling
of faba beans by farmers. In Spain, for example, farmers used to choose
as mother plants (for the next years seeds) the earliest ones in their
plots and even, in some cases, the earliest tillers of the earliest
plants (this might be considered as useless, but any somatic mutations
would be detected). This way of selecting the reproducers allows, in a
mild climate such as the Mediterranean winter, for selection of the
plants which start flowering before the period of intense pollinator
activity; "self-fertile" genes will then increase in frequency.

This is not a general rule, of course. We have found extreme self-
fertile types in some Indian and Ethiopian accessions. The progenies of
the crosses between them and commercial cultivars show that self-fertility
is a character which is easily transmitted. To explain how these
accessions are so highly self-fertile would also require some knowledge
of their history and use. In particular, the case of the accession
VF 172 (paucijuga) (in our opinion the most primitive material, from a
morphological point of view, in present collections) has been explained
by Cubero and Suso (1981) as being a consequence of human selection. These
plants, with 1-2 flowers/node, few flowering nodes per tiller and less
than 3 ovules/ovary (on average), had only a few possibilities for yield:
namely to be highly fertile and to increase the number of tillers; and
this they did.

The methods discussed here are by no means definitive; Lawes (1973)
and Poulsen (1975) have described other methods. The object of the
present paper is to present our current work in breeding faba beans for
self-fertility.

COMPONENTS OF SELF-FERTILITY. GENETIC STUDIES

To provide a definition of self-fertility is easier than to split this character and then to study it from a genetic point of view. We will be concerned here only with absolute self-fertility, that is, the plant will produce as much in the absence of pollinators as in open pollination. We do not propose to deal with conditioned self-fertility, that is, those plants producing as indicated but only after tripping. The response to tripping in faba beans is an interesting and as yet unresolved question (the floral structure seems to be the most important feature; Kambal et al 1976) but our aim is to study absolute self-fertility, since it is the only one which may allow for the reproduction of commercial self-fertile cultivars. Conditioned self-fertility would also require insect pollination. There is also a differential response to tripping in self-sterile plants, conditioned self-sterile plants showing some response, whilst absolutely self-sterile plants show none.

A plant or an inbred line can be fully self-fertile but produce a very poor yield, due to such causes as a low number of seeds/pod, very low number of flowering nodes and abortion of young pods owing to physiological constraints. We began our studies by examining what happens in the floral node. Several indices have been described in the literature for studying self-fertility (weight/node, number of seeds or pods/node, tripped v. non-tripped nodes etc.). We prefer to use a number of characters to describe the stages from real fertilization to seed production considering the transformation of flowers to pods and that of ovules to seeds. Thus the characters used were: flowers/node, fertilized ovaries/flower, fertilized ovaries/node, mature pods/fertilized ovary, pods/flower and pods/node, and, according to the transformation ovules/seeds: ovules/ovary, seeds/ovule and seeds/pod. These nine characters do not exhaust the components of self-fertility, but they may help in elucidating the problem.

The study was performed in a triple way: the correlation between characteristics (phenotypic, genotypic, environmental and maternal effects), the heritabilities and, finally, the genetic control of these characters.

a) Correlation between characteristics.

It is not possible to provide a complete picture of the results in a short report of this nature but those from two different F_2s indicated high phenotypic, genotypic and environmental correlations (all of them positive) for:

> pods/node - pods/flower
>
> pods/flower - pods/fertilized ovary
>
> fertilized ovaries/node - pods/fertilized ovary
>
> pods/node - pods/fertilized ovary.

In the latter, one F_2 produced intermediate values. Medium to high genotypic correlations were obtained in both crosses for:

> fertilized ovaries/node - pods/node
>
> pods/flower - fertilized ovaries/flower
>
> fertilized ovaries/flower - pods/fertilized ovary

With other materials (F_1, pure lines and outbreds) high values for the correlations coefficient were obtained for pods/flower-pods/node.

Other high values were obtained for:

> pods/flower - ovules/ovary (genotypic, negative)
>
> pods/flower - flowers/node (gen., neg.)
>
> seeds/ovule - seeds/pod (phenotypic, gen., environmental
>
> > all positive)
>
> ovules/ovary - flowers/node (gen., pos.)

In different material, the coefficient for pod/flower-seeds/ovule showed intermediate but generally significant values (gen., phen.; both pos.).

Of course, correlation values are dependent on the material; we are more confident of the F_2 results than those obtained from inbreds or F_1s (Cubero & Martin 1981) and more crosses will have to be analysed in order to obtain a better understanding of the behaviour of different lines. Even the F_2 generation is not perfect material because of its residual heterozygosity. The results expressed above can be tentatively summarized as follows: the higher (lower) the number of selfed flowers, the higher (lower) the number of young pods produced, and the higher (lower) the probability of the latter reaching maturity. Thus pods/node is a good index of the number of flowers transformed to pods. In a similar way, seeds/pod is a good index of the proportion of fertilized ovules (in this case in F_1 and also in a large collection of land races both without any inbreeding and with one inbreeding generation). Finally, if confirmed, the correlation between pods/flower and seeds/ovule (positive) could mean that the barrier found by the pollen tube in the stigma and style is found again in the micropyle. That is: floral fertility corresponds roughly to ovular fertility. This correlation could also be explained by the existence of a "correct" or an "incorrect" vascular system in the plant: if both flowers and ovules are well fed (see Gates and Boulter, this volume), the double transformation of flowers to pods and ovules to seeds will be high. If the vascular system does not allow for an adequate uptake of nutrients of both flowers and ovules, these indices will be low. Both hypotheses (i.e., incompatibility and vascular system) are not incompatible to each other.

At the level of maternal effects, pods/flower was positively correlated with pods/node. Again the latter is a good index of the former. But pods/node was also highly but negatively correlated with seeds/ovule, that is, the mother plant could compensate for the increase in pods/node by reducing the number of ovules reaching the seed stage.

b) Heritabilities.

Repeatabilities (in F_2) were around 0.50 or higher for flowers/node, fertilized ovaries/node, pods/flower and (almost) pods/node. The repeatibility is a lower limit for heritability, and it was confirmed by estimates of broad and narrow sense heritabilities (F_1, inbreds): high (broad and narrow) for flowers/node and pods/flower and medium to high (idem) for pods/node. Heritability was found to be medium to high in the broad sense but low in the narrow sense for seeds/pod. To interpret these results, we must consider again the dependence of these parameters on the material studied. Tentatively, they indicate a fair degree of genetic transmission of self-fertility, better for the transformation of flowers to pods than for that of ovules to seeds. Our experimental results support this hypothesis, as well as those of other authors (Adcock and Lawes 1976, Poulsen 1975, Lim Eng Siong and Knight 1979). Rowlands (1961, 1964) did not find a positive response after three generations, but found that self-fertility was dominant over self-sterility (Rowlands 1960).

c) Genetic control.

It was studied in two ways: by mean of F_1 diallel crosses and by mean of F_2 segregations. A wide range of variation was present in the parental lines. This variation produced on the one hand, an inconvenient effect, because of the problems of interactions. But on the other hand, in some cases, the results indicated the existence of different genetic systems (Table 1).

TABLE 1 Genetic systems controlling fertility characters in Vicia faba.

	Diallel F_1		F_2 segregations	
	Inheritance	sense	Inheritance	sense
Flowers/node	NO DOM.	-	NO DOM.	-
	PART. DOM.?	NEG.	PART. DOM.	NEG.
Pods/flower	PART. DOM.	POS.	PART. DOM.	POS.
F.O./flower	-		PART. DOM.?	NEG. POS.
Ovules/ovary	PART. DOM.	POS.	DOM.?TRANSG.	NEG.?
Pods/F.O.	-		PART. DOM.	NEG.
Pods/node	OVERDOM.	POS.	PART. DOM.	NEG. POS.
F.O./node	-		TRANSG.	POS.
Seeds/pod	OVERDOM.	POS.	-	
Seeds/ovule	PART. DOM.?	POS.	PART. DOM.?	POS.

For example, in the case of pods/flower crosses between two paucijuga lines showed partial dominance of positive sense (i.e., self-fertility dominant), but the F_2 between the self-fertile and a self-sterile minor produced dominance of negative sense.

Ovules/ovary showed a bias (but not general) in F_2 towards the negative sense of dominance, but in an F_1 diallel the conclusion was that there was no dominance or partial dominance of positive sense. Crosses between minor and paucijuga accessions in all combinations also showed a high degree of heterogeneity for pods/node, as in the case of pods/flowers, with a large proportion of transgressive inheritance.

When comparing the results obtained from F_1 diallel crosses with the F_2 derived from them, it was obvious that many cases of dominance and even overdominance inferred from the former could be seen as partial dominance (for example, pods/node) after F_2 results. In the case of ovules/ovary, the sense of the dominance changed from positive (F_1) to negative (F_2) always within partial dominance; perhaps, the conclusion is that there is no dominance. The case of number of seeds/ovule moved from positive partial dominance in F_1 to no dominance or partial dominance of negative sense in F_2. The main conclusion to be drawn from these results is that F_1 diallel crosses alone are probably not the best means of studying the genetics of these characters since the heterotic effects are strongly reinforced in faba bean F_1's. The "Mendelian way" for genetics -i.e., Parentals, F_1, F_2, F_3 and Backcrosses- seems to be a much more realistic approach. It was not in vain that Mendel did his experiments working on a close relative of faba beans!

REFERENCES

Adcock, M.E. and Lawes, D.A. 1976. Self-fertility and the distribution of seed yield in Vicia faba L. Euphytica, 25, 89-96.

Cubero, J.I. and Martin, A. 1981. Factorial analysis of yield components in Vicia faba L. In "Vicia faba L.: Physiology and Breeding" (Ed. R. Thompson). (Martinus Nijhoff, The Hague). pp. 139-151.

Cubero, J.I. and Suso, M.J. 1981. Primitive and modern forms of Vicia faba. Kulturpflanze 29, 137-145.

Kambal, A.E., Bond, D.A. and Toynbee-Clarke, G. 1976. A study on the pollination mechanism in field beans (Vicia faba L.). J. Agric. Sci., Camb., 87, 519-526.

Lawes, D.A. 1973. The development of self-fertile field beans. Annual Report of the Welsh Plant Breeding Station, 19, 1972, 163-176.

Lim Eng Siong and Knight, R. 1979. Fertility studies on some Vicia faba L. population. Sobrao Journal, 11, 133-146.

Poulsen, M.H. 1975. Pollination, seed setting, cross-fertilization and inbreeding in Vicia faba L. Z. Pflanzenzuchtung, 74, 97-118.

Rowlands, D.G. 1960. Fertility studies in the field bean (Vicia faba L.) I. Cross and self-fertility. Heredity, 15, 161-173.

Rowlands, D.G. 1961. Fertility studies in the field bean (Vicia faba L.) II. Inbreeding. Heredity, 16, 497-508.

Rowlands, D.G. 1964. Fertility studies in the broad bean (Vicia faba L.) Heredity, 19, 271-277.

A DESIGN FOR TESTING SOME GENOTYPE x ENVIRONMENT INTERACTIONS

J. Picard and G. Duc

Institut National de la Recherche Agronomique
Station d'Amelioration des Plantes
B.V. 1540   21034 Dijon Cedex,   France

ABSTRACT

With faba beans (*Vicia faba* L.), breeding for better yield stability is a very important aim and testing for this character is difficult to achieve.

At the Plant Breeding Station in Dijon we are testing the possibility of estimating yield stability through estimation of stability of some of the more prevalent yield components early in the course of the breeding programme i.e. using limited amounts of seed.

Compared with classical experiments, in 1982, the Nelder design seems valuable. Results have to be confirmed with data from the 1983 harvest.

INTRODUCTION

The first EEC meeting devoted to *Vicia faba* and other crops was held in Padova in 1976. From this emerged the joint field bean test which, as part of its objectives, had the appraisal of yield and yield stability over locations and years.

Since that time, much more work has been done on yield stability than in the past but the problem still remains for the farmer to decide what level of confidence he can put on the expected yield when he sows his field. For the breeder the problem is how to get reliable information with the minimum use of resources. The following factors have been suggested as causes of yield instability:

Flower drop - 95% or over has been reported - is always impressive but frequently high percentages are due to the very high number of flowers produced by vigorous genotypes which also give the highest yields.

Deficient pollination is often proposed as a possible cause of yield instability. Studies by Kambal (1969), Chapman *et al.* (1979), Rowland *et al.* (1983) and Rowland and Duc (to be published, 1983) have shown that inadequate pollination is of limited importance as most abscised flowers have been pollinated and fertilized.

P.D. Hebblethwaite, T.C.K. Dawkins, M.C. Heath and G. Lockwood (eds.)
Vicia faba: Agronomy, Physiology and Breeding. ISBN 90-247-2964-5.
© 1984, Martinus Nijhoff/Dr W. Junk Publishers. Printed in The Netherlands.

If external factors (temperature, drought, diseases, pests) are to be excepted, intraplant competition seems to be the most important factor controlling yield stability, as shown through experiments by Hodgson and Blackman (1957), Jacquiery (1977) and Gehriger (1978). Explanations of direct use for breeders or growers can be expected from more physiological or biochemical experiments and such work is to be actively supported.

Whatever the factors which control yield stability the breeder is concerned to quantify this character as early as possible in the breeding cycle. The method of doing this must allow use of a limited amount of seed and the possibility of testing a large number of genotypes.

In this paper we describe the experimental approach we used in 1982 and again in 1983 to test at one site a number of genotypes under different environmental conditions.

MATERIALS AND METHODS.

In spring of 1982, eight genotypes (12 in 1983) were sown in two different experiments. The plant material was chosen to represent a large diversity of genotypic variability : breeding lines, commercial varieties and an experimental hybrid, and phenotypic variability; monostem or no tillering types, high or low number of flowers per raceme or per plant etc..

Two experiments were set up with these genotypes in 1982 and 1983 :
- a classical split-plot design with 2 replicates, the first order subplots being devoted to irrigation (none and about 60 mm given in the middle of June), the second order to densities (respectively 25, 50, 75 plants per square metre at sowing and the third order to genotypes.
- a Nelder design experiment (Nelder 1962) with 2 replicates in which the positioning of plants gave densities from 10 to 100 plants per square metre (10, 20, ... 90, 100).

In the classical design yield was estimated by harvesting 7 m^2 plots (border rows discarded) and 10 plants were used throughout the growing period for observations on number of stems, number of flowers per node, number of pods etc..

In the Nelder experiment the first flowering node was located on each plant. Twelve plants per density were harvested when ripe and data were collected on yield components.

Results given here are limited to those from 1982.

TABLE 1   Nelder experiment - Statistical significances

	Stems/plant	Podded nodes/plant	Pods/podded node	Pods/plant node	Seeds/pod	Seeds/plant	Seed weight	Yield	Protein content
Density	HS	HS	NS	HS	NS	HS	NS	HS	NS
Genotype	HS	HS	HS	HS	HS	HS	HS	HS	HS
Density x genotype	HS	HS	NS	HS	NS	HS	NS	S	NS

TABLE 2   Classical density experiment - Statistical significances

	Stems/plant	Flowers/plant	Young pods/flower (1)	Pods/flower nodes/plant (1)	Podded nodes/plant	Pods/podded node	Pods/plant	Seeds/pod	Seed weight	Yield	Protein content
Irrigation	NS	NS	NS	HS	NS	NS	HS	NS	NS	HS	HS
Density	HS	HS	HS	HS	HS	NS	HS	NS	NS	HS	S
Genotype	HS	HS	HS	HS	HS	HS	HS	HS	HS	HS	HS
Density x Genotype	HS	HS	NS	S	NS	NS	HS	NS	NS	S	NS
Irrigation x Density	NS	NS	NS	HS	NS	NS	NS	NS	NS	HS	NS
Irrigation x Geno-type	NS	NS	HS	NS	NS	NS	NS	NS	NS	HS	NS

(1) Estimate based on flowering nodes number 3 and 4 on the main stems.

NS : non significant   S : significant at P= 0,05   HS : significant at P= 0,01

RESULTS.

Tables 1 and 2 give a brief summary of the significance of the variance ratios obtained in statistical analyses of the results from the
two experiments. Some general remarks can be made :
- over the two experiments the genotype effect is highly significant for
every character studied, indicating that there is substantial genetic
variability in the sample of material studied.
- some characters were analysed in both trials, giving rise to common columns in table 1 and 2. The significances of the variance ratios are in good
agreement and among 18 results, only two differ.
- some characters are very responsive to density, including stems/plant,
flowers/plant, pods/flower, podded nodes/plant and pods/plant. Others are
largely stable, including seeds/pod and seed weight.

In a multiple correlation analysis of the two experiments, the three
characters appeared in the same order accounting for the largest part of
yield per unit area (Table 3).

TABLE 3 R values for correlations with yield in the two
experiments

	Density exper.	Nelder exper.
Number of tillers/plant	0.623	0.566
Seed weight	0.659	0.649
Seeds/pod	0.714	0.674

DISCUSSION.

We do not intend in this paper to discuss in detail the biological
or agronomic significance of the results. It is clear, for instance, that
watering at the mid-flowering stage can hardly influence tillering.

Two points must be emphasised when discussing the results.
- Firstly, the multiple correlation analysis explains only a limited part
of the yield variation, 59 and 55% respectively in the classical and Nelder
experiments. This is probably due to the fact that the responses observed
were not linear, and this will be examined when the 1983 results are analysed statistically.
- Secondly, the range of individual genotypes included in the trials determined the intensity of the different interactions. The results are

strongly influenced by this effect and therefore only apply to the specific genotypes examined.

It would seem surprising to have the number of stems per plant as the most important component of yield. However, if in the Nelder experiment, we look at the multiple correlation analysis for each genotype separately, very often (5 out of 8) the most important character is the number of pods per plant.

We want to focus here on the good agreement we obtained between the two types of experiments. If this agreement is confirmed by the 1983 results, the Nelder-type experiment can be recommended as a way of testing genotypes for yielding ability, yield characteristics and yield stability. Some more work is needed to test the validity of this type of experiment with other series of genotypes with reduced variability for characters such as seed weight, seeds per pods and stems per plant.

Using a more restricted number of plant densities this type of experiment could be used to test genotypes through densities, locations and other agronomic factors such as nutrition and irrigation. Compared to clasical experiments, it requires a limited amount of seed and therefore can be used earlier in a breeding programme. Observations may be restricted to a limited number of characters and specially pods per plant. Nevertheless, many of the cultural operations are difficult to mechanize and therefore the system as a whole is laborious.

REFERENCES.

Kambal, A.E. 1969. Flower drop and fruit set in field beans *Vicia faba* L. J. Agric. Sci., Camb. 72, 131-138.
Chapman, G.P., Fagg, C.W. and Peat, W.E. 1979. Parthenocarpy and internal competition in *Vicia faba* L. Z. Pflanzenphysiol. 94. 247-255.
Hodgson, G.L. and Blackman, G.E. 1977. An analysis of the influence of plant density on the growth of *Vicia faba* II. The significance of competition for light in relation to plant development at different densities. J. Exp. Bot. 8. 195-219.
Jacquiery, R. 1977. Etude de la chute des fleurs chez la fèverole (*Vicia faba* L.) - Relations avec la disponibilité en assimilats marqués au 14C - Thèse n° 5893 - Ecole Polytechnique Fédérale - Zurich.
Gehriger, W. 1978. Influence de la température et de l'écimage sur le développement de la fèverole (*Vicia faba* L.) et étude de la nutrition des fleurs en assimilats marqués au 14C. Thèse n° 6133 - Ecole Polytechnique Fédérale - Zurich.
Nelder, J.A. 1962. New kinds of systematic design for spacing experiments. Biometries 18. 283-301.

Rowland, G.G., Bond, D.A. and Parker, M.L. 1983. Estimates of the frequency of fertilization in field beans. J. Agric. Sci. Camb. <u>100</u>, 26-33.

Rowland, G.G., Duc, G. and Picard, J. The effect of environment, apex excision and flower removal on fertility components of fababeans (*Vicia faba* L.) (in publication).

ANNEXE I

Genotypes used in the two experiments.

1- Ascott - Variety on the French national list.
2- Experimental single cross hybrid.
3- Wierboon - Old Dutch variety.
4- 123 - Breeding line from a Leningrad collection.
5- 240 - Breeding line from a Leningrad collection.
6- 319 - Breeding line from a Leningrad collection.
7- 370 - Breeding line from a collection from Iran.
8- 795 - Breeding line from a collection from Pullman (USA).

MICRODENSITOMETRIC ANALYSIS OF VICIA FABA L. CHROMOSOME IMAGES [+)]

E. Filippone*, L.A. Smaldone** and L.M. Monti***

*Centro Miglioramento Genetico Ortaggi C.N.R.
Portici (Napoli)
**Osservatorio Astronomico Capodimonte - Napoli
***Istituto di Agronomia, Miglioramento Genetico,
Università di Napoli

ABSTRACT

An analysis of photographed chromosome images (1250x) of Vicia faba L. was performed by a microdensitometer scanner and by FORTRAN programmes.

Data on length and arm-ratio of each chromosome showed an error definitely lower than that obtained by usual techniques.

Differences among the chromosomes were found with respect to the number and the position of density areas inside the densitometric profiles.

INTRODUCTION

A method was set up in order to utilize photographic images of plant mitotic chromosomes for identifying the chromosomes by means of photometric parameters (Filippone et al., 1982).

This paper reports data concerning the application of this method to Vicia faba L. ($2n = 12$), in the framework of a research program carried out in Portici on the cytogenetics of this species.

MATERIALS AND METHODS

Mitotic metaphase chromosomes (1250x), stained with Feulgen, were photographed using an Agfa panchromatic film. Twenty photographic plates were analysed by a digital microdensitometer (Perkin-Elmer PDS 1010 A), connected to a minicomputer (Digital PDP 8/A); a scan slit of 25x25 μm was used. This analysis converts the continuous photographic density distribution into a matrix recorded on a magnetic tape (Roberti et al., 1982). Two hundred readings per second were achieved.

For each plate a matrix of 600 x 500 density data was obtained in about 25 minutes. The matrices were analysed by a minicomputer (Digital PDP 11/34

+) Contribution no. 1 from Centro di Studio per il Miglioramento Genetico degli Ortaggi - C.N.R. - Portici (Napoli) - Italy

P.D. Hebblethwaite, T.C.K. Dawkins, M.C. Heath and G. Lockwood (eds.)
Vicia faba: Agronomy, Physiology and Breeding. ISBN 90-247-2964-5.
© 1984, Martinus Nijhoff/Dr W. Junk Publishers. Printed in The Netherlands.

with 128 Kbytes of main memory) and FORTRAN programmes were utilised to obtain for each plate the number of the chromosomes (the satellite of chromosome no. 1 was considered as an independent structure), their perimeter, area and position on the plates.

On each chromosome, the densitometric profiles of the chromatids were drawn. A semi-automatic method was followed: the digitized chromosome images were re-built on a computer video-graphic terminal and the profiles were drawn manually. The elaboration of data was automatically performed by means of appropriate programmes. The arm-ratio of each chromosome was calculated on the basis of the centromere position. The length of each chromosome was the average of the chromatids and identification was based on the relative length, with the longest chromosome (no. 1) equal to 100.

For each chromosome type, a mean profile was obtained as a sum of the profiles of their chromatids. The mathematical analysis of the profiles showed that they are a linear combination of Gaussian curves; therefore each chromosome profile could be defined by the parameters of these Gaussian curves (i.e. their number, positions, heights and widths).

All the dimensional parameters and the densitometric profiles for each chromatid were permanently recorded on the computer in order to obtain a data bank of the species.

RESULTS

Table 1 reports the length and the arm-ratio of the chromosomes of Vicia faba L. numbered according to decreasing length (Errico, 1980); values of standard deviations are reported and an average 5% coefficient of variation was found.

Using the method of microdensitometric analysis, areas with different optical densities were found in all the chromosomes. Table 2 reports the position of these areas in each of the six chromosome types plus the satellite; the position is expressed as percentage of each chromosome length. The frequency with which each density is present in the indicated area is also reported in table 2.

TABLE 1   Length (as % of the longest chromosome) and arm-ratio of chromosomes in Vicia faba L.   Data from 20 processed metaphase   plates

Chromosome number	Length	Arm-ratio
1	100	1.6 ± 0.2
2	53.4 ± 1.8	5.3 ± 1.2
3	50.6 ± 1.9	5.8 ± 1.7
4	48.1 ± 1.9	6.0 ± 1.7
5	45.8 ± 1.7	6.1 ± 1.5
6	42.9 ± 1.7	5.4 ± 1.1

TABLE 2   Position (as % of the chromosome) of density areas and their frequency in each chromosome of Vicia faba L.

	Chromosome number						
	1 without satellite	satellite chromosome no.1	2	3	4	5	6
processed chromatids	24	70	51	42	58	52	50
position	8	28	8	18	14	16	21
frequency %	96	84	100	100	100	73	84
position	25	62	19	67	29	49	49
frequency %	92	94	100	93	81	100	98
position	44		33	86	58	84	83
frequency %	100		100	43	88	4	2
position	63		60		87		
frequency %	100		94		86		
position	81		83				
frequency %	100		98				
position	92		90				
frequency %	79		98				

A different number of density areas is present in the different chromosome, with the highest number in chromosome 1 and 2; the satellite shows
only two density areas. Differences among the areas were also found as far as
their parameters are concerned.

Optical density profiles of the six chromosome types based on data from
table 2 are shown in fig. 1, where the satellite is not reported; the centromere of each chromosome is also indicated.

DISCUSSION

Karyotypic analyses in plants are usually performed on the basis of dimensional parameters (i.e. length and arm-ratio) and in some cases on the basis of simple photometric parameters (i.e. Giemsa banding); these analyses
are done by manual and visual measurements.

Some techniques were introduced in the recent past for more accurate
measurement by the use of a computerized analysis of chromosome images obtained by light microscopy and recorded on photographic film (Andronico et al.,
1978); records on magnetic tape by a TV camera have been also used in cyto-
photometric analysis (Van Oostveldt, 1980).

Improvement of the microdensitometer scanner and of the computer facilities offered the possibility of analysing photographic plates with a resol-
ving power higher than that at present obtainable by the above methods.

Some attempts have been made to perform a fully automatic analysis by
these means (Senf, 1979), but the procedure works only when the chromosomes
in the plates are straight or little curved; to overcome this limitation, we
utilized a semi-automatic procedure, which allowed the measurements of all the
chromosomes, whatever their shape.

In comparison with traditional karyotypic analysis, our measurements
show an error at least 5 times lower. It means that this method can be particulary used for those species in which the chromosomes show little difference in length.

Furthermore, this analysis allows differentiation of the chromosomes
for their densitometric profiles: differences in the density areas were
found among the chromosomes of Vicia faba L. both for their number and for

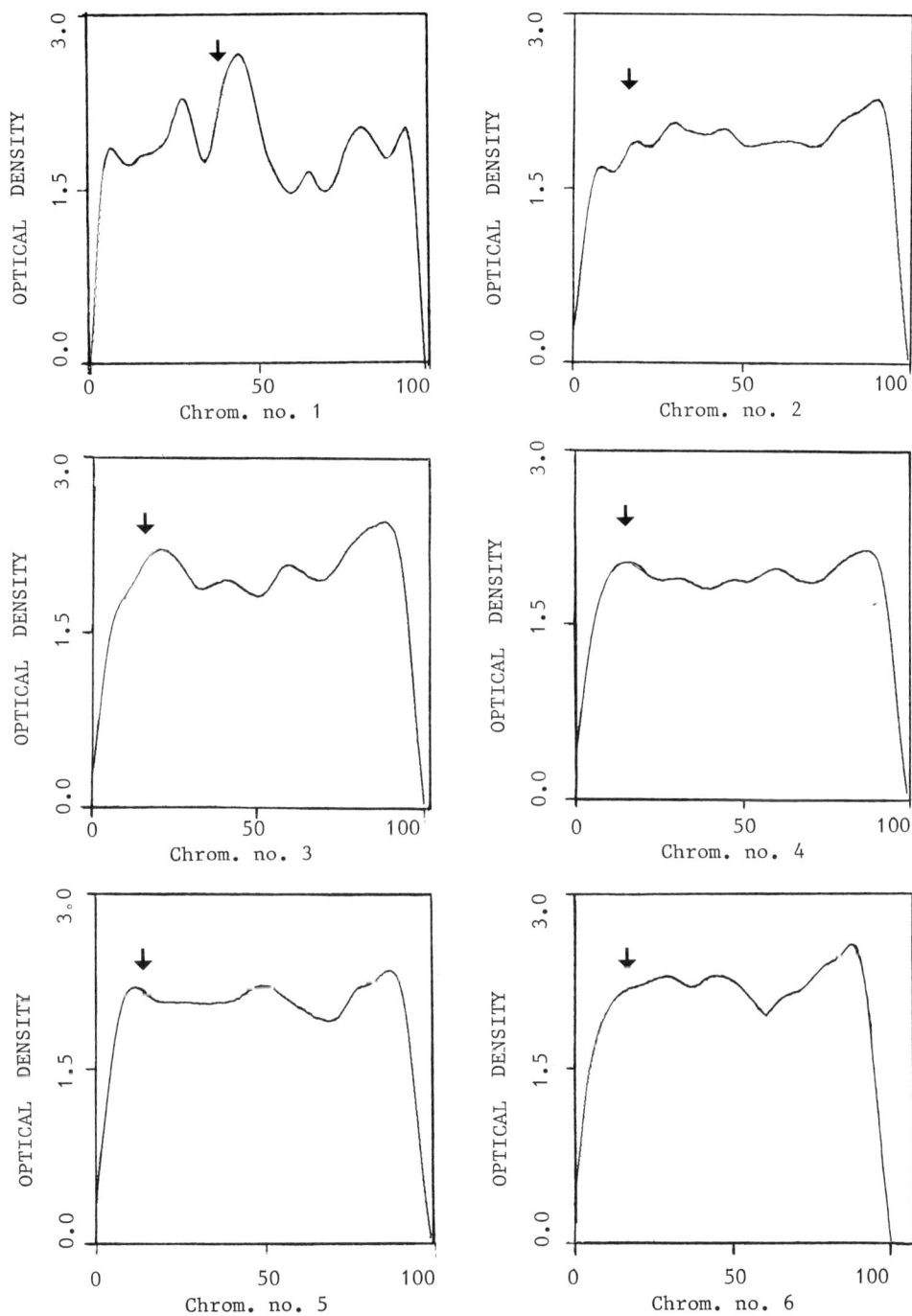

Fig. 1 Densitometric profiles of the six chromosome types of
Vicia faba L.   Centromeres are indicated by arrows

their position. We think that these differences are reliable because of the high frequency with which these density areas are present in the same position in each chromosome.

An analysis is now in course on lines of Vicia faba L. characterized by known translocations among chromosomes (Dobel et al., 1978) in order to check the density areas in the standard and in the modified karyotypes; the densitometric data of the chromosomes of the different karyotypes can be easily allocated in data banks of the species.

Densitometric analysis of chromosome profiles can be very useful when a fine description is desired of the banding patterns obtained by specific staining methods, such as the Giemsa staining procedure or Quinacrine fluorescence (Vosa et al., 1972; Dobel et al., 1973).

REFERENCES

Andronico, A., Bernardi, F., Casalini, P.C. and Marchetti, G. 1979. Digital filtering for image processing. Mem.Soc.Astronomica Ital.,50, 483-495

Döbel, P., Schubert, I. and Rieger, R. 1978. Distribution of heterochromatin in a reconstructed karyotype of Vicia faba as identified by banding and DNA late replication patterns. Chromosoma, 69, 193-209

Errico, A. 1980. Karyotype analysis in Vicia faba L. In "Annali Facoltà di Agraria di Portici (Napoli)", 14, 102-107

Filippone, E., Smaldone, L.A., Errico, A. and Monti, L.M. 1983. Analisi microdensitometrica di immagini fotografiche di cromosomi di Vicia faba. Genetica Agraria, 37, 169-170

Roberti, G., Russo, G., Russo, P., Segré, G., Smaldone, L.A. 1982. Un sistema a minicomputer per l'analisi automatica di immagini biomediche: descrizione, prestazioni e problematiche. In "Atti I Conv.Intern. I Sistemi Informativi Sanitari", 445-467

Senf, G. 1979. Automatische Chromosomenanalyse. Akad.Landwirtsch.-Wiss DDR, Berlin, 168, 71-79

Van Oostveldt, P. 1983. Quantitative and qualitative microspectrophotometric techniques for the analysis of seed varieties at different stages of development. In "EC Perspectives for Peas and Lupins as protein crops-Sant'Agnello (Napoli) 19-22/10/1981", 241-246

Vosa, C.G. and Marchi, P. 1972. On the quinacrine fluorescence and Giemsa staining patterns of the chromosomes of Vicia faba. Giorn.Bot.Italiana, 106, 151-159

VARIABILITY INDUCED BY A DICENTRIC CHROMOSOME
IN VICIA FABA L. [1]

C. Conicella*, A. Errico**, F. Saccardo*** and N. La Gioia***

*Centro Miglioramento Genetico Ortaggi C.N.R.
Portici (Napoli)
**Istituto di Agronomia, Miglioramento Genetico,
Università di Napoli
***E.N.E.A, Casaccia, Roma

ABSTRACT

M_2, M_3 and M_4 plants of Vicia faba derived from an M_1 variegated plant characterized by a dicentric chromosome were analysed by cytologic and phenotypic methods. Genomic structures with dicentric, telocentric, ring and metacentric chromosomes, and triploid plants were found. Mutants for sterility, earliness, plant height and leaf shape were also identified. The results show that in this species a wide range of variation can be induced by a dicentric chromosome.

INTRODUCTION

X-irradiation of Vicia faba pollen induced a variegated plant in M_1 characterized by the presence of a dicentric chromosome. An account of the origin and the behaviour of this unstable chromosome was given previously (Errico et al., in press). This paper reports the analysis carried out in successive generations of material derived from the M_1 variegated plant, with the aim of studying the cytologic and phenotypic variability determined by the dicentric chromosome.

MATERIALS AND METHODS

Cytologic analysis was performed on plants grown in greenhouses. M_2, M_3 and M_4 plants derived from M_1 plant number 59 were grown in fields under cages to avoid cross-pollination.

Fifty fertile M_2 plants gave 716 M_3 plants for analysis; 5 M_3 variegated plants and 145 M_3 non-variegated plants gave a total of 1212 M_4 plants. Biometric analysis of the mutants was limited to those progenies which showed good phenotypic stability. Control plants of the mother line were also analysed.

1) Contribution no. 2 from Centro di Studio per il Miglioramento Genetico degli Ortaggi - C.N.R. - Portici (Napoli) - Italy

P.D. Hebblethwaite, T.C.K. Dawkins, M.C. Heath and G. Lockwood (eds.)
Vicia faba: Agronomy, Physiology and Breeding. ISBN 90-247-2964-5.
© 1984, Martinus Nijhoff/Dr W. Junk Publishers. Printed in The Netherlands.

On each plant, the dimensions of the first two leaflets at the 13th node were measured at the beginning of the flowering, and number of days from the sowing until anthesis of the first flower was recorded. After harvesting each plant, measurements were taken of total plant height and length of the internode between the 12th and 13th nodes and the number of pods and the weight of dry seeds were recorded. In each generation, the plants with a variegated phenotype and the plants with no seeds were separated, the latter ones being considered as sterile.

Mitotic analysis was done on squashes of 1.5 - 2 cm long root tips pretreated with a 0.05% solution of colchicine for 7 hours at 3°C, fixed in ethanol-acetic acid 3 : 1 for at least 2 hours and stained with the Feulgen technique (hydrolysis in 1N/HCl at 60°C for 7 min.). Meiotic analysis was carried out on squashes of anthers fixed for 48 hours in ethanol : propionic acid 3 : 1, to which a small amount of ferric chloride had been added as a mordant, and stained with acetocarmine. Pollen fertility was estimated by staining with acetocarmine.

RESULTS

Table 1 summarizes the cytologic analysis of the chromosome mutants isolated in three generations derived from M_1 plant 59. Different genomic structures were identified with dicentric (Fig. 1), telocentric, ring and metacentric (Fig. 2) chromosomes, and in addition two triploid mutants were

TABLE 1   Cytologic analysis of chromosome mutants found in the M_2, M_3 and M_4 generations of M_1 variegated plant 59.

2 n =	Number of plants analysed	Diakinesis
11+1 Dicentric	30	6II;4II+1IV
10+1 Dicentric+1 Telocentric	11	6II;5II+2I;4II+1IV
12+1 Ring	4	6II+1 Ring
10+1 Dicentric+1 Metacentric	1	6II;4II+1IV;5II+2I
17+1 Dicentric	2	6III;5III+1II+1I;4III+2II+2I; 3III+3II+3I

found. Some of these mutants were first identified in M_4.

The meiotic analysis of these mutants in diakinesis is summarized in Table 1. The percentage of four-chromosome associations, univalents, and ring chromosomes (Fig. 3) differed according to the mutant. From 3 to 6 trivalents were found in the triploid plants (Fig. 4).

The percentages of the plants with a variegated phenotype (that is a phenotype similar to that of M_1 plant 59) in the different generations are reported in Table 2: lower rates (1 - 3%) were found in M_3-M_4 in comparison with the higher frequency found in M_2 (23.5%); variegated plants were also found in progenies of non-variegated plants.

TABLE 2   Incidence of variegated and normal plants is successive generations of variegated and normal plants derived from variegated M_1 plant 59.

	M_2	M_3	M_4
Variegated plant progenies (n°)	1	12	5
Plants analysed (n°)	51	152	29
Variegated plants (%)	23.5	1.3	3.4
Normal plant progenies (n°)	-	39	145
Plants analysed (n°)	-	564	1183
Variegated plants (%)	-	1.6	0.2

Table 3 summarizes the recurrence of sterile plants in successive segregating generations, showing that the percentage increased from 2% in M_2, to 13% in M_3 and to 31% in M_4.

Biometric data concerning four mutants with a stable phenotype in M_4 are reported in Table 4. Mutant 7 is characterized by leaves with a reduced width/length ratio, described as "oblong leaf".

234

TABLE 3   Frequency of sterile plants in the M_2, M_3 and M_4 generations of M_1 variegated plant 59.

	M_2	M_3	M_4
Progenies analysed (n°)	-	50	150
Progenies segregating for sterile plants (%)	-	40.0	29.3
Plants analysed in the segregating progenies (n°)	51	313	440
Sterile plants (%)	1.9	13.1	30.7

Mutants 22 and 24 are respectively 16 and 18 days earlier to flowering than control, mutant 24 is also shorter in stature than control (123 cm compared with 174 cm). Mutant 20 is semi-sterile and some completely sterile plants were found in its progeny.

TABLE 4   Main traits of four mutant M_4 lines derived from M_1 plant 59.

Mutant line Phenotype	7 oblong leaves	20 semi - sterile	22 early	24 short and early	Control
M_4 plants analysed (n°)	10	11*	8	6	9
Leaflet width/length ratio	0.40±0.02	0.52±0.04	0.48±0.04	0.49±0.02	0.52±0.07
Internode length (cm)	7.8 ±1.0	6.3 ±0.4	6.8 ±0.8	3.3 ±0.7	6.5 ±0.7
Plant heigth (cm)	188±9	166±7	184±6	123±4	174±16
Days from sowing to flowering	143±2	140±2	122±1	120±1	138±2
Pods per plant (n°)	14	3	33	28	24
Seed yield per plant (g)	12.85	2.15	29.23	36.98	26.93

* Some plants sterile

DISCUSSION

A breakage at the centromere of one chromosome and a deletion in another chromosome, both induced by irradiation treatment of the pollen grain participating in the formation of M_1 variegated plant number 59, were considered responsible for the occurrence of dicentric and telocentric chromosomes (Errico et al.,in press). The dicentric chromosome was shown to be heterodicentric giving rise to "breakage - fusion - bridge" cycles, but with the secondary centromere partially inactive.

In peas, the presence of unstable chromosomes always causes a clear variegated pattern of the plant (Monti and Saccardo,1969); this does not hold true for V. faba, where dicentrics were found in non-variegated plants and where variegated plants were found in the progenies of non-variegated ones; this can be related to partial inactivity of the secondary centromere, which causes a parallel type of disjunction, and a low rate of bridge formation at anaphase and in consequence a limited mixochimeric structure.

The analysis performed on M_3 and M_4 plants showed the extensive cytologic variability which is present in progenies of M_1 variegated plant 59, due probably to persistent bridges in telophase II formed by the dicentrics; diploid gametes were also formed (Errico et al.,1982), which gave rise to two triploid plants in which dicentric chromosomes were also recognised. Triploid plants had 50% pollen sterility and some pollen grains were triangular-shaped, similar to those found by Poulsen and Martin (1977) in tetraploid mutants. Ring chromosomes found in our material also come from dicentrics, as found in other species (Darlington and Wylie,1953; Saccardo,1971).

Further cytologic variability was determined by the telocentric chromosome induced by the mutagenic treatment in our M_1 plant. The metacentric chromosome which was first found in M_4 was probably derived from the telocentric chromosomes by means of mis-division, as has been found already in maize, Triticum and peas (Rhoades,1940; Sears,1954; Saccardo and Monti,1974).

There is no doubt that further chromosome rearrangements will be revealed in later generations of the M_4 plants. In peas, for example, many aneuploids and some mutants with duplicated chromosome segments were found in the progenies of plants with dicentric chromosomes (Saccardo and Monti,1971).

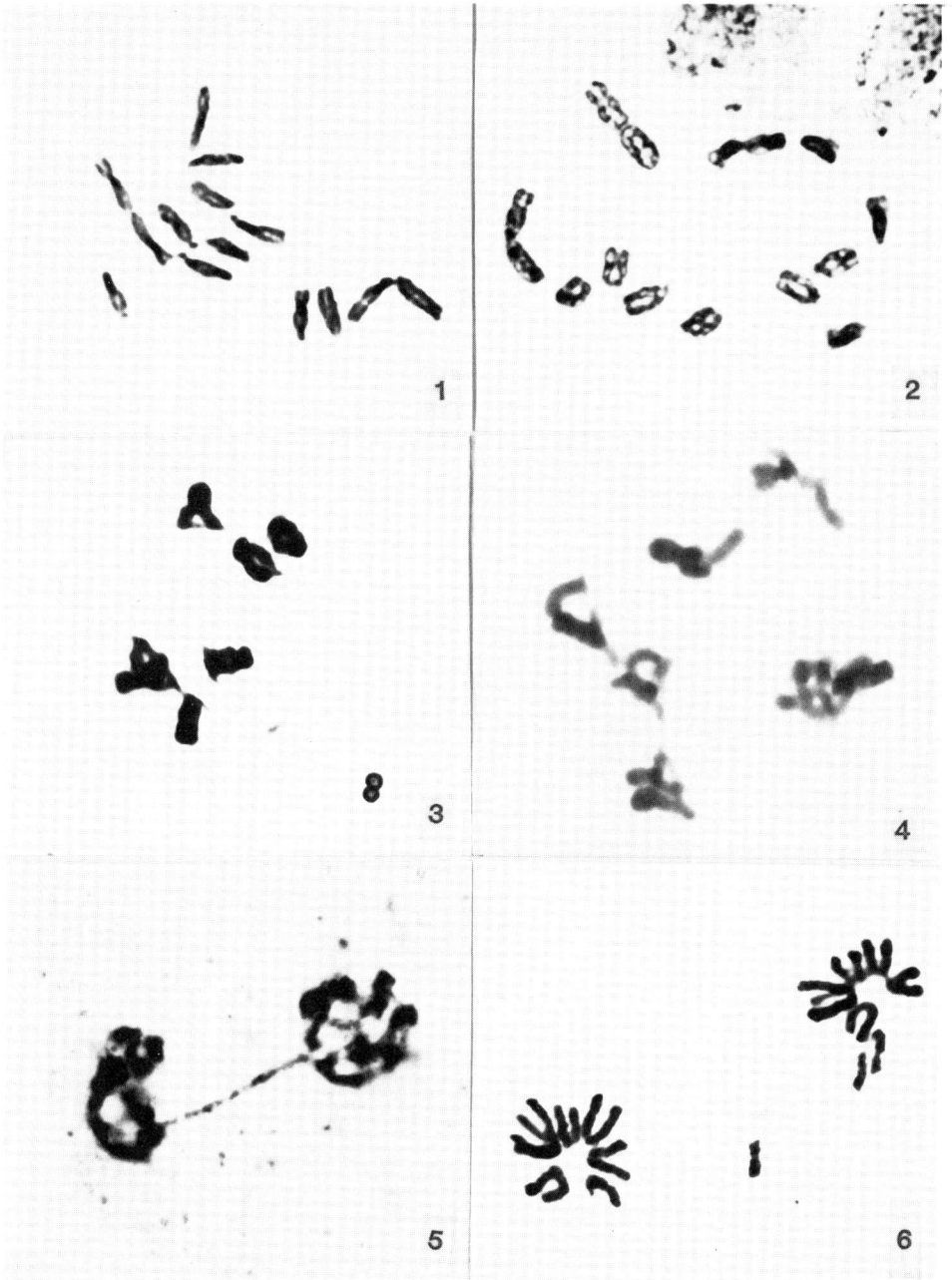

Figs. 1-2 Mitotic metaphases from variegated plants with 2n=11+1 dicentric
and 11+1 metacentric chromosomes. Fig. 3 Diakinesis with 6 bivalents plus a
duplicate ring chromosome. Fig. 4 Diakinesis from a triploid plant with 6
trivalents. Fig. 5 Telophase II with persistent bridge. Fig. 6 Anaphase I
showing disjunction of one chromosome.

This paper also shows that unstable chromosomes induce extensive phenotypic variability. Mutants for sterility, earliness, short plant height and leaf-shape were identified in the M_2 generation of M_1 variegated plant 59. Such mutants were confirmed in M_3 and M_4 where one semi-sterile mutation was also found. The early flowering and the short straw mutations may be of agronomic value. A genetic and caryotipic analysis of these mutations should be carried out in order to understand whether some or all of these mutations are due to gene-mutations induced independently of the dicentric chromosome or whether they are a consequence of the chromosome rearrangements due to the unstable induced chromosome mutation.

At present, the second hypothesis seems the more probable, firstly because chromosome aberrations were found in many of the M_2 plants and secondly because it is rather difficult to hypothesise that so many point mutations could be induced in the pollen grain which gave origin to the M_1 variegated plant, together with the induction of the dicentric chromosome.

In conclusion, it seems that induction of unstable chromosomes is a very important tool for induction of further new variability in Vicia faba, as has been shown already in Nicotiana (Saccardo and Devreux,1967) and in peas (Saccardo,1971).

REFERENCES

Darlington, C.D. and Wylie, A.P. (1953). A dicentric cycle in Narcissus. Heredity, Suppl., 6, 197-213.

Errico, A., Saccardo, F., Conicella, C. and Monti, L.M. (1982). Variabilità indotta da un cromosoma dicentrico in Vicia faba L. XXVI Convegno S.I.G.A., Bordighera.

Errico, A., Saccardo, F., Monti, L.M. and Conicella, C. (1983). Induction and behaviour of a dicentric chromosome in Vicia faba L. Zeitschrift für Pflanzenzüchtung, in press.

Monti, L.M. and Saccardo, F. (1969). Mutations induced in pea by X-irradiation of pollen and the significance of induced unstable chromosomes in mutagenic experiments. Caryologia, 22, 81-96.

Poulsen, M.H. and Martin, A. (1977). A reproductive tetraploid Vicia faba L. Hereditas, 87, 123-126.

Rhoades, M.M. (1940). Studies of a telocentric chromosome in maize with reference to the stability of its centromere. Genetics, 25, 483-520.

Saccardo, F. and Devreux, M. (1967). Caracteristiques morphologiques et cytogenetiques de quinze mutants de tabac. Caryologia, 20, 239-256.

Saccardo, F. and Monti, L.M. (1970). Isolation of trisomics in peas. Caryologia, 23, 347-358.

Saccardo, F. (1971a). Behaviour of dicentric chromosomes in peas. Caryologia, 24, 71-84.

Saccardo, F. (1971b). Comportamento e trasmissibilità di cromosomi ad anello radio-indotti. Genetica Agraria, 25, 99-111.

Saccardo, F. and Monti, L.M. (1974). Isolation of aneuploids originating from induced unstable chromosome aberrations. IAEA - PL - 503/37.

Sears, E.R. (1952). The behaviour of isochromosomes and telocentrics in wheat. Chromosoma, 4, 551-562.

FABA BEAN CYTOGENETICS AT CORDOBA, SPAIN

A. Martin, J.A. Gonzalez and P. Barceló
Escuela Técnica Superior de Ingenieros Agrónomos
Apdo. 3048, Córdoba, Spain.

ABSTRACT

Recent advances in the breeding of a tetraploid Vicia faba are reported. No differences in meiotic behaviour between inbred lines and F_1s were found, but seed setting was superior in F_1s.
Production, phenotypic characterization, meiotic behaviour and transmission rates of V. faba trisomics are presented. Leaflet shape is the most discriminatory character between trisomics.

INTRODUCTION

Apart from academic interest, cytogenetics may allow the plant breeder to overcome problems that would be difficult or impossible to solve by other means. It is not unusual, therefore, to find crops such as wheat, maize, barley and tomato among those studied in depth by cytogeneticists.

The Faba bean (Vicia faba) possesses qualities that make it very attractive to the cytogeneticist. No fertile variants in chromosome number were available, however, until recently when tetraploids and trisomics (Poulsen and Martin, 1977; Martin, 1978) emerged from work with Sjödin's (1971) PO-1 mutant.

We decided, at Córdoba, to work on the cytology of V. faba with a dual aim namely to increase fertility of the tetraploid and to produce the complete set of trisomics.

The influence of date of sowing, temperature, decapitation and meiotic behaviour on the fertility of the V. faba tetraploid has been studied (Martin and Gonzalez, 1981). Date of sowing was the only important factor, apart from genetic differences among lines.

Selection of the smallest seeds among the tetraploids produced the first trisomics (Martin, 1978). We attempted to produce triploids by crossing tetraploid and diploid V. faba, in both directions, with the eventual aim of obtaining trisomics in the progeny of the triploid or backcrossing it with diploids. Nevertheless all attempts were unsuccessful.

Sjödin's material was also found to be the most suitable for producing V. faba aneuploids. When asynaptic mutants (Sjödin, 1970) were

P.D. Hebblethwaite, T.C.K. Dawkins, M.C. Heath and G. Lockwood (eds.)
Vicia faba: Agronomy, Physiology and Breeding. ISBN 90-247-2964-5.

crossed with standard diploids, trisomics and double trisomics were produced (Gonzalez and Martin, 1983).

We report here the most recent advances made in our work at Córdoba on tetraploidy and trisomy in V. faba.

MATERIAL AND METHODS

PO-1 and the asynaptic mutants originated from Dr. Sjödin in Sweden.

For cytological observations the following standard procedures were used. Somatic chromosome counts were made from root-tips, treated in a 0.05% colchicine solution, fixed in 1:3 acetic alcohol and stained by the Feulgen procedure. Meiotic observations were carried out on anthers fixed and stained by the same procedure.

Pollen viability was estimated from slide preparations stained with aceto-carmine.

RESULTS AND DISCUSSION

Tetraploidy

Four generations of selfing and selection did not give a reliable gain in seed setting. However the best lines had not decreased in fertility. Crosses between lines differing in fertility and length of life cycle were carried out to obtain recombinants.

In order to test relationships between degree of inbreeding, meiotic behaviour and fertility, 32 inbreds which had been selfed for at least four generations and 11 F_1s were studied. The meiotic behaviour was studied by only one observer to minimize errors.

Table 1 shows the results obtained in glasshouse conditions. It is difficult to reach a definite conclusion from these results given the size of the samples and the standard deviation of some of the parameters. Nevertheless, some trends can be inferred.

There was little difference in meiotic behaviour, ovules per ovary, pollen sterility and flowers per node between inbred lines and F_1s, but F_1s produced more seeds than inbred lines. It can be concluded that differences in fertility may be due to heterosis independently of meiotic behaviour. Nine of the 32 inbreds and 2 of the 11 F_1s did not yield any seed.

Samples from two high yielding F_1s were sown in the glasshouse and they produced a mean of 10.88 $\pm$ 0.77 and 14.95 $\pm$ 0.94 seeds respectively with 6 out of 17 plants with no seeds, in both crosses.

These data suggest that genetic variability for fertility is present in the V. faba tetraploids. We will continue working on this long term project, following the scheme already devised: selection of more productive lines, crossing them to each other, selfing and selecting for at least four generations, and crossing again.

TABLE 1   Values of meiotic and fertility variables of two populations of Vicia faba tetraploids.

| | Inbred lines | | F_1s | |
	mean	cv*	mean	cv*
Bivalents per cell	0.59	0.40	0.70	0.16
Univalents per cell	0.25	0.24	0.20	0.30
Micronuclei (diad stage)	0.17	0.24	0.16	0.38
Micronuclei (tetrad stage)	0.15	0.33	0.14	0.29
Pollen sterility (%)	44.02	0.24	31.54	0.18
Seeds per pod	0.54	1.24	0.99	0.76
Seeds per plant	2.55	1.63	7.20	0.93
Ovules per ovary	4.10	0.04	4.20	0.06
Flowers per node	6.05	0.17	6.53	0.09
* cv coefficient of variation %				

Trisomy

We have obtained V. faba trisomics by crossing asynaptic mutants (Sjodin, 1970) with a white flowered autofertile line. These two characters are important because it will be easier to maintain trisomics carrying genes for self fertility, and the presence of a marker will help when using the trisomic for mapping genes.

In 1981 we used two asynaptic lines differing in frequency of chromosome pairing. In the progeny of the crosses using the mutant with the highest asynaptic level, from 75 seeds 9 were trisomics and 1 a double trisomic. When the line with lower level of asynapsis was used none out of 96 seeds obtained were trisomic.

In 1982 one plant selected for high asynaptic level was crossed with the autofertile white flowered line. We obtained 126 plants (23 of them were trisomics) all with similar genetic background, differing only in the effect of the extra chromosome when present.

This material was sown in the glasshouse under standard environmental conditions. Twelve diploid and the 23 trisomic plants were used to assess which characters should be scored to differentiate between diploids and trisomics, and if possible, among trisomics.

Chromosome numbers were scored using root tips that emerged from the bottoms of the pots.

Unfortunately a fungal attack on the roots affected the plants differentially, killing some of them and decreasing the value of the results.

In Table 2, vegetative characters from the 12 diploid and 23 trisomic plants are presented. The third leaf stage, flowering and first pod occurred later for trisomics than for diploids, indicating that trisomics develop more slowly than diploids. The coefficient of variation (cv) is similar for these three characters in trisomics and diploids, and for this reason they have no discriminatory value. Numbers of flowers and pods per node are lower on trisomics and cvs higher. These two characters have to be taken into account when classifying trisomics. The mean number of ovules per ovary and its cv are similar in trisomics and diploids. The mean height of the first flower scored by node number is higher, and there is more variation in trisomics than in diploids. This character may have value in classification.

TABLE 2   Values of vegetative characters from <u>Vicia</u> <u>faba</u> euploids (top) and trisomics (bottom).

| | Number of days to | | | | Number of | | |
	Third leaf	First flower	First pod	Flowers per node	Pods per node	Ovules per ovary	Height first flower
mean	11.9	29.18	41.18	6.04	1.91	3.43	5.45
cv	0.16	0.14	0.09	0.12	0.29	0.06	0.13
mean	14.09	40.17	53.77	3.87	1.20	3.25	7.34
cv	0.16	0.16	0.13	0.27	0.72	0.08	0.32

In Table 3, we present data from 12 diploids and 20 trisomics. L is the length of leaflet; A_1 is the width of the leaflet measured at 1/4 of L; A_2 the same in the middle, and A_3 at 3/4 of L. The first three leaves were used to obtain a mean value for a plant. For every parameter there are differences between diploids and trisomics in mean value or cv, or both. The leaflet shape is the most discriminatory phenotypic character.

TABLE 3 Values of some leaflet parameters from <u>Vicia faba</u> euploids (top) and trisomics (bottom) (explanation in text).

	L	A_1	A_2	A_3	A_1/L	A_2/L	A_3/L
mean	6.21	3.04	4.20	3.57	0.49	0.68	0.57
cv	0.08	0.15	0.17	0.16	0.06	0.07	0.07
mean	4.69	2.01	2.73	2.44	0.46	0.61	0.54
cv	0.33	0.23	0.24	0.25	0.34	0.25	0.21

Some seed parameters are presented in Table 4. From this first trial, width, thickness and mean seed weight are the most discriminatory characters between diploids and trisomics.

TABLE 4 Values of some seed parameters from <u>Vicia faba</u> euploids (top) and trisomics (bottom).

	Number of seeds	Length cm.	Width cm.	Thickness cm.	Mean weight
mean	42	1.32	0.96	0.81	0.75
cv (%)	72	3	2	3	4
mean	31	1.36	0.96	0.86	0.70
cv (%)	69	4	10	7	19

We have classified the trisomics into four phenotypic groups.

The first group includes trisomics which are very similar to diploids, differing from them in leaflet shape. The A_3 value is higher for trisomics than for diploids. On the basis of Giemsa banding pattern, we think that trisomics for chromosomes II and VI are in this group.

The second group includes trisomics with low A/L values, weak stems, few flowers per node and larger seeds than other trisomics. We have obtained this trisomic four times, but only once produced a good number of seeds.

The trisomics in the third group were obtained from asynaptics. They have A/L values higher than the average, strong stems and few flowers per node. The height of the first flower is lower than normal. We think that the extra chromosome is number V.

The trisomics in the fourth group also appeared in crosses with asynaptics. They develop more slowly than other trisomics. They are similar to the group 3 trisomics, differing in having fewer veins on the leaflets. These trisomics are the least fertile, and frequently the pods are cracked.

Some values of meiosis and pollen fertility for each trisomic group are shown in Table 5. There is no relationship between either value and trisomic fertility.

TABLE 5 Micronuclei at the tetrad stage and stained pollen from Vicia faba trisomics and euploids %.

Trisomic group	Number of plants	Cells with 0 micronuclei	Normal pollen
1	9	80.7 ± 12.6	38.4 ± 1.7
2	2	88.6 ± 1.3	43.85± 3.2
3	4	81.2 ± 3.1	64.1 ± 0.2
4	3	97.7 ± 1.7	58.2 ± 0.1
Diploid	2	99 ± 0	81.7 ± 1.2

For group 1 trisomics the female rates of transmission of trisomy to F_1 and F_2 are about 22 and 17% respectively, while for group 3 it is only 4.4%. No trisomic was obtained from 71 seeds tested from male transmission. We do not have values of transmission of trisomy for the other groups.

Our efforts are now directed in four areas:

1. Identification of the extra chromosome in each group, using banding techniques.
2. Crossing trisomics with mutants to obtain the genetic map of V. faba.
3. Studying the effect on quantitative characters of the extra chromosome.
4. Producing trisomics by crossing asynaptic mutants with diploids of high combining ability and resistance to root diseases.

REFERENCES

Gonzalez, J.A. and Martin, A. 1983. Development, use and handling of Vicia faba trisomics. FABIS 6, 10-11.

Martin, A. and Gonzalez, J.A. 1981. Factors influencing fertility of a tetraploid in Vicia faba. In "Vicia faba: Physiology and Breeding" (Ed. R. Thompson). (Martinus Nijhoff Publishers, The Hague, Boston, London) pp. 129-145.

Martin, A. 1978. Aneuploidy in Vicia faba. The Journal of Heredity, 69, 421-423.

Poulsen, M.H. and Martin, A. 1977. A reproductive tetraploid Vicia faba L. Hereditas, 87, 123-126.

Sjödin, J. 1970. Induced asynaptic mutants in Vicia faba L. Hereditas, 66, 215-232.

Sjödin, J. 1971. Induced morphological variation in Vicia faba L. Hereditas, 67, 155-180.

THE INCIDENCE OF OVULE FERTILIZATION IN FABA BEAN FLOWERS FROM COMMERCIAL
CROPS AND FROM EXPERIMENTAL PLOTS OF CONTRASTING GENOTYPES

F.L. Stoddard and G. Lockwood

Plant Breeding Institute, Maris Lane, Trumpington, Cambridge

ABSTRACT

Epifluorescence microscopy was used to evaluate pollen deposition and germination, and fertilization of ovules, in over 16000 flowers from commercial crops and experimental plots of winter and spring faba beans in the 1982 flowering season. When sufficient pollen tubes were present, all ovules were fertilized in almost every ovary: few ovules were passed or not reached by the pollen tubes. In spring bean crops and in most experimental plots, 80 to 95% of flowers were pollinated, but only 50% in winter bean crops, 65% in spring long-podded types and 45% in a spring closed flowered type in experimental plots. In winter bean crops a further 5 to 13% of flowers had ungerminated pollen on their stigmas. As few as 14% of ovules were fertilized in some parts of some winter bean fields, and there was no association between incidence of fertilization and distance from the margins of the field. In winter bean crops the presence of honeybee hives had no discernible effect on any component of the process of fertilization. It is concluded that more reliable fertilization of winter bean crops is desirable.

INTRODUCTION

Yields of faba beans fluctuate more widely than those of many other agricultural crops (Bond, 1977) under the influence of interacting edaphic, climatic, pathological and zoological factors. Observations that up to 90% of flower buds fail to develop into pods on winter bean plants (Soper, 1952) have been interpreted as 'wastage of flowers and the potential yield they represent' (Chapman 1983). At a sowing density of 58 plants/m^2 individual plants of the spring variety Cockfield produced an average of 80 flowers with 3.74 ovules; with an average seed weight of 0.58 g (G. Lockwood and F.L. Stoddard, unpublished), the yield potential was over 100 t/ha. It is inevitable that a high proportion of ovules will be lost; thus securing a high yield depends partly on minimizing this proportion.

Flowers may be shed without being fertilized or, by analogy to temperate fruit crops, the interval between anthesis and fertilization may be so great that abscission is already inevitable (Williams, 1966). Loss of developing pods as a consequence of internal competition is a separate process which will not be considered further here.

In most studies pollination and fertilization have not been separated because responses to experimental treatments have been measured indirectly by observing floral abscission or final yield. Williams (1973) showed that

P.D. Hebblethwaite, T.C.K. Dawkins, M.C. Heath and G. Lockwood (eds.)
Vicia faba: Agronomy, Physiology and Breeding. ISBN 90-247-2964-5.

fluorescence microscopy can be used to observe pollen and pollen tubes on and within bean pistils. Gates et al. (1981) used such methods to examine flowers which had abscissed after being tripped in a glasshouse, finding pollen tubes in 42% of 462 flowers. In an intensive study of flowers from complete racemes taken from field-grown spring beans, Rowland, Bond and Parker (1983) found that only 48% of flowers had one or more fertilized ovules, with 33% of ovules fertilized overall. The incidence of fertilization was lower (25%) in plants which were heavily irrigated during the flowering season than it was in those which were not irrigated (41%). In the experimental variety Ticol, which has Sjodin's (1971) gene for the terminal inflorescence character, only 17% of ovules were fertilized compared to 26 to 43% in conventional varieties. Rowland et al. (1983) concluded that in their trial inadequate fertilization may be an important cause of floral abscission. If failure of fertilization does limit yields of faba beans, then the problem may be most acute in autumn-sown crops which flower when effective activity of pollinators is less assured.

Successful fertilization depends on a chain of events:

> deposition on stigma of sufficient numbers of viable pollen grains
> pollen germination
> growth of pollen tubes
> penetration of the micropyle and fusion of gametes

In this study, these processes were examined in both autumn and spring sown crops in large fields as well as experimental plots.

MATERIALS AND METHODS

The breeding material studied is briefly described in Tables 1 and 2.

Racemes bearing at least three flowers were considered suitable for collection when the proximal flowers had wilted and the distal ones had started to wilt and therefore had been open for at least two days (Synge, 1947). Only one raceme was taken from any plant.

Commercial crops

Four fields of winter beans and one of spring beans were surveyed three times at weekly intervals during their flowering seasons (Table 1). Two transects were worked at right angles to each other along the long and short axes of the fields. Collections, each of two racemes, were taken every 5m for the first and last 25m of each transect, then every 25m across the body of the field.

Experimental plots

    Two trials at the Plant Breeding Institute, Cambridge (Table 2) each
had four replicates. The winter trial had 4-row plots 25m long with 12 cm
between plants within rows, 30cm between the rows and 60cm between plots.
In the spring trial there were six 2m rows per plot with all rows 50cm apart
and an average sowing density of 58 seeds/m^2.

    In the winter trial, four racemes were taken from flowering nodes 1 to
7 on each of five days. In the spring trial, six racemes were taken from
flowering nodes 1, 3 and 4 on each of five days.

Preservation and examination of flowers

    Racemes were brought to the boil in 70% ethanol, cooled, stripped of
petals and then stored in the ethanol in a cool and dark place until
needed. Racemes for examination were transferred to 70% lactic acid, heated
in a boiling water bath for 30 seconds, cooled for 10 minutes and washed
with three changes of water. The pH of the tissue was raised by immersion
in 0.1M K_3PO_4 overnight or longer. The pistils were split on a microscope
slide and stained with 0.2% aniline blue (water soluble, batch 15574,
E. Gurr Ltd.) in 0.1M K_2HPO_4 for 30-90 minutes. Excess stain was blotted
from the preparations, a coverslip was mounted with glycerol, and material
was examined using epifluorescence microscopy. Presence, amount and
germination of pollen and the entry of pollen tubes into ovules were
scored. Such entry is taken as evidence of fertilization.

RESULTS

Commercial crops

    Losses of potential seeds due to lack of pollination or fertilization
are shown in Table 1. Up to 41.5% of flowers were not pollinated in the
fields of winter beans, and in 4.6-12.9% more, the pollen had not
germinated. A separate analysis (not given) showed that the percentage of
flowers with ungerminated pollen did not change with position of the flower
on the raceme.

    In winter beans, a small proportion of flowers, 2.5-3.0%, was damaged
by insects and no pollen tubes were evident; a further 1.1-3.6% had pollen
tubes arrested in the style. There was no discernible association between
the incidence of arrested pollen tubes and flower position within the
raceme.

In those flowers of winter beans where at least the first ovule was reached by a pollen tube, 4.6-7.1% of ovules remained unfertilized because insufficient pollen grains had germinated. Pollen tubes failed to reach another 2.3-6.1% of ovules, and passed very few, 0.6-1.3%, without entering them. In flowers with at least one fertilized ovule, between 85 and 90% of all ovules were fertilized.

Spring beans had a much smaller proportion (6.1%) of flowers not pollinated, and almost 85% of flowers had at least one fertilized ovule; in such flowers 94.3% of ovules were fertilized. The most important factor limiting fertilization was damage done by insects to 6.9% of flowers.

The mean incidence of fertilization of ovules in the field of spring beans was, at 80%, much higher than the mean of 43% from the four fields of winter beans. There were marked fluctuations from week to week in the incidence of fertilization in the winter fields but not in the spring field.

Within all fields, the incidence of pollination varied with position, but it was not related to the distance from the edge, since minima and maxima were found near the ends of the transects. In the winter fields some collection sites consistently showed low incidences of pollination, with as few as 14% of ovules being fertilized.

Experimental plots

The incidence of pollination of flowers of winter beans was much greater (75.9-89.9%) than in commercial crops (cf. Tables 1 and 2), as was the incidence of fertilization of ovules (65.7-83.0%). The proportion of ovules fertilized in fertilized flowers (88.1-94.6%) was only slightly greater than in the commercial crops (85.9-89.5%).

Varieties of spring beans with conventional flowers gave similarly high incidences of pollination of flowers (82.3-94.1%) and fertilization of ovules (62.4-81.3%) as those obtained in the commercial crop. The proportion of ovules fertilized in fertilized flowers was also high (87.3-94.9%). The incidences of pollination (82.3%) and fertilization (62.4%) were underestimated in Maris Bead; the first racemes were collected early in the flowering season when few were suitable and it was necessary to take some younger material. Varieties with long pods or closed flowers showed substantial depression in the incidences of pollination (44.8-68.6%) and fertilization (21.3-38.2%), and in the proportion of ovules fertilized in flowers with fertilized ovules (59.6-64.3%) (Table 2).

DISCUSSION

The survey of farm crops and of experimental plots has shown that if a sufficient number of pollen grains is deposited on the stigma and germinates, then it is almost certain that all ovules will be fertilized. The principal factor limiting fertilization is failure of pollination and, in the case of winter beans, the next most important factor is pollen germination. Poor germination and growth of pollen in winter bean flowers may occur when relatively low temperatures prevail at flowering time (Williams, 1973). Once pollen grains have germinated, the main hazard to successful fertilization is damage done to the style by insects, especially in spring beans. When pollen tubes are plentiful, few ovules escape fertilization by being passed. The higher incidence of pollination in experimental plots compared to fields could be attributed to smaller areas involved, wider row spacings and numerous pathways facilitating access by bees.

There was little difference among the five winter bean genotypes in the incidences of pollination and fertilization, and the ti-type was not at a disadvantage. Similarly, in the spring beans, the ti-types TC38 and Ticol, and the 3 conventional varieties Maris Bead, Cockfield and Deiniol, had comparable incidences of fertilization. Hylon and LG5/14, which were less well pollinated, flowered earlier in the season and had larger flowers, which may make them relatively unattractive to bees. The closed-flower type had the lowest incidences of pollination, and of fertilization in flowers with at least one fertilized ovule, results which suggest that this genotype is insufficiently autofertile.

The weather during the winter bean flowering season in 1982 was relatively warm and dry, so it seemed likely to have favoured bee activity. Nonetheless the incidence of pollination was markedly lower than in the spring beans. Through the winter bean flowering season, the incidence of fertilization rose steadily in one of the fields with supplementary beehives and decreased in the other. In one of the two fields without hives the incidence of fertilization was lower in mid-season than earlier or later, and the other showed a slight decrease before a large increase. It is clear from the results that provision of the hives did not give a clear advantage in any component of the process of fertilization. In contrast, the commercial crop of spring beans, which did not have supplementary hives, had consistently high incidences of pollination and fertilization. The relative incompleteness of pollination and fertilization in winter beans compared to spring ones suggest that enhanced autofertility would be most beneficial to yields of winter crops.

ACKNOWLEGEMENTS

We are grateful to D.A. Bond and G. Toynbee-Clarke for thoughtful discussions of results, to M.L. Parker for providing microscope facilities, and to M. Pope for assistance with field trials. FLS acknowledges studentships from Imperial Oil (Canada) Ltd. and St. John's College, Cambridge.

REFERENCES

Bond, D.A. 1977. A breeder's approach to stabilising production in field beans (Vicia faba L.). In "Proceedings of the symposium on the Production, Processing and Utilisation of the Field Bean (Vicia faba L.) held at SHRI on 9 March 1977". (Ed. R. Thompson). Scottish Horticultural Research Institute Symposium Bulletin Number 15, pp. 10-16.

Chapman, G.P. 1983. The modernisation of field beans. Span, 26, 62-64.

Gates, P., Yarwood, J.N., Harris, N., Smith, M.L. and Boulter, E. 1981. Cellular changes in the pedicel and peduncle during flower abscission in Vicia faba. In "Vicia faba: Physiology and Breeding" (Ed. R. Thompson) (Martinus Nijhoff, the Hague). pp. 299-312.

Poulsen, M.H. 1977. Obligate autogamy in Vicia faba L. J. agric. Sci., Camb., 88, 253-256.

Rowland, G.G., Bond, D.A. and Parker, M.L. 1983. Estimates of the frequency of fertilization in field beans (Vicia faba L.). J. agric. Sci., Camb., 100, 25-33.

Sjodin, J. 1971. Induced morphological variation in Vicia faba L. Hereditas, 67, 155-180.

Soper, M.H.R. 1952. A study of the principal factors affecting the establishment and development of the field bean (Vicia faba). J. agric. Sci., Camb., 52, 335-346.

Synge, A.D. 1947. Pollen collection by honeybees (Apis mellifera). J. Anim. Ecol., 16, 122-138.

Williams, R.R. 1966. Pollination studies in fruit trees: III. The effective pollination period for some apple and pear varieties. Long Ashton Res. Stn. Univ. of Bristol, Rep. for 1965, 136-138.

Williams, R.R. 1973. Pollination of field beans. Long Ashton Res. Stn., Univ. of Bristol, Rep. for 1972, p. 49.

TABLE 1. Commercial crops of faba beans growing near Cambridge in the 1982 flowering season: characteristics of the fields, pollination, pollen tube growth and number of seeds per pod.

	Bulldog Winter Hardwick		Bulldog Winter Harlton		Bulldog Winter Oakington		Throws M.S. Winter Hildersham		Danas Spring Arrington	
Variety	Bulldog		Bulldog		Bulldog		Throws M.S.		Danas	
Type	Winter		Winter		Winter		Winter		Spring	
Nearest village	Hardwick		Harlton		Oakington		Hildersham		Arrington	
Honeybee hives	Yes		Yes		No		No		No	
Length x width, m.	535 x 230		285 x 270		485 x 240		585 x 460		485 x 320	
Collecting sites per week	50		41		48		61		51	
Collecting season	25/5-7/6		28/5-10/6		24/5-7/6		26/5-9/6		15/6-29/6	
	%	No.	%	No.	%	No.	%	No.	%	No.
All flowers examined	100	1368	100	1020	100	1321	100	1648	100	1775
not pollinated	38.2		35.1		32.2		41.5		6.1	
pollen not germinated	9.2		4.6		12.9		8.4		1.6	
pollen tubes arrested	1.1		1.9		2.0		3.6		0.7	
style damaged by insects	3.0		3.0		2.3		2.5		6.9	
first ovule reached by tubes	48.5		55.7		50.8		44.1		84.7	
total ovules where first one reached	100	2065	100	1864	100	2067	100	2414	100	5540
ovules not reached by tubes	2.3		4.5		5.0		6.1		3.3	
ovules passed by pollen tubes	0.6		1.3		0.9		0.8		0.4	
insufficient pollen tubes	6.8		6.7		4.6		7.1		2.0	
ovules fertilized	87.4		87.2		89.5		85.9		94.3	
All ovules examined		4241		3297		4066		5497		6512
mean fertilization, week 1	54		38		49		31		79	
week 2	41		45		38		28		83	
week 3	33		64		52		53		77	
all weeks	42.5		49.3		45.5		37.4		80.1	
in pollinated flowers	71.8		78.3		68.9		66.7		92.2	
Flowers with at least 1 ovule fertilized	48.3	660	55.7	568	50.8	671	43.8	722	84.5	1500
ovules per flower	3.10		3.23		3.08		3.34		3.67	
fertilized ovules per fertilized flower	2.73		2.86		2.76		2.85		3.48	
seeds per pod	2.17		2.36		2.55		-		3.03	
Total pods examined		822		1273		953		0		1291

Bulldog : synthetic variety
Throws M.S. : inter-pollinated mixture of four populations
Danas : large tick bean population

TABLE 2. Pollination and fertilization of faba beans in experimental plots at the Plant Breeding Institute, Cambridge.

Variety	Number of flowers examined	Percentage pollinated	Percentage of flowers fertilized	Percentage of ovules fertilized in fertilized flowers	Percentage of all ovules fertilized	Number of ovules per flower	Number of fertilized ovules per fertilized flower
Winter trial							
Maris Beagle	1255	89.9	86.9	94.6	83.0	3.25	3.12
Polar	996	80.4	77.0	88.1	68.0	3.71	3.28
T18BSv	850	81.5	75.7	90.2	69.3	3.87	3.55
Line 194	1275	75.9	72.7	89.5	65.7	3.60	3.25
Line 669	1088	89.4	85.1	92.5	79.6	3.92	3.67
Spring trial							
Maris Bead	621	82.3	71.7	87.3	62.4	3.95	3.44
Deiniol	710	91.3	82.7	89.2	74.0	3.94	3.53
Cockfield	699	94.1	88.0	92.3	81.3	3.74	3.45
Ticol	369	83.5	76.4	91.5	69.6	3.74	3.40
TC38	290	89.2	81.0	94.9	76.9	3.26	3.09
Hylon	432	68.6	60.9	62.7	38.2	8.11	5.09
LG5/14	520	66.5	58.3	64.3	37.5	6.73	4.34
Closed	416	44.8	35.6	59.6	21.3	3.89	2.32

Maris Beagle : synthetic variety
Polar : white-flowered population
T18BSv : population derived from first backcross of Sjodin's ti mutant to English winter line T18
Line 194 : English winter line
Line 669 : vigorous English winter line
Maris Bead : tick bean population
Deiniol : population with fixed autofertility
Cockfield : high yielding horse bean population with heterotic autofertility
Ticol : population derived from crosses between ti and four English winter lines
TC38 : autofertile line selected from (ti x Sudanese autofertile)
Hylon : broad bean variety with up to 10 seeds per pod
LG5/14 : partially inbred line derived from (Imperial longpod x English tick bean) with intermediate seed size
Closed : line with Poulsen's (1977) cf gene and <2% outcrossing

NEW DEVELOPMENTS IN CYTOPLASMIC MALE STERILITY IN *Vicia faba* L.

G. Duc[1], R. Scalla[2], A. Lefebre[1]

Institut National de la Recherche Agronomique

(1) Station d'Amelioration des Plantes

(2) Laboratoire des Herbicides

B.V. 1540 21034 Dijon Cedex, France.

ABSTRACT

The instability of the two determinisms of nucleo-cytoplasmic male sterility which are available in *Vicia faba* is described. In the case of the 447 cytoplasm, the cytoplasmic spherical bodies were characterised in biochemical studies and it was confirmed that they are directly linked with male sterility. Increased particle content in pistils seems to be a good predictor of greater stability of male sterile lines.

INTRODUCTION

The heterosis which has been observed in *Vicia faba* could, if exploited in hybrid cultivars, provide a significant improvement in the yields of this crop (Picard et al. 1982).

Two sources of nucleocytoplasmic male sterility (CMS) termed "447"[1] and "350"[2] appeared spontaneously in this species. Both express instability leading to fertile revertants in progenies of backcrosses with maintainer lines. This instability has until now hindered utilisation of CMS in large scale production of hybrid seed. Thus the answer to the problem is either to isolate new forms of CMS or to stabilize the existing ones.

Following the first purpose, mutagenesis work was developed by members of ACVF[3] and led to the induction of several genic male sterilities and one new form of CMS, the stability of which is as yet unknown.

Following the second purpose, new data was obtained which provided new methods of screening for stability of the 447 type of CMS.

The present paper focusses on these recent developments.

---

(1) Discovered by Bond D.A. P.B.I. Cambridge U.K.

(2) Discovered by Berthelem P. I.N.R.A. Rennes France.

(3) Association des Createurs de Varietes Fourrageres composing INRA, BLONDEAU and CLAUSE.

P.D. Hebblethwaite, T.C.K. Dawkins, M.C. Heath and G. Lockwood (eds.)

Vicia faba: Agronomy, Physiology and Breeding. ISBN 90-247-2964-5.

© 1984, Martinus Nijhoff/Dr W. Junk Publishers. Printed in The Netherlands.

1 - CHARACTERISTICS OF THE INSTABILITY OF THE 447 CYTOPLASM.

This instability leads to production of viable pollen grains in the progeny of a maintainer cross which is expected to be fully sterile. The proportion of fertile pollen grains varies both between flowers within a plant and between plants within a progeny. This may be expressed progressively, leading to juxtaposition on the same plant of male sterile and completely male fertile stems. Studying the intermediate levels of male sterility, plant breeders have shown that intra-mother plant variation is maternally transmitted to the progenies (Thiellement, 1977 ; Duc, 1976). Concerning the ultimate state of complete male fertility, which is called reversion, genetic analysis indicates that the cytoplasmic factor causing sterility is lost, in the same way as in the case of nuclear restoration.

The factors which induce this instability have been sought, in order to develop screening tests. Two of them have been identified, and the causes of the difficulties met by breeders trying to select against these characters are becoming clear :

a) Some minor recessive genes for restoration seem to act beside the major maintainer gene. They were detected by using Hayman's method to analyse a diallel cross of male sterile lines and their isogenic maintainer counterparts (Duc et al. 1983). This result is in agreement with the fact that instability is usually not expressed in the first backcross generations and that homozygosity seems to favour it (Berthélem and Le Guen,1974). The consequence is that a large amount of work has to be done before an unstable male sterile line can be eliminated. Another consequence comes from the genetic variability that was noticed among these minor restorer genes which opens possibilities of breeding for the best recombinations.

b) Environmental factors influence instability : the study in this field started with the observation that there were good years and bad years concerning the stability of CMS. Temperature and light have been shown to induce the production of fertile pollen grains on a male sterile plant (Letouzey, 1981 ; Duc, 1976). These environmental factors directly influence the fertility of the growing plant and since they also influence the embryos produced by this plant, they affect its progeny too (Duc, 1980). From these results, it follows that instability may not always be expressed in field conditions and that artificial conditions for screening for stability should be developed.

2 - IMPROVED UNDERSTANDING OF THE 447 CYTOPLASM COULD COME FROM STUDY OF
   MOLECULAR GENETICS :

Two approaches have been taken in molecular genetic studies of the
447 cytoplasm :

a) the first relates to mitochondria where Boutry and Briquet (1982
and personal communication) analyzed mitochondrialDNA and characterized
in vitro synthesized polypeptides from maintainers, sterile and restored
lines. Nothing was found at the level of the mt DNA which would indicate
that mitochondria are involved in CMS. One polypeptide seems to be synthesized only when male sterility is expressed.

b) the second line of research concerns the virus-like particles
(CSB) discovered by Edwardson *et al*(1976). These particles, about 70 nm in
diameter, are spherical in shape and are made up of a peripheral unit
membrane enclosing a dark central zone. They are found in all tissues of
sterile plants.

Scalla *et al*.(1981) devised a biochemical method of preparing these
particles, and were able to verify the close relationship between the presence of the particles and the existence of the sterility factor (Table 1).

TABLE 1 Correlation of genetic data with the presence of the CSB peak in
       sucrose gradients.

Génotypes		Nuclear genes	Pollen phenotype	Detection of CSB
Ad 23	Male sterile	rf.rf	Sterile	Yes
Ad 23	Maintainer	rf.rf	Fertile	No
Ad 23 x HG 115	Restored hybrid	rf.rf	Fertile	No
159	Male sterile	rf.rf	Sterile	Yes
159	Maintainer	rf.rf	Fertile	No
159	Reversion	rf.rf	Fertile	No
135	Male sterile	rf.rf	Sterile	Yes
135	Maintainer	rf.rf	Fertile	No
123	Male sterile	rf.rf	Sterile	Yes
123	Maintainer	rf.rf	Fertile	No

Analysis of these particles enabled us to demonstrate that they contain a double-stranded RNA of high molecular weight (about $12-10^6$). The
presence of this RNA in sterile plants was simultaneously demonstrated by
Grill and Garger (1981). Recent results from Scalla and Lefebvre (unpublished) indicate that RNA polymerase activity is associated with the par-

ticles, which thus perhaps possess their own system of replication.

Are these particles a cause or a consequence of CMS? Simultaneous transmission of the large RNA molecule and CMS through dodder bridges was claimed by Grill and Garger (1981). We have not succeeded in repeating this experiment despite about 50 attempts. We have also tried graft transmission with negative results, as in Bond's (1966) experiments. In consequence, conclusive evidence that the particles are the basis of CMS is still lacking. However, several facts indicate that this is so.

Firstly, the concentration of the particles in pollen grains increases simultaneously with degeneration of their cytoplasmic contents (Moussel *et al.*, 1982).

Secondly, fractionation of particles through sucrose gradients enabled us to estimate the number of particles present in pistils (see Table 2). The results show that there is high genetic variability in the number of particles. More striking, there is a link between the quantity of particles and the stability of the male-sterile lines that we have examined. The difficulty in the establishment of this correlation is the low number of stable male sterile lines available for examination. A strong supporting argument for such a link follows from comparison of two segregating lines derived from the same backcross (Ad 23 and Ad 23 E) : the more stable line has the higher particle content.

TABLE 2   Study of the link between pollen fertility (as measured by Alexander's pollen staining technique) expressed as the percentage of fertile pollen grains among a sample of 1500 pollen grains and the area of the peak observed when extracting particles on a sucrose gradient.

Genotypes	Mean pollen fertility of previous BC generation (%)		EXP I		EXP II	EXP III
			Pollen fertility of analysed plant (%)	area of peak (cm^2)	area of peak (cm^2)	area of peak (cm^2)
Ad 23*	0	(BC 11)	0.06	0.40	0.46	0.45
NY	0.04	(BC 4)	0.02	0.24	0.71	0.5
123	0	(BC 11)	0	0.15	0.30	0
159	32.7	(BC 4)	0.3	0.14	0.27	0.22
249	9.7	(BC 4)	2.7	0.06	0.19	0.20
135	0.9	(BC 5)	0.9	0.02	0.10	0.15
I	7.7	(BC 5)	1.7	0	0.07	0
241	20.8	(BC 4)	31.9	0	0	0
93			70.2	0	0.08	0
Ad 23 E*			15.3	0	0	0

* Selected by P. Berthélem.

This result indicates that we have a quantitative method of estimating the stability of our lines, which could be performed on early backcross generations and thus save labour. However, our biochemical method
of assay of the amount of particles is too time-consuming and we intend
to develop a serological test to measure it more readily.

Finally, to explain the relationship between the stability of CMS
and the amount of particles, it can be assumed that, in order to be distributed in all parts of the plant, the particles must be in adequate number in the cells, otherwise they could be lost in the growing organs
through successive mitoses.

Besides the short term implications of such discoveries for the selection of more stable CMS in faba bean, the cytoplasmic particles also
set interesting problems in the fields of Virology and Molecular Biology.
They appear to belong to a new type of viruses, and similar particles have
already been described in other plants (*Tropaeolum majus* : Ie, 1972 ;
Epilobium : Anton Lamprecht, 1965) and in fungi (Albouy, 1973). Their mode
of transmission, if any, is unknown, and they raise at least two other
questions : how is their RNA organized ? and do they really possess an
autonomous system of replication, as suggested by our experiments ?

REFERENCES.

Albouy, J., Lapierre, H., Molin, G. 1973. Mise en évidence d'un nouveau
     type de particules dans les hyphes de carpophore d'*Agaricus bisporus*.
     C.R. Acad. Sci. Paris Série D, 276, 2805-2807.
Anton-Lambrecht, I. 1965. Electron microscopical evidence of unusual struc-
     tures in the cytoplasm of some plasmotypes of *Epilobium* hybrids.
     J. Ultrastruct. Res. 12, 624-633.
Berthélem, P., Le Guen, J. 1974. Rapport d'activité de la Station INRA.
     Amélioration des Plantes de Rennes.
Bond, D.A., Fyfe, J.L., Tonynbee-Clarke, G. 1966. Male sterility in field
     beans (*Vicia faba* L.) III Male sterility with cytoplasmic type of
     inheritance. J. Agric. Sci. Cambridge 66, 359-367.
Boutry, M., Briquet, M. 1982. Mitochondrial modifications associated with
     cytoplasmic male sterility in faba bean. Eur.J. Biochem. 127,
     129-135.
Duc, G. 1976. Modalités d'expression et hypothèses explicatives du manque
     de stabilité de la stérilité mâle cytoplasmique chez la féverole
     (*Vicia faba* L.) Thèse n°369 de l'Université de Paris Sud Centre
     d'Orsay.

Duc, G. 1980. Effect of environment on the instability of two sources of cytoplasmic male sterility in faba beans. Fabis Newsletter 2,29–30.

Duc, G., Huglo, B. 1983. La stérilité mâle nucléo-cytoplasmique 447 chez la fèverole (*Vicia faba* L.) Etude du déterminisme paternel de l'instabilité phénotypique de la stérilité mâle (to be published in Agronomie).

Edwardson, J.R., Bond, D.A., Christie, R.G. 1976. Cytoplasmic sterility factors in *Vicia faba* L. Genetics, 82, 443–449.

Grill, L.K., Garger, S.J. 1981. Identification and characterization of double stranded RNA associated with cytoplasmic male sterility in *Vicia faba*. Proc. Nat. Acad. Sci. USA 78, 7043–7046.

Ie, T.S. 1972. Cytoplasmic particles in *Tropaeolum majus*. Planta 106, 227–236.

Letouzey, D. 1981. Influence de la température sur les différents aspects phenotypiques de la stérilité mâle cytoplasmique chez la fèverole (*Vicia faba* L.) Thèse de Docteur Ingénieur ENSA Rennes 81/4/C/7.

Moussel, B., Moussel, C., Audran, J.C., Bouillot, J., Duc, G. 1982. La stérilité mâle nucléo-cytoplasmique chez la fèverole (*Vicia faba* L.) III. Répartition au sein des tissus de l'anthère et évolution quantitative au cours de l'ontogenèse pollinique de particules associées à la stérilité mâle nucléo-cytoplasmique de type 447. Rev. Cytol. Biol. Végét. Bot. 5, 81–93.

Picard, J., Berthélem, P., Duc, G., Le Guen, J. 1982. Male sterility in *Vicia faba*, future prospects for hybrid cultivars. G. Hawtin and C. Webb (eds.) ICARDA. ISBN 90247 259 33.

Scalla, R., Duc, G., Rigaud, J., Lefebvre, A., Meignoz, R. 1981. RNA containing intracellular particles in cytoplasmic male sterile fababean (*Vicia faba* L.) Plant Science Letter 22, 269–277.

Thiellement, H. 1977. La stérilité mâle cytoplasmique chez *Vicia faba* L. III. Hérédité des phénotypes polliniques. Ann. Amélior. Plantes 27, 555–562.

INSTABILITY OF CYTOPLASMIC MALE STERILITY IN VICIA FABA :
ROLE OF TEMPERATURE AT MEIOSIS

J. Le Guen, P. Berthelem and F. Rousselle

Plant Breeding Station, Le Rheu, France

ABSTRACT

The results presented arise from a programme of research developed as part of an E. E. C programme at Rennes in 1980. Two important causes of instability were shown '(a) increased reversion from sterile to fertile pollen when meiosis occurs above 17°C and (b) increased reversion on the upper nodes of the plants. These findings provide a basis for both the practical utilisation of existing types of cytoplasmic male sterility and development of improved male sterility.

INTRODUCTION

Male sterility created considerable interest in France and Great Britain in the 1960s and 1970s when it was shown that heterosis in *Vicia faba* is associated with high yield.

Further interest was generated when two different sterile cytoplasms were identified : PBI447 discovered by Bond in Cambridge (Bond et al, 1966) and REN350 discovered a few years later by Berthelem in Rennes (Berthelem, 1966). Important research programmes aiming to combine different nuclear genotypes with these cytoplasms, especially the PBI447 one, were developed in Great Britain and France. These programmes resulted in the creation of experimental hybrid varieties which were tested in trials and proved to have high yielding ability. Multiplication of the parental lines of some hybrid varieties (F1 and three-ways hybrids) was commenced in France and Great Britain.

However, hybrid production on a commercial scale has never been achieved because it is not possible to produce the female parent of the hybrid varieties in pure form on a large scale. This is due to a phenomenon which has been termed "instability of cytoplasmic male sterility in faba bean".

The general features of this instability have been described by many authors . The phenomenon is expressed at the phenotypic level, in two ways :

- Firstly, by sudden reversions from sterility to fertility, the best example being what Bond termed "tiller-fertile plants".

- Secondly, by slow evolution of plants from a sterile state to a fertile one, with various intermediate forms (plants which are partly sterile and partly fertile) being recognizable during this evolution.

P.D. Hebblethwaite, T.C.K. Dawkins, M.C. Heath and G. Lockwood (eds.)
Vicia faba: Agronomy, Physiology and Breeding. ISBN 90-247-2964-5.

When considering this phenomenon in relation to possible breeding strategies, two characteristics can be isolated and noted ;

- Firstly, there is an interaction between cytoplasm and different
recurrent genotypes, which is expressed as different rates of segregation
into sterile and fertile forms in the progenies of some sterile plants during the backcrossing process.

- Secondly, in the course of successive generations of multiplication
of the few lines which have been stabilized for sterility after 9 or 10
backcross generations, fertility increases as an exponential function. This
observation means that, starting for example with 1 % of fertile plants in
the first generation of multiplication, at the fifth 40 % to 60 % of the
plants may be fertile.

Many of these traits are well known and the aim of our work was to
try to understand and explain how and why male sterility in faba beans
is unstable.

MATERIALS AND METHODS

Most of our experiments were conducted using two French winter faba
bean lines (245.17 A and 972 A) the first one being able to maintain the
PBI447 cytoplasm, the second being a rather poor maintainer. Both materials
were at backcross generation 10 when used. (Results reported in this note
relate in their totality to the 972 A line).

Plants were grown either in completely controlled conditions in cabinets where light and temperature were maintained at a constant level or, in
semi-controlled conditions, in glasshouses in which temperatures were
controlled by a cooling system.

Plants were sown in containers, and at flowering one flower was removed from each flowering node just before anthesis. By collecting buds at
various times from young inflorescences of control plants, it was found
that meiosis of these flowers occurred 150 to 180 accumulated degree-days
before anthesis (F. Rousselle, Personal communication).

Analysis of sterility was done by microscopic examination of pollen
grains stained by Alexander's method (Alexander, 1969) and dividing the
grains into five classes according to their staining densities. These ranged from sterile grains devoid of cytoplasm and non-stained, to fertile
grains which contained cytoplasm and were fully stained. Each pollen class
was expressed per thousand of all the pollen grains from a given flower.

At the same time other characteristics were noted such as the number of the tiller from which the flower was collected and the position of the flowering node on the tiller. Temperatures were recorded continuously in every experiment.

The data were analysed by a multivariate method. From this computation a "Sterility index" has been defined (Le Guen and Berthelem, 1983) corresponding to the projection on a plane, expressed in non-Euclidian distances, of the mean values of all the cytoplasmic densities of the pollen grains observed in a flower. It is necessary to use such computed values to express sterility because the phenomenon of reversion appears to be a quantitative one, and all the pollen grain phenotypes were significant and had to be taken into account.

RESULTS

When sterile faba bean plants are grown in field conditions pollen phenotype fluctuates in many different ways ; this phenomenon appears to be completely random for a given progeny. When plants are grown in controlled conditions the fluctuations appear less random as shown in Figure 1 (Le Guen and Berthelem, 1983).

This figure represents the projection, as explained previously, of three sister progenies, that is to say the flowers of three plants derived from the same parent. Each symbol corresponds to a single flower. The upper left quarter of the plane contains the most sterile flowers, and the lower left the most fertile ones. The two right quarters contain intermediate sterile or fertile flowers.

Figure 1 demonstrates clearly that there are three different progenies emanating from the same mother plant.

The flowers represented by a closed circle are, in the main, concentrated in the upper part of the graph and belong to a sterile plant, whereas most of the flowers symbolized by an open circle are in the lower part of the figure and were produced from a more fertile plant.

The flowers characterized by a closed square are scattered on the plane, with a greater concentration in the right hand part, and were produced from an intermediate plant. The figure also shows that for a given progeny there is considerable dispersal of the flowers on the plane.

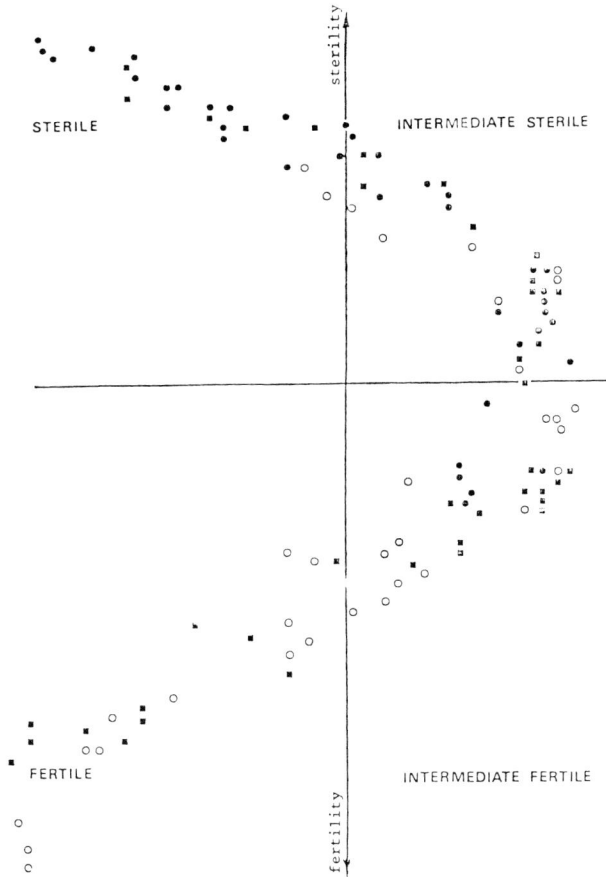

Figure 1

Plant 1 ●
2 ○
3 ▣

Figure 2 provides an initial explanation of this problem. The values of the "Sterility index" are shown on the Y-axis while the X-axis represents the time when the flowers were sampled, expressed in days from the beginning of the year.

On two dates (98 and 108) the "Sterility index" indicates reduced reversion from sterility to fertility with the sterility of the flowers on day 98 being even greater than that of the flowers collected on day 77.

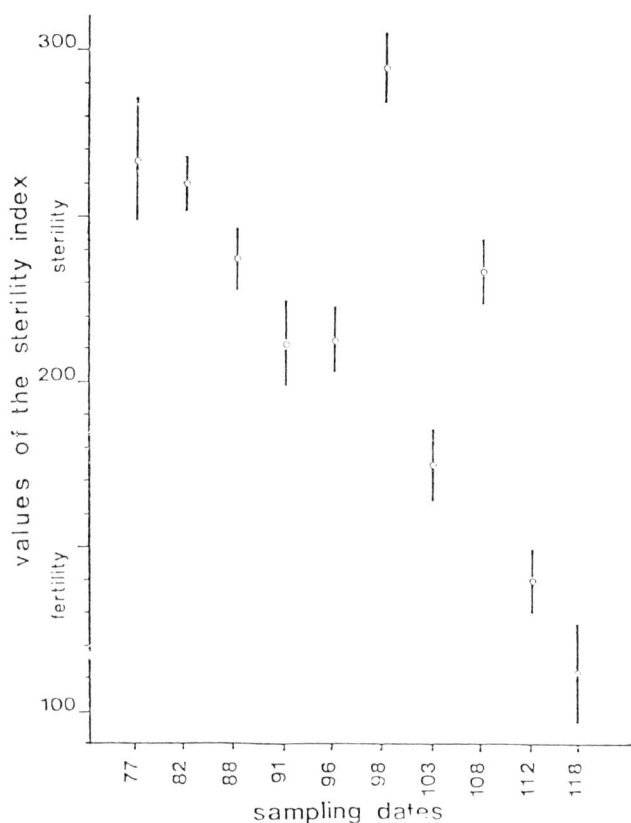

Figure 2

Further explanation of this phenomenon can be obtained by examination
of Figure 3, which illustrates three parameters ;

        - dates of sampling

        - sterility index

        - temperatures at meiosis

Two main characteristics can be observed in the figure ;

    - Firstly, there is a very close relationship between sterility index
and temperature at meiosis, the two curves being of similar shape. At sampling dates 98 and 108 there was unexpectedly little or no reversion from
sterility to fertility corresponding to rather low temperatures at meiosis (less than 18°C).

    - Secondly, the phenomenon of reversion seems to be amplified at the
end of the sampling time, as if an additive or a cumulative effect of tem-

perature takes place. It is clear that low meiotic temperatures permit the expression of the sterile phenotype of pollen, whereas high temperatures at meiosis lead to a fertile expression of this pollen.

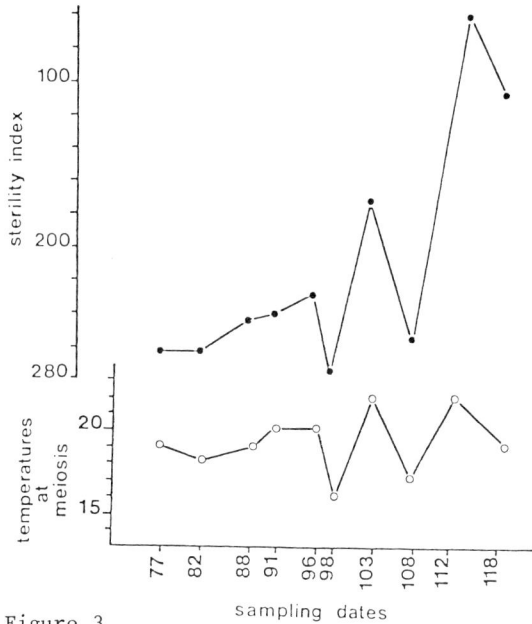

Figure 3

Complementary information was obtained by observing what happened to flowers of different flowering nodes when meiosis occured at a given temperature. Four temperatures (16°C, 19°C, 20°C and 22°C) were chosen and flowers were taken only from the even helical flowers series, i.e. from nodes 2, 4, 6 etc...

Figure 4 represents the relationship between pollen fertility, flowering node and temperature at meiosis. In this representation the expression of pollen fertility is the mean values of the two most fertile classes (according to their staining densities) already described in the Materials and Methods section.

This figure shows that there is a rapid increase in fertility from the second to the fourteenth flowering node and that the progression has the shape of an exponential function. There is an important inversion in the relationship at the tenth and twelfth nodes and we are currently looking at this phenomenon because it has been observed for other characteristics, such as in particular susceptibility to chocolate spot (Le Guen and

Lenormand, manuscript in preparation).

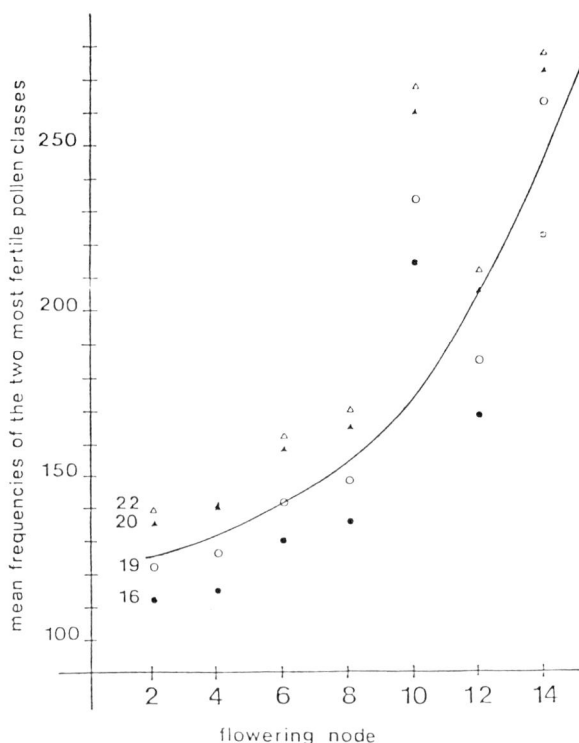

Figure 4

Figure 5 represents the same phenomenon in a different form. Meiotic temperatures have the same type of effect on every flowering node and, in general, the critical temperature for increasing fertility is between 19°C and 20°C. The inversion of the tenth and twelfth flowering nodes may also be seen in this figure.

The most important point from these observations is that temperature at meiosis affects the lower nodes and the upper ones in different ways. This observation was the basis of a new breeding method which we have termed "zonal selection", consisting of selecting seed for further breeding only from the lower pod setting nodes in order to avoid the great increase in fertility which occurs at the upper ones. This method has been shown to be utilizable in controlled temperature conditions but is practically inadequate in field conditions because temperatures do not always increase regularly from the bottom to the top of the plants, and so inversions in the theoretical profile often occur.

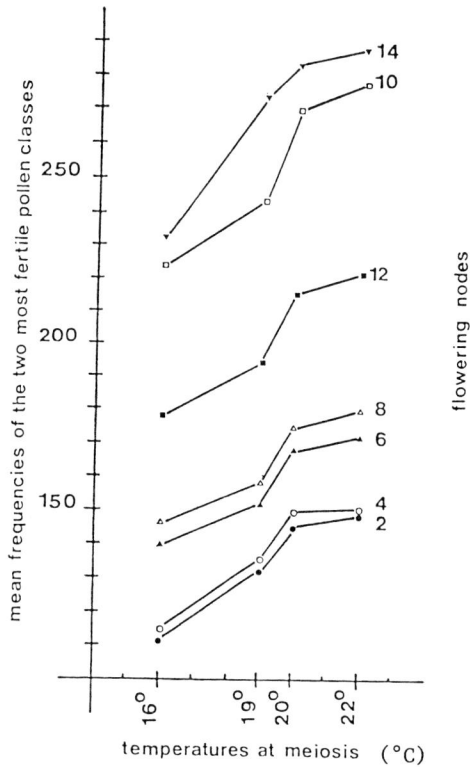

Figure 5

The effect of increased temperature at meiosis on pollen phenotype was assessed more precisely by considering only the fertile pollen class (phenotype F) which was divided into 10 equiprobable subclasses based on a sample of pollen frequencies as can be seen from Figure 6. The way in which the distribution of pollen frequencies over these subclasses changed when temperatures at meiosis fluctuated was examined.

At 16°C, 17°C and 18°C, the great majority of the pollen grains belonged to subclasses 0 to 4, and thus the mean pollen fertility of the plants was very low at these temperatures. The higher the meiotic temperature the greater were pollen frequencies in the high F subclasses and the transition from rather sterile to rather fertile plants was observed to take place between 19°C and 20°C.

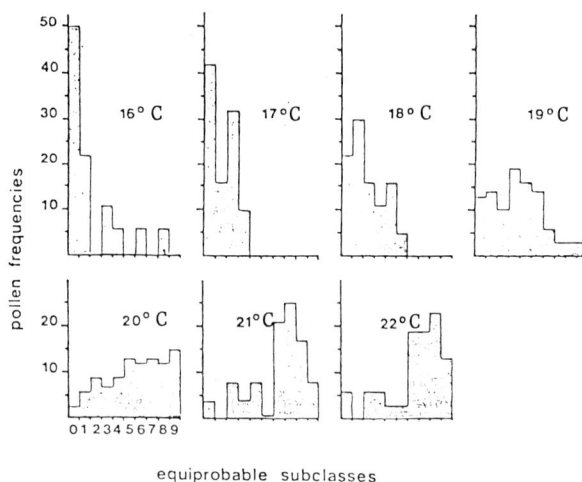

equiprobable subclasses

SUBCLASS	EXTREME VALUES OF POLLEN F FREQUENCIES		SUBCLASS	EXTREME VALUES OF POLLEN F FREQUENCIES	
0	0.0 to	1.8	5	7.7 to	8.9
1	1.9	3.0	6	9.0	12.0
2	3.1	4.5	7	12.1	14.5
3	4.6	5.8	8	14.6	19.7
4	5.9	7.6	9	19.8	58.7

Figure 6

DISCUSSION

The results presented here constitute a part of the breeding programme for male sterility which has been developed at Rennes over several years. Simultaneous studies have been concerned with other aspects of this work such as the biochemical processes involved, and these have been reported elsewhere.

It seems clear, from the results reported here, that instability of cytoplasmic male sterility in *V. faba* is not a random phenomenon, but is influenced by two separate characteristics ;

- Firstly, there is a strong correlation between temperature at meiosis and pollen phenotype, and the higher the temperature the more fertile the pollen. It is also clear that temperatures below 17°C produce a high degree of sterility. This point has been confirmed in other respects in complementary experiments.

- Secondly, at a given temperature, an increase in fertility occurs with increasing number of the flowering node. This observation is related to the fertility gradient (Le Guen and Berthelem, 1983). The cause of this phenomenon in unclear, but it may arise from the effect of temperature units accumulated by the plants during their growth.

Whatever the cause the fact remains that under field conditions temperatures during the flowering period exceed the critical value of 17°C and are liable to variation thus preventing the use of male sterility under natural conditions, at least for the type of material tested.

Several experiments have been designed with the hope of improving cytoplasmic male sterility or to use it in particular conditions. These experiments, which constitute another aspect of this programme, are based on mutagen treatments made in co-operation with the plant breeding station at Dijon and French private breeders. In this programme, several genetic male sterilities have been found and one, which seems to be dominant, is being utilized at Rennes in a recurrent selection scheme.

Another aspect developed at Rennes is the use of low controlled temperatures (between 15°C and 16°C) to produce the first two of five generations of multiplication of male sterile lines.

In this way we hope to reduce the slope of the exponential evolution of sterility to fertility below the level which is expressed when all the multiplications are made under natural conditions. Using this method, we hope to obtain seed of the fifth generation which is sufficiently sterile to produce commercial seeds in a hybrid scheme.

REFERENCES

Alexander, M. P. 1969. Differential staining of aborted and non aborted pollen. Stain technol., 44, 117-122.

Berthelem, P. 1966. Rapport d'activité de la Station d'Amélioration des Plantes de Rennes., 1965-1966.

Bond, D.A., Fyfe, J.L. and Toynbee-Clark, G. 1966. Male sterility with a cytoplasmic type of inheritance in field bean. J. agric. Sci., 66, 359-367.

Le Guen, J. and Berthelem, P. 1983. Instabilité phénotypique de la stérilité mâle nucléocytoplasmique chez Vicia faba L. Hétérogénéité inter-plantes et intraplante. Agronomie, 3, 265-272.

AN ASSESSMENT OF THE POTENTIAL FOR IMPROVING THE NUTRITIVE VALUE OF

FIELD BEANS (VICIA FABA) - A PROGRESS REPORT

D. Wynne Griffiths

Welsh Plant Breeding Station,
Aberystwyth, Dyfed, Wales, UK

ABSTRACT

The amino acid contents of various varieties, populations and selections of field beans (Vicia faba) with widely differing protein contents have been determined. Highly significant statistical correlations were found between protein content and six nutritionally essential amino acids. All with the exception of arginine were negatively correlated, indicating that selection for improved protein content would result in a slight decrease in the biological value of field bean protein.
Studies have also been undertaken to evaluate the variation and nutritional significance of various anti-nutritive factors; in particular protease inhibitors, phytates and tannins. Considerable inter- and intra-varietal variation has been found for trypsin inhibitor activity but the relative trypsin and chymotrypsin inhibitor activities were considerably less than that found for soyabean. Phytate content has also been found to vary considerably both within and between varieties and by determination of the solubility profiles of these compounds, it has been possible to prepare various residues and protein precipitates with similar protein content but with widely differing phytate content. Field bean testa tannins have been shown to inhibit in vitro a wide range of digestive enzymes and the results of rat feeding experiments have shown that these polyphenolic compounds significantly reduce live weight gain and the availability of protein and other nutritionally essential dietary constituents. The significance of these results in relation to breeding for improved nutritive value are discussed.

INTRODUCTION

The field bean (Vicia faba) probably represents one of the best

sources of indigenous plant protein for both the United Kingdon and much

of western Europe. However compared to the oilseed crop residues current-

ly available to the animal feed processing industry, its protein content

is comparatively low and the results of previously reported feeding trials

(e.g. Palmer and Thompson, 1975) have consistently reported that the

nutritive value of the field bean is considerably lower than that pre-

dicted by its amino acid composition. The objective of this work has been

to assess the potential for and consequences of improving the nutritive

value of the crop.

PROTEIN QUANTITY AND QUALITY

Early investigations carried out at the Welsh Plant Breeding Station

P.D. Hebblethwaite, T.C.K. Dawkins, M.C. Heath and G. Lockwood (eds.)
Vicia faba: Agronomy, Physiology and Breeding. ISBN 90-247-2964-5.
© 1984, Martinus Nijhoff/Dr W. Junk Publishers. Printed in The Netherlands.

(Griffiths and Lawes, 1978) indicated that considerable variation in crude
protein content existed both within and between varieties. In an examination of some thirty varieties, genotypes and populations of diverse geographical origin protein content varied by over 15 units, the values found
ranging from 22.9% in an Ethiopian selection to 38.5% in a mutant variety
originating from Svalöf, Sweden. Considerable variation was also found
within existing varieties with values ranging over 8-10 units of protein
within each variety. Numerous single plants with a protein content excee-
ding 36% were found and preliminary heritability studies indicated that
much of the observed variation was of genetic origin.

In order to assess the possible effect of selection for increased
protein content on nutritive value, the amino acid composition of various
varieties, genotypes and single plant selections whose protein content
spanned the range expected to be found was determined (Griffiths, 1983).
The results of correlating crude protein content with the nutritionally
essential amino acid content (Table 1) indicated that five of these acids
were negatively correlated but with the exception of lysine the calculated
rate of decline was relatively low. In contrast arginine was positively
correlated with protein content, the magnitude of the rate of increase
being clearly demonstrated by the increase from 8.6 g/100 g in the low
protein (24.2%) line to 11.4 g/100 g in the high protein (37.5%) line.

TABLE 1   Correlation and regression coefficients between amino
acid content (g/100 g protein) and crude protein content.

Amino acid	r-value	Significance	b-value
Threonine	-0.280	NS	-
Cystine	-0.461	**	-0.013
Valine	-0.051	NS	-
Methionine	-0.433	**	-0.014
Iso-leucine	-0.129	NS	-
Leucine	0.283	NS	-
Phenylalanine	-0.650	***	-0.076
Histidine	-0.670	***	-0.041
Lysine	-0.592	***	-0.118
Arginine	0.742	***	0.196

Attempts to relate the data to changes in the ratio of the two major
storage proteins, vicilin and legumin were not entirely successful, indicating that increased protein content may also affect albumin protein
ratios or as suggested by Barratt (1982) the amino acid content may be

altered by the presence of more free amino acids in the high protein selections.

Nutritionally it would appear that selection for increased protein content would result in a small but significant reduction in the nutritive value of the protein, mainly due to the decreased levels of the already nutritionally limiting sulphur amino acids, methionine and cystine. Particular attention should also be given to lysine content, since too great a reduction in the relative concentration of lysine could lead to a diminishing of the value of field bean protein as an ideal supplement to the lysine deficient cereals.

ANTI-NUTRITIVE FACTORS

The field bean, in common with other grain legumes has the capacity to synthesise various compounds, which, when consumed by man or other animals may reduce the nutritive value of the crop, as predicted by protein content and quality. Many such compounds have been identified as being present in the field bean and our investigations have been restricted to three namely, the tannins, protease inhibitors and phytates.

Protease inhibitors

Since their first discovery in 1938 (Read and Haas) protease inhibitors have been found to be widely distributed throughout the plant kingdom. These proteinic compounds have been shown to significantly retard growth rates and to induce pancreatic hypertrophy (Liener and Kakade, 1969) but their mode of action *in vivo* has not been completely elucidated, although it has been suggested that the observed reduction in growth rates may be related to the increased secretory activity of the pancreas. This would result in an increased demand for amino acids by this organ and consequently reduce their availability for growth.

A number of trypsin inhibitors have been isolated and characterised from field beans (Warsy, Norton and Stein, 1974) and in our investigation of the variation in trypsin inhibitor activities of a wide range of differing genotypes, varieties and populations (Griffiths, 1979) a two-fold range in values was found. However, none of the values found approached that for soyabean and no significant correlation was found between trypsin inhibitor activity and protein content. Typical values for trypsin inhibitor content of beans are given in Table 2 and also included are values for some common pea varieties.

TABLE 2   The trypsin and chymotrypsin inhibitor content of
various pea and bean varieties.

Variety	Trypsin inhibitor (sbe*)	Chymotrypsin inhibitor (sbe*)	Variety	Trypsin inhibitor (sbe*)	Chymotrypsin inhibitor (sbe*)
Vicia faba			Pisum spp.		
Dacre	5.0	5	Minerva	0.5	4
Danas	4.5	4	Marathon	1.2	12
Maris Bead	4.5	5	Rosakrone	4.3	33
White flower	5.5	4	Progretta	11.1	64
Minden	4.5	4	Maro	12.5	76

*mg soyabean containing the same quantity of
inhibitor as 100 mg of the given variety

Comparison of the trypsin inhibitor activities of peas and beans
revealed that the former appeared to contain on average slightly less inhibitor than beans. However, two pea varieties namely Progretta and Maro
had significantly greater trypsin inhibitor activity but again the levels
present in these varieties were almost 8 times less than that found in
soyabean. Also included in the above table are values for chymotrypsin
inhibitor content, in contrast to the result for trypsin inhibitor, peas
appear to contain significantly more chymotrypsin inhibitor activity than
beans and indeed the levels found in Maro and Progretta were over 15 times
higher than those for field beans.

A study of the stability of both trypsin and chymotrypsin inhibitors
revealed that for both peas and beans the inhibitor activities remained
unchanged after heating samples in ovens up to 100°C but were readily in-
activated on autoclaving.

Nutritionally it would appear that little benefit would result for
selection for low protease inhibitor varieties of field beans, since the
levels of both trypsin and chymotrypsin inhibitors are very low as compared with soyabeans. Indeed this appears to be confirmed from the results of Abbey, Neale and Norton (1979) who found no statistically significant reduction in growth parameters resulted on feeding purified trypsin
inhibitors to rats at levels commonly found in field beans. However, it
should be pointed out that more recently Sjodin et. al. (1981) in a comparison of high and low trypsin inhibitor selections found a significant
improvement in biological value on feeding the low trypsin inhibitor lines
to rats.

Phytates

The results of macro element analysis would suggest that the field bean represents an excellent source of phosphorus and other nutritionally essential metal ions. However, more detailed examinations have shown (Griffiths and Thomas, 1981) that between 40-60% of the total phosphorus is present as phytate, the mixed calcium and magnesium salts of myoinositol 1, 2, 3, 4, 5, 6-hexakis dihydrogen phosphate. These phosphorus rich compounds have the ability to form insoluble chelates with various trace and macro elements and thus may significantly reduce the amount of phosphorus and associated trace or macro element available for absorbtion in vivo.

An examination of a range of genotypes, selections and populations revealed that considerable inter-varietal variation with regard to both phosphorus and phytate content existed within field beans and as has been found for other nutritionally important characteristics a large variation in phytate phosphorus content (1.93 - 4.03 g/kg) has been found within existing varieties. However, although sufficient variation may exist for the commencement of selection programmes for reduced phytate content, comparatively little evaluation of the possible nutritional advantages to be gain by such a programme have been undertaken. Consequently in order to achieve this objective the possibility of preparing residues of both high and low field bean phytate content has been studied (Griffiths, 1982). An examination of the solubility of field bean phytate at various pH values revealed that as for field bean protein both reached minimum solubility at around pH 3. However, as the pH of the extracting solvent was increased, phytate solubility increased rapidly whilst protein solubility remained low, up to pH 5. Consequently extracting at pH 3 was found to a leave a residue with 38.5% protein and 2.6% phytate whilst at pH 5 the residue contained similar protein content but only 0.8% phytate.

A similar response was obtained on examining protein precipitates prepared from a pH 12 extraction of Minor cotyledons. Precipitation of the protein at pH 3 yielded a precipitate containing over 90% protein and 6.1% phytate, whilst at pH 5 the precipitate again contained over 90% protein and only 1.1% phytate. Thus by selection of either extracting or precipitating pH samples with similar protein but widely differing phytate contents may be prepared and could be used in feeding trials to evaluate more precisely the nutritive significance of these compounds.

Tannins

The presence of condensed tannins in the seed coats of coloured flowered varieties of field beans has been demonstrated by several workers including Bond (1976) and Griffiths and Jones (1977). In vitro studies
(Griffiths, 1979) showed that extracts from coloured flowered testa not
only reduced protein solubility but also inhibited the activities of
several digestive enzymes including trypsin, $\propto$ -amylase and lipase. White
flowered varieties showed no such activity and it is therefore probable
that the protease inhibitor reported to be present in field bean testa
(Wilson et al., 1972) was not a specific enzyme inhibitor as is found in
the cotyledons but a non-specific enzyme inhibitor such as tannin.

In order to assess the nutritive significance of these polyphenolic
compounds a feeding trial in which diets containing 10% seed coats from
both a coloured flowered and a white flowered variety were fed to rats
(Moseley and Griffiths, 1979). The high tannin diets (i.e. those containing testa from coloured flowered varieties) were seen to reduce live
weight gain by 15% and to significantly reduce the true digestibility of
nitrogen as well as the apparent digestibilities of both carbohydrates and
lipids. An examination of intestinal enzyme activity of the rats revealed
that those receiving the high tannin diets had significantly reduced
trypsin and $\propto$ -amylase but also increased lipase activity (Griffiths and
Moseley, 1980). It may be therefore that the nutritional significance of
field bean tannins is not solely restricted to the formation of insoluble
dietary protein-tannin complexes but may also reduce the availability of
other nutrients due to the inhibition of digestive enzymes.

From these results it is clear that a significant improvement in nutritive value, resulting from an increase in net protein utilization would
result in selecting for tannin-free, white flowered varieties. However,
the consequences of such selections on both yield and disease resistance
require close examination before such a programme could be confidently recommended.

CONCLUSION

The extent of the variation with respect to various chemical parameters of nutritional significance to be found both within and between
field bean varieties would suggest that a substantial improvement in the
nutritive value of the crop could be achieved by plant breeding methods.
An increase in protein content would appear to be nutritionally desirable,

although selection for such an improvement would be associated with a slight reduction in protein quality. The elimination, or at least reduction in the levels of the anti-nutritional factors, particularly the tannins could also lead to a significant improvement in nutritive value. However, since many of these compounds may play vital roles in natural plant defence mechanisms or in the physiology of the developing plant and seed, further basic research is required to determine more precisely the significance of these anti-nutritional compounds.

REFERENCES

Abbey, B.A., Neale, R.J. and Norton, G. 1979. Nutritional effects of field bean (Vicia faba L.) protease inhibitors fed to rats. Br. J. Nutr. 41, 31-38.

Barratt, D.H.P. 1982. Chemical composition of mature seeds from different cultivars and lines of Vicia faba L. J. Sci. Fd Agric. 33, 603-608.

Bond, D.A. 1976. In vitro digestibilities of the testa in tannin-free field beans (Vicia faba L.). J. agric. Sci., Camb. 86, 561-566.

Griffiths, D.W. 1979. The inhibition of digestive enzymes by extracts of field beans (Vicia faba). J. Sci. Fd Agric. 30, 458-462.

Griffiths, D.W. 1982. The phytate and iron binding capacity of various field bean (Vicia faba) populations and extracts. J. Sci. Fd Agric. 33, 847-851.

Griffiths, D.W. 1983. The amino acid composition of high and low protein faba bean (Vicia faba) varieties and selections. Fabis 6 (in press).

Griffiths, D.W. and Jones, D.I.H. 1977. Cellulase inhibition by tannins in the testa of field beans. J. Sci. Fd Agric. 28, 983-989.

Griffiths, D.W. and Lawes, D.A. 1978. Variation in the crude protein content of field beans in relation to the possible improvement of the protein content of the crop. Euphytica 27, 487-495.

Griffiths, D.W. and Moseley, G. 1980. The effects of diets containing field beans (Vicia faba) of high or low polyphenolic content on the activity of digestive enzymes in the intestines of rats. J. Sci. Fd Agric. 31, 255-259.

Griffiths, D.W. and Thomas, T.A. 1981. Phytate and total phosphorus content of field beans (Vicia faba). J. Sci. Fd Agric. 32, 187-192.

Liener, I.E. and Kakade, M.L. 1969. Protease inhibitors. In: "Toxic Constituents of Plant Foodstuffs" (Ed. I.E. Liener). (Academic Press). pp. 293-318.

Moseley, G. and Griffiths, D.W. 1979. Varietal variation in the anti-nutritive effects of field beans (Vicia faba) when fed to rats. J. Sci. Fd Agric. 30, 772-778.

Palmer, R. and Thompson, R. 1975. A comparison of the protein, nutritive value and composition of four cultivars of field beans (Vicia faba). J. Sci. Fd Agric. 26, 1577-1583.

Read, J.W. and Haas, L.W. 1938. Studies on the baking quality of flour as affected by certain enzyme systems. V. Further studies concerning potassium bromate and enzyme activity. Cereal Chem. 15, 59-68.

Sjödin, J., Martensson, P. and Magyarosi, T. 1981. Selection for anti-nutritional substances in field beans (Vicia faba). Z. Pflanzenzuchtg. 86, 231-247.

Warsy, A.S., Norton, G. and Stein, H. 1974. Protease inhibitors from
    broad beans, isolation and purification. Phytochem. 13, 2481-2496.
Wilson, B.J., McNab, J.M. and Bentley, H. 1972. Trypsin inhibitor
    activity in the field bean (Vicia faba). J. Sci. Fd Agric. 23,
    679-684.

QUANTITATIVE MEASUREMENT OF QUALITY DETERMINING CONSTITUENTS IN SEEDS
OF DIFFERENT INBRED LINES FROM A WORLD COLLECTION OF VICIA FABA

M. Frauen, G. Röbbelen, E. Ebmeyer

Institut für Pflanzenbau und Pflanzenzüchtung,
Georg-August-Universität, Von-Siebold-Str. 8,
D-3400 Göttingen, Federal Republic of Germany

ABSTRACT

Investigations in a sample of 125 inbred lines of Vicia faba were
carried out to elaborate analytical screening methods for characters of
nutritional quality and to estimate variation in these traits. The results
encourage more intensive breeding work in the field of nutritional quality
of faba beans.

INTRODUCTION

The primary objectives of most breeding programmes with faba beans

are improvement of yield, stability of yield and disease resistance.

Attention, however, should also be paid to characters determining nutritio-

nal quality. In contrast to many other leguminous species, including soy-

bean, faba beans do not contain harmful concentrations of toxic constitu-

ents and therefore, they have a high potential for utilization in animal

feeding. Nevertheless,there are indications, supported by experimental

results, that some components of faba bean meal impair its feeding value

to monogastric animals under certain conditions.

In view of these findings investigations were carried out in order

(a) to devise and evaluate efficient analytical screening methods for the

quantitative measurement of quality traits and (b) to estimate the varia-

bility of these traits in a broad sample of Vicia faba inbred lines. The

results were expected to indicate the extend to which nutritional quality

might be improved by application of appropriate selection procedures.

EXPERIMENTAL MATERIAL

The material examined consisted of 125 more or less homozygous inbred

lines of V.faba originating from a world wide collection of cultivars as

well as less improved material, which showed extensive variation in plant

and seed characteristics. Ten plants of each line were grown in 1979 and

1980 in beeproof cages and a sample of 20 seeds per line was taken for

biochemical analysis.

P.D. Hebblethwaite, T.C.K. Dawkins, M.C. Heath and G. Lockwood (eds.)
Vicia faba: Agronomy, Physiology and Breeding. ISBN 90-247-2964-5.
© 1984, Martinus Nijhoff/Dr W. Junk Publishers. Printed in The Netherlands.

QUALITY CHARACTERISTICS AND METHODS OF THEIR DETERMINATION

The characters which were analyzed and affect the nutritional quality of faba bean seed can be classified into three groups:

1. Quantity and quality of proteins
   - crude protein
   - methionine
   - cyst(e)ine
   - tryptophan

2. Quantity and quality of carbohydrates
   - hulls
   - crude fibre
   - saccharose
   - α-galactosides

3. Contents of toxic components
   - tannins
   - vicine
   - trypsin inhibitors

(1) It is well known that the protein content of faba beans varies considerably. Yet it is not only the quantity but also the quality of the seed protein which define the proportion of faba beans which can be offered in feeding rations including, e.g., cereal grains. The relatively high lysine content of the faba bean protein compensates favourably for the lysine deficiency of, e.g., barley. On the other hand, the contents of sulphur-containing amino acids (methionine and cyst(e)ine) is low in faba beans, thus limiting the biological value of their proteins. Tryptophan has also been analyzed in our collection being the next limiting amino acid after methionine.

Automated analysis with volumetric N_2 measurement at constant temperature was used to determine total N (x 6.25 = crude protein) of the faba bean meals (Merz, 1970). Among the sulphur-containing amino acids, initially only methionine was determined by BrCN treatment of seed meal and gas-liquid-chromatography of the resulting methylthiocyanate (Paul, 1977). This method allows a good first screening of early generation material but does not give quantitatively accurate estimates of the methionine content. With this objective material was oxidized with performic acid and then hydrolized with HCl before the sulphur-containing amino acids were separated on the short column of an automated amino acid analyzer.

Correctness of the measurements was checked by enzymatic protein digestion with pronase E/pankreatin in the presence of SDS and reducing agents
(Winkler and Schön, 1979). This was followed by photometric determination
of the liberated cystine and cysteine by means of dithio-bis-nitrobenzoic
acid and of the methionine as a platinum complex by means of an autoanaly-
zer (Holz, 1982). The correlation of results from both methods with those
from the amino acid analyses was higher than 0.9. Sulphur content per se
was estimated by microanalysis using combustion in pure oxygen and separa-
tion of sulphate from other oxidation products by ionic chromatography.
The strong correlation of these data with the amino acid content indicated
that 80-95% of the total sulphur in faba bean meal is bound in the amino
acid fraction. Since, however, the relative amounts present in methionine
and cysteine cannot be distinguished by sulphur measurement, this method
can at best be used for preselection in screening programmes. A detailed
comparison of the different methods of sulphur amino acid determination
will be published elsewhere (Schön, in preparation). Finally, for tryptophan determinations the reaction of the oxidized hydrolysate with N-1-
(naphthyl)-ethylene diamine dihydrochloride and its photometric measurement
(Basha and Roberts, 1976) proved to be useful, as was shown by parallel
analyses on the amino acid analyzer.

(2) The amount of carbohydrates in V.faba seeds is inversely proportional to the protein content and their composition is decisive for the
energy value of faba bean meal. Crude fibre content was determined by
Weender analysis. Seed coat (hull) content was measured by simple dehull-
ing. It was established that hull content is strongly correlated with
cellulose content, but much easier to measure. Within the soluble sugars,
α-galactosides comprise a large fraction (70 up to 90%). Since they are
not digested by enzymes in the stomach and intestine of animals but only
by microbial activity, they are of no nutritional value and cause nothing
but flatulence upon ingestion. They were estimated by calculation of the
difference between the toal soluble sugars and the saccharose content. The
total soluble sugars were determined by anthrone reaction in an 80% ethanol
extract (Pape et al., 1969) after enzymatic hydrolysis with invertase as
its glucose moiety, which was measured using the hexokinase/glucosedihy-
drogenase test.

(3) Digestibility and availability of proteins can be affected by

secondary plant metabolites. Tannins are active by binding soluble proteins and thus blocking certain enzyme systems. Quantitative estimates of tannin content were made photometrically after methyl extraction from the meal and reaction with vanillin (Burns, 1971). Very low tannin content is known to be strongly correlated with the white flower character in V.faba (Crofts et al., 1980). The content of the toxic seed constituent vicine was estimated by UV-photometry at 274 nm (Collier, 1976). Trypsin inhibitor content, although generally low in faba beans, was determined by measuring the inhibition of trypsin digestion which arose from addition of a faba bean extract to a synthetic peptid (benzoyl-arginin-nitroanilid). The result was expressed as trypsin unit inhibition (TUI) by 100 mg seed meal (Kakade et al., 1974).

EVALUATION OF INBRED LINES

Mean values ranges, standard deviations and coefficients of variation of all the characters examined are summarised in Table 1. The largest

TABLE 1   Mean value, range, standard deviation (S.D.), coefficient of variation and heritability (h^2) of 11 characters relating to nutritional quality in a collection of 125 Vicia faba inbred lines.

Character			Mean $\bar{x}$	Range min. – max.	S.D. s	C.V. s%	h^2
Crude protein	(%)	[1]	29.8	25.5 – 38.1	3.39	11.4	0.51
Methionine	(%)	[4]	0.16	0.09 – 0.22	0.03	16.0	–
Cysteine	(%)	[4]	0.35	0.23 – 0.58	0.06	16.2	–
Tryptophan	(%)		0.31	0.24 – 0.42	0.04	12.9	0.22
Hulls	(%)		13.3	9.5 – 17.8	1.21	9.1	0.60
Crude fibre	(%)		7.3	5.1 – 9.9	0.98	13.4	0.18
Saccharose	(%)		1.65	1.27 – 2.56	0.19	11.6	0.40
α-galactosides	(%)		2.39	1.66 – 3.58	0.30	12.7	0.30
Tannins	(C.E.)	[2][5]	36.7	6.5 –156.5	19.3	52.6	–
Vicine	(%)		0.44	0.26 – 0.63	0.07	16.6	0.59
Trypsin inhibitors (TUI/mg)		[3]	12.6	7.0 – 19.2	2.54	20.1	0.56

[1] Percent of air dried seed meal

[2] Catechin equivalent

[3] Trypsin unit inhibition in mg meal

[4] 1979 only

[5] 1980 only

variation was found in tannin content, ranging from 6.5 to 156.5 with a
coefficient of variation of 53%. This is due to the fact that white flowering as well as coloured flowering genotypes were included in the collection. All other characters exhibited smaller amounts of variability with
coefficients of variation between 10 and 20%.

Estimates of heritability, as given in the last column of Table 1,
were calculated by correlating the measurements of the two years in which
the plant material was grown at different locations. Low heritabilities of
between 0.18 and 0.30 were determined for the contents of saccharose, α-
galactosides, tryptophan and crude fibre. All the other traits showed heritabilities ranging from 0.40 to 0.60, indicating good possibilities for
improvement by selection.

TABLE 2   Correlation coefficients between different characters of nutritional quality. Coefficients absolutely greater than 0.18 are
significant at P = 0.05.

Character	Methionine	Cysteine	Tryptophan	Hulls	Crude fibre	Saccharose	α-galactosides	Tannins	Vicine	Trypsin inhibitors
Crude protein	0.40	0.26	0.25	−0.20	−0.41	0.10	0.20	−0.20	−0.08	0.19
Methionine		0.23	0.15	−0.10	−0.13	0.06	0.10	−0.17	−0.05	0.26
Cysteine			−0.01	−0.04	−0.19	0.00	0.05	−0.07	0.03	0.13
Tryptophan				0.47	0.22	−0.05	0.01	0.31	−0.08	0.16
Hulls					0.56	0.09	0.08	0.56	0.15	−0.12
Crude fibre						−0.14	−0.09	0.34	0.09	−0.03
Saccharose							0.08	−0.06	0.28	−0.05
α-galactosides								−0.28	0.08	−0.01
Tannins									−0.05	−0.04
Vicine										0.16
Trypsin inhibitors										

The correlations between quality traits are given in Table 2. As anticipated, the highest positive correlation found was the coefficient of 0.56 between hull and crude fibre content. The high positive correlation between hull and tannin content confirms the fact that tannins are mainly located in the seed coat. According to expectation, a negative correlation (-0.41) was established between crude fibre and protein content. Weak, but significant, positive correlations were found between the three amino acids and protein content. The correlations between all other traits of nutritional quality were generally rather weak. In particular, variation in contents of trypsin inhibitors may be regarded as being more or less independent of all the other characters investigated.

On the basis of the present investigations, nine inbred lines were selected as exhibiting the most divergent quality characteristics (Table 3). Feeding trials will be used to evaluate their value for future breeding.

TABLE 3   Inbred lines selected for extreme expressions of nutritional quality characteristics.

Inbred line	Crude protein (%)	Crude fibre (%)	Hulls (%)	Tannins (C.E.)	Vicine (%)	Trypsin inhibitors (TUI/mg)	General quality
2	31.6	6.6	14.8	28.0	0.52	9.1	average
12*	31.9	4.8	11.4	8.5	0.59	13.9	average
16	26.6	7.6	13.9	50.5	0.63	18.3	poor
27	36.0	6.6	11.9	17.8	0.28	15.6	average-good
32	33.3	6.0	11.5	22.0	0.35	8.7	good
41	27.5	8.7	16.8	156.5	0.34	14.0	poor
46*	32.2	6.9	11.7	7.5	0.46	8.9	good
64	25.9	9.3	14.8	87.5	0.54	12.3	poor
122	28.9	6.8	13.0	59.0	0.35	8.7	average

* White flowering lines

CONCLUSIONS

The investigations carried out on a sample of 125 inbred lines of V. faba were aimed firstly to elaborate efficient analytical mass-screening methods and secondly to get information on the amount of variability available to plant breeders in characters of nutritional quality. In all traits enough variability was observed to encourage intensification of breeding work in the field of nutritional quality. Regarding relationships between different characters, no conclusively negative correlations were found. If some of the good expressions of quality traits found in this investigation could be made available in commercial varieties then the nutritional quality of faba beans would show definite improvement.

REFERENCES

Basha, S.M.M. and Roberts, R.M. 1976. A simple and colorimetric method for the determination of tryptophane. Anal.Biochem. 77, 378–386.

Burns, R.E. 1971. Method for estimation of tannin in grain sorghum. Agron.J. 63, 511–512.

Collier, H.B. 1976. The estimation of vicine in faba beans by an ultraviolet spectrophotometric method. J.Inst.Sci.Technol.Aliment. 9, 155–159

Crofts, H.J., Evans, L.E. and McVetty, P.B. 1980. Inheritance, characterization and selection of tannin-free faba beans (Vicia faba). Can.J. Plant Sci. 60, 1135–1140.

Holz, F. 1982. Automatisierte, photometrische Direktbestimmung des Methionin- und Cyst(e)ingehaltes von Samenproteinen und ihre Anwendung als analytische Schnellmethode bei der Selektion von Ackerbohnen. Z.Pflanzenzüchtg. 88, 103–117.

Kakade, M.L., Rackis, J.J. and McGhee, J.E. 1974. Determination of trypsin inhibitor activity of soy products: A collaborative analysis of an improved procedure. Cereal Chem. 51, 376–383.

Merz, W. 1970. Neuer Automat zur Stickstoffschnellbestimmung. G-I-Z Fachz. Lab. 14, 617–625 .

Pape, G., I-Shun-Shen und Schön, W.J. 1969. Zur quantitativen Bestimmung Reservekohlenhydrate in Pflanzenmaterial. Z.Acker- u. Pflanzenbau 130, 16–32.

Paul, Chr. 1977. GLC method for determination of methionine after cyanogen bromide reaction and its use in mass screening of field beans (Vicia faba L.). Z.Pflanzenzüchtg. 78, 97–112.

Winkler, U. und Schön, W.J. 1979. Analysis of amino acid composition of plant proteins following enzymatic hydrolysis. An expressive view to prove their nutritive value. Z.Acker- u. Pflanzenbau 148, 430–445.

ANTINUTRITIONAL AND FAVISM INDUCING FACTORS IN VICIA FABA L.:
NUTRITIONAL VALUE OF FABA BEANS; METABOLISM AND PROPERTIES OF
VICINE, CONVICINE AND DOPA-GLUCOSIDE

B. Bjerg*, B.O. Eggum**, O. Olsen* and H. Sørensen*

*Chemistry Department
Royal Veterinary and Agricultural University
40 Thorvaldsensvej, DK-1871 Copenhagen V, Denmark
**National Institute of Animal Science
Department of Animal Physiology and Chemistry
25 Rolighedsvej, DK-1958 Copenhagen V, Denmark

ABSTRACT

Comprehensive investigations of antinutritional and favism inducing factors in faba beans (Vicia faba L.) have been performed. The results are presented in this and allied papers and include studies of the nutritional value of faba beans in relation to the protein quality and the presence of antinutritional and/or toxic constituents in the seeds. N-balance trials with growing rats have revealed that fortification with threonine and methionine improves the biological value and consequently the nutritional value. True protein digestibility was markedly higher for faba beans with relatively low content of tannins and crude fibre (varieties with white flowers) compared to faba beans with higher content of tannins and crude fibre (varieties with coloured flowers). However, it has been revealed that other low molecular weight faba bean compounds have appreciable antinutritional and/or toxic effects. The favism inducing factors - vicine, convicine, dopa-glucoside and their aglucones - have been assigned special attention and it is shown that especially convicine is responsible for a reduced biological value of faba beans. Methods of analysis suitable for quantitative determination of these glucosides individually as well as isolation of sufficient quantities of these compounds to perform feeding experiments are presented. Properties of vicine, convicine, dopa-glucoside and their aglucones; metabolism studies; and results from plant breeding programmes, devoted to the elimination of antinutritional and favism inducing factors are briefly discussed.

INTRODUCTION

Amino acid composition and properties of proteins in faba beans (Vicia faba L.) and its subspecies or botanical varieties, e.g., major, equina, minor and paucijuga are dominated by vicilin and legumin as for other legumes (Derbyshire et al., 1976; Casey and Short, 1981). Nutritional studies have shown that faba beans contain antinutritional and/or toxic constituents in addition to tannins, haemagglutinins and protease inhibitors (Griffiths and Moseley, 1980; Sjödin et al., 1981; Griffiths, 1981; Muduuli et al., 1982; Bjerg et al., 1983a). The deleterious effects can in part be related to low molecular weight constituents of faba beans suspected of causing favism in man (Mager et al., 1980). In this respect, vicine, convicine

P.D. Hebblethwaite, T.C.K. Dawkins, M.C. Heath and G. Lockwood (eds.)
Vicia faba: Agronomy, Physiology and Breeding. ISBN 90-247-2964-5.
© 1984, Martinus Nijhoff/Dr W. Junk Publishers. Printed in The Netherlands.

and dopa-glucoside are considered to produce specific and serious problems.

This paper describes results obtained from a comprehensive study of the nutritive value of faba beans. The structures and properties of vicine, convicine, dopa-glucoside and their a-glucones are presented as well. The aglucones divicine, isoura-mil and dopa were produced by β-D-glucosidase catalysed hydrolysis. Oxidation of the aglucones were followed by the use of different redox cofactors in the assay mixtures. The coupled redox reactions have been adapted to specific and sensitive methods of vicine and convicine detection as well as to studies of possible toxic effects caused by aglucones of the compounds.

The results presented reveal promising possibilities of improving the nutritive quality of faba beans by breeding or selecting for a low content of inherent antinutritional compounds.

MATERIALS AND METHODS

The plant material consisted of <u>Vicia faba</u> L. cultivars/lines with origin in different parts of Europe, Africa and the Middle East. The plants were grown for at least one year under identical conditions in Denmark and some of them were highly inbred as described previously (Bjerg et al., 1983b).

General methods and equipment used for isolation, purification and qualitative determination of the low molecular weight compounds, including ion-exchange chromatography, gel filtration, PC, HVE, UV, ^{1}H- and ^{13}C-NMR-spectroscopy, HPLC, GC/MS, amino acid analysis and other chemical analyses have been described elsewhere (Bjerg et al., 1983a; Bjerg et al., 1983b, and refs. cited therein).

Vicine, convicine and dopa-glucoside (Fig. 1) were isolated from faba bean meal in appreciable amounts (Bjerg et al., 1983a). Dopa and other chemicals were commercially available preparations of analytical-reagent grade.

Glucosidase (EC 3.2.1.21) was a preparation from almonds (Boehringer-Mannheim). The catalytic activities of the enzyme towards the synthetic substrate PNPG (4-nitrophenyl-β-D-glucopyranoside) and the glucosides from V. <u>faba</u> (Fig. 1) were measured in terms of the rate of the reactions catalysed in McIlvaine buffer (pH = 5.0) at 25°C. The PNPG assay has been described previously (Bille et al., 1983); when other substrates were used (vide infra) determinations of time dependent changes in substrate and/or product concentration were performed spectrophotometrically. 1 Unit = 1 μmole of substrate transformed into products/min/mg of protein.

Animals, diets and methods used in the N-balance trials with growing rats have been described in detail elsewhere (Bjerg et al., 1983a).

RESULTS AND DISCUSSION

Structure and properties of vicine, convicine and dopa-glucoside

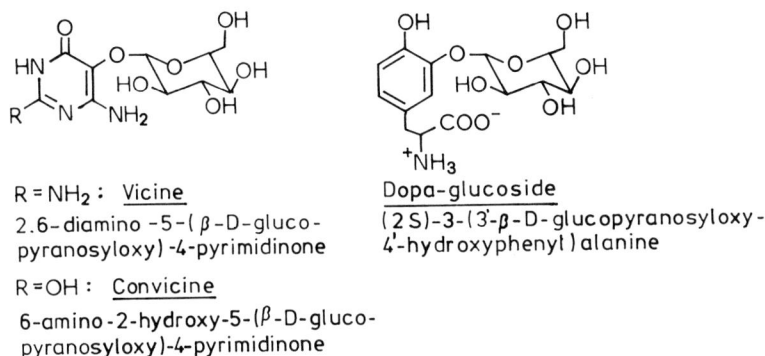

R = NH$_2$: Vicine
2.6-diamino-5-(β-D-gluco-
pyranosyloxy)-4-pyrimidinone

R = OH : Convicine
6-amino-2-hydroxy-5-(β-D-gluco-
pyranosyloxy)-4-pyrimidinone

Dopa-glucoside
(2S)-3-(3'-β-D-glucopyranosyloxy-
4'-hydroxyphenyl)alanine

Fig. 1.   Structure of vicine (Bendich and Clements, 1953), convicine (Bien
et al, 1968) and dopa-glucoside (Andrews and Pridham, 1965).

UV-spectra (Fig. 2) of the three glucosides demonstrate
intense absorptions which can be utilised in connection with
commonly applied methods of analysis: vicine, $\varepsilon_{275,H_2O} \approx 13220$
$l \cdot mole^{-1} \cdot cm^{-1}$ (Bendich and Clement, 1953); convicine, ε_{272,H_2O}
$\approx 14500 \ l \cdot mole^{-1} \cdot cm^{-1}$ (Bien et al., 1968); dopa-glucoside,
$\varepsilon_{294,0.1M \ NaOH} \approx 2000 \ l \cdot mole^{-1} \cdot cm^{-1}$.

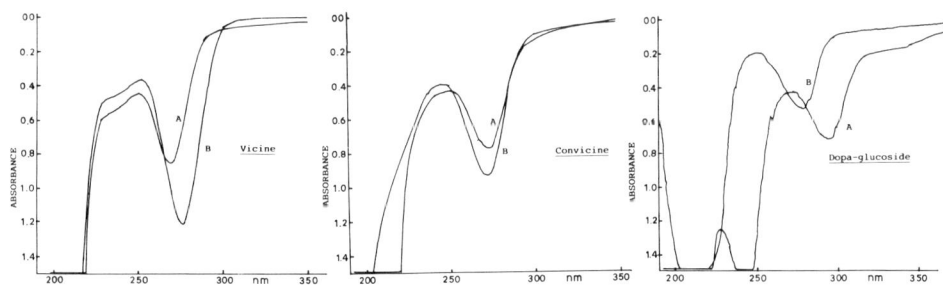

Fig. 2.   UV-spectra of vicine, convicine and dopa-glucoside.
A = in 0.1 M NaOH; B = in H$_2$O.

NMR spectra (Fig. 3) of vicine, convicine, dopa and dopa-
glucoside confirm the structures of these compounds. ^{1}H-NMR
spectra of vicine and convicine in heavy water (D$_2$O) solution
show absorption peaks only for the β-D-glucoside part of the
molecules. A ^{1}H-NMR spectrum of dopa-glucoside in D$_2$O solution

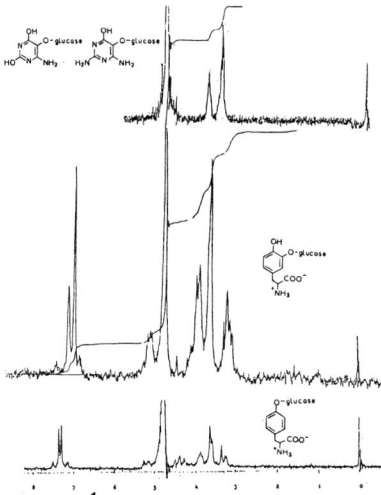

Fig. 3. ^{1}H-NMR spectra

is compared with a corresponding spectrum of tyrosine-glucoside in Fig. 3. The spectra show peaks for all non-exchangeable protons at the expected positions with the anomeric proton close to the HOD signal. ^{13}C-NMR spectra of these compounds (Fig. 4) are especially valuable for structure confirmation. However, chemical shift values for the pyrimidine carbon atoms are very sensitive towards changes in solvent/solvent pH as shown for vicine, especially the value for carbon atom No. 6. Comparison of the spectra for dopa and dopa-glucoside confirms the 3'-O-β-D-glucopyranoside structure by the expected change in chemical shifts for carbon atoms o- and p- to the glucosylation, Nos. 2', 4', 6'. The spectrum of vicine dissolved in DMSO is in accordance with previously reported data (Dutta et al., 1981).

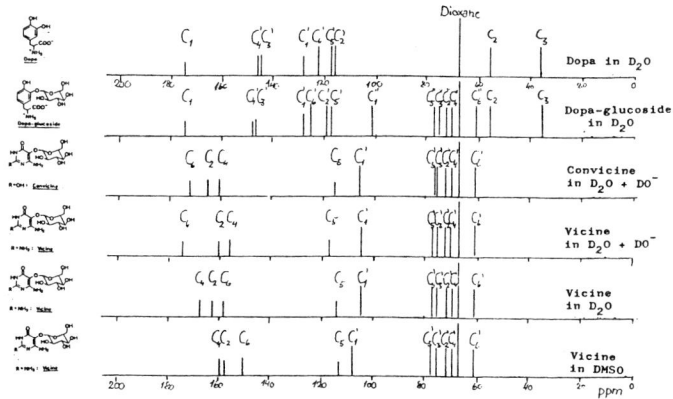

Fig. 4. ^{13}C-NMR spectra of dopa, dopa-glucoside, convicine and vicine in different solvents; dioxane as internal standard.

Methods of quantitative analysis of individual faba bean glucosides

Gas liquid chromatography (GLC) determination of per-trime-

thylsilylated vicine and convicine has been described previously (Pitz and Sosulski, 1979). Using conditions as described for
other glucosides (Olsen and Sørensen, 1981) (except for a modified temperature programme: initially 250°C and 2°C/min) the
results obtained were as presented in Fig. 5. The GLC method is
valuable for identification purposes, but for quantitative
determinations it is more expensive, experimentally difficult
and time consuming than the gel filtration/UV method (Fig. 5)
described elsewhere (Bjerg et al., 1980; Bjerg et al., 1981).
However, the most simple, fast and easy method for quantitative
analysis of all of the individual glucosides and dopa is the
high performance liquid chromatography (HPLC) method originally
described by Marquardt and Frolich (1981) and since then used
by others with minor changes (Bjerg et al., 1983b) (Fig. 5).

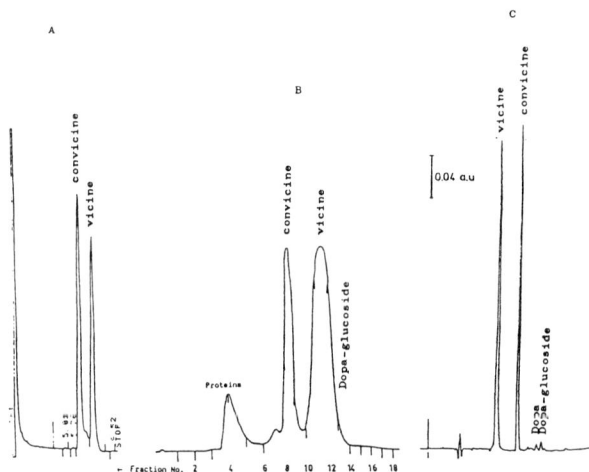

Fig. 5.   Different methods used in quantitative analysis of faba bean
          constituents: A = gas chromatography; B = gel filtration/UV;
          C = HPLC.

HVE; PC; TLC; glucosidase reaction; aglucone instability

A thin layer chromatography (TLC) determination of the
glucosides has been described previously (Olsen and Andersen,
1978). However, paper chromatography (PC) and high voltage
electrophoreses (HVE) appear to be less sensitive to impurities
in the extracts. Table 1 shows the Rf-values obtained by PC and
mobilities obtained by HVE.

TABLE 1. Rf-values and ionic mobilities of vicine, convicine, dopa-
glucoside and some reference compounds.

Compound	Rf-values from PC in solvent*			HVE-mobilities (cm) at pH 1,9**
	1.	2.	3.	
Vicine	0.16	0.36	0.13	18.4
Convicine	0.18	0.28	0.15	2.0
Dopa	0.27	0.30	(0.10)	20.4
Dopa-glucoside	0.18	0.31	0.10	16.3
Tyrosine-glucoside	0.12	0.45	0.14	16.6
Glu/Asp	0.28/0.22	0.26/0.16	0.15/0.12	25.0/22.6

* Solvent 1: n-BuOH-HOAc-H_2O (12:3:5); solvent 2: PhOH-H_2O-13M NH_4OH
(120:30:1); solvent 3: i-PrOH-H_2O-13M NH_4OH (8:1:1).
** pH 1.9 buffer: HOAc-HCO_2H-H_2O (4:1:45).

Detection of the amino acids are performed with ninhydrin;
pyrimidine derivatives are detectable in short wave UV light
(dark spots) if alkaline/phenolic solvents are avoided. More
specific and sensitive detection of vicine and convicine are
obtained by spraying the chromatograms with the glucosidase
assay mixture[1] (vide infra; without vicine and convicine). White
spots on a light blue background are formed where vicine or
convicine are present when DCIP is used as redox cofactor.
Other redox cofactors with appropriate properties (E_o' values)
are also usable instead of DCIP; thus, PMS + INT form orange
spots when divicine or isouramil are produced.

In the presence of oxygen, both of these AH_2-type aglucones are transformed into unstable oxidation products and hy-
drogen peroxide (Chevion et al., 1982)

[1] Assay mixture: 200 µl glucosidase (10 mg/ml buffer; 2.4 U/mg protein in the
PNPG assay) + 1.0 ml DCIP (15.3 mg/100 ml H_2O) + 50-1400 µl vicine or convicine (1.1 mg/5 ml) + McIlwaine buffer in a total volume of 3.2 ml.

$$AH_2 + O_2 \longrightarrow A + H_2O_2$$

This reaction is not surprising if we compare the structures of divicine and isouramil with structures, properties and redox potentials of flavins, folic acid and pterin coenzymes/cofactors. It was expected and found that these aglucones easily reacted with DCIP and PMS + INT in the following types of reactions:

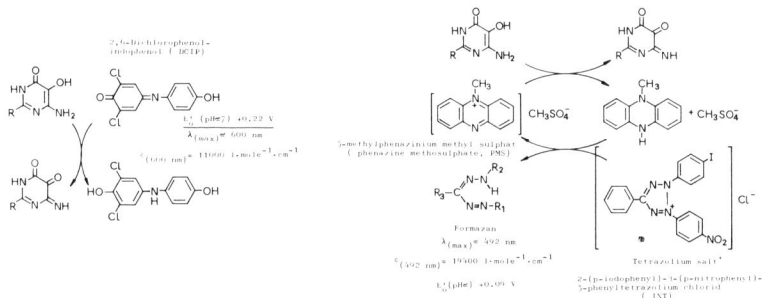

The oxidised aglucones of vicine and convicine are quite unstable, but owing to differences in their structures, the final reactions/reaction products and possible physiological/toxic effects are probably different as found in the N-balance trials with rats. Consideration of the above mentioned reactions/ problems is also important in studies of favism (Arese et al., 1981).

The use of plant breeding to eliminate antinutritional and favism factors

Plant breeding programmes to improve sources of plant protein which include chemical and biochemical aspects have been used in developing faba bean cultivars with low content of vicine, convicine and dopa-glucoside (Bjerg et al., 1983b and ref. cited therein). The content of vicine and convicine in faba beans on a dry matter basis range between 0.1-1.1% for vicine and 0.1-0.7% for convicine and dopa-glucoside. It is, therefore, possible to select cultivars with a low content of these glucosides. Such quality improvement of faba beans, especially combined with selection for low content of tannins and eventually crude fibers (white flowers) seems to be very promising.

Nutritional value of faba beans as influenced by antinutritional factors

The glucosides have been isolated in crystalline state and added singly, in varying amounts, to a standard rat diet which was used in N-balance trials comprising studies of antinutritional and toxic effects caused by the glucosides or products thereof.

Net protein utilization of faba beans has been improved by about 20% by addition of essential amino acids to the diets. True protein digestibility is about 10% lower for seeds of normal coloured faba beans compared to those with white flowers. Convicine reduces the biological value of proteins in standard diets much more than vicine and dopa do (Bjerg et al., 1983a).

Metabolism of vicine, convicine and dopa-glucoside

Biosynthesis and/or accumulation in the seeds of vicine, convicine and dopa-glucoside occurs during maturation and development of the seeds (Jamalian and Bassiri), 1978; Bjerg et al 1983b). During germination, the glucosides disappear from the seeds. Green plant parts accumulate dopa (but not the glucosides to a quite high concentration (unpublished results).

Acknowledgement

Support from DANIDA, The Danish Council on Development Research is gratefully acknowledged.

REFERENCES

Andrews, R.S. and Pridham, J.B. 1965. Structure of dopa glucoside from Vicia faba. Nature, 205, 1213-1214.

Arese, P., Bosia, A., Naitana, A., Gaetani, S., D'Aquino, M. and Gaetani, G.F. 1981. Effect of divicine and isouramil on red cell metabolism in normal and G6PD-deficient (mediterranean variant) subjects. Possible role in the genesis of favism. The Red Cell: Fifth Ann Arbor Conference, Alan R. Liss, Inc., New York, 725-744.

Bendich, A. and Clements, G.C. 1953. A revision of the structural formulation of vicine and its pyrimidine aglucone, divicine. Biochim. Biophys. Acta, 12, 462-477.

Bien, S., Salemnik, G., Zamir, L. and Rosenblum, M. 1968. The structure of convicine. J. Chem. Soc. (C), 496-499.

Bille, N., Eggum, B.O., Jacobsen, I., Olsen, O. and Sørensen, H. 1983.The effects of processing on antinutritional constituents and value of double low rapeseed meal. Z. Tierphysiol., Tierernähr. u. Futtermittelkde., 43, 148-163.

Bjerg, B., Eggum, B.O., Jacobsen, I., Olsen, O. and Sørensen, H. 1983a. Protein quality in relation to antinutritional constituents in faba beans (Vicia faba L.). The effects of vicine, convicine and dopa added to a standard diet and fed to rats. Z. Tierphysiol. Tierernähr. u. Futtermittelkde. submitted.

Bjerg, B., Knudsen, J.C.N., Olsen, O., Poulsen, M.H. and Sørensen, H. 1983b. The possibility of elimination of antinutritional and favism releasing factors in Vicia faba L. by plant breeding. Quantitative estimation of vicine, convicine and dopa-glucoside. Z. Pflanzenzüchtg. submitted.

Bjerg, B., Knudsen, J.C.N., Poulsen, M.H. and Sørensen, H. 1981. Elimination of antinutritional and favism releasing factors in Vicia faba L. by plant breeding. Pulse Crops Newsletter, 1, 36-38.

Bjerg, B., Poulsen, M.H. and Sørensen, H. 1980. Quantitative elimination of favism releasing factors in Vicia faba seeds. Fabis Newsletter, 2, 51-53.

Casey, R. and Short, M.N. 1981. Variation in amino acid composition of legumin from Pisum. Phytochem., 20, 21-23.

Chevion, M., Navok, T., Glaser, G. and Mager, J. 1982. The Chemistry of favism-inducing compounds. The properties of isouramil and divicine and their reaction with gluthathion. Eur. J. Biochem., 127, 405-409.

Derbyshire, E., Wright, D.J. and Boulter, D. 1976. Review. Legumin and vicilin, storage proteins of legum seeds. Phytochem., 15, 3-24.

Dutta, P.K., Chakravarty, A.K., Chowdhury, U.S. and Pakrashi, S.C. 1981. Vicine, a favism-inducing toxin from Momordica charantia Linn. seeds. Ind. J. Chem., 20B, 669-671.

Griffiths, D.W. 1981. The polyphenolic content and enzym inhibitory activity of testas from bean (Vicia faba) and pea (Pisum spp.) varieties. J. Sci. Food Agric., 32, 797-804.

Griffiths, D.W. and Moseley, G. 1980. The effects of diets containing field beans of high or low polyphenolic content on the activity of digestive enzymes in the intestines of rats. J. Sci. Food Agric., 31, 255-259.

Jamalian, J. and Bassiri, A. 1978. Variation in vicine concentration during pod development in broad bean (Vicia faba L.). J. Agric. Food Chem., 26, 1454-1456.

Mager, J., Chevion, M. and Glaser, G. 1980. Favism. In "Toxic Constituents of Plant Foodstuffs", second edition (Ed. L.J. Liener). (Academic Press, New York). pp. 265-294.

Marquardt, R.R. and Frohlich, A.A. 1981. Rapid reversed-phase high-performance liquid chromatography method for quantitation of vicine, convicine and related compounds. J. Chromatogr., 208, 373-379.

Muduuli, D.S., Marquardt, R.R. and Guenter, W. 1982. Effect of dietary vicine and vitamin E supplementation on the productive performance of growing and laying chickens. Br. J. Nutr., 47, 53-60.

Olsen, H.S. and Andersen, J.H. 1978. The estimation of vicine and convicine in faba beans (Vicia faba L.) and isolated faba bean proteins. J. Sci. Food Agric., 29, 323-331.

Olsen, O. and Sørensen, H. 1981. Sinalbin and other glucosinolates in seeds of double low rape species and Brassica napus cv. Bronowski. J. Agric. Food Chem., 28, 43-48.

Pitz, W.J. and Sosulski, F.W. 1979. Determination of vicine and convicine in faba bean cultivars by gas-liquid chromatography. J. Can. Inst. Food Sci. Techn., 12, 93-97.

Sjödin, J., Mårtensson, P. and Magyarosi, T. 1981. Selection for increased protein quality in field bean (Vicia faba L.). Z. Pflanzenzüchtg., 86, 210-220.

FABA BEAN RESEARCH IN EUROPE

P.D. Hebblethwaite

Department of Agriculture and Horticulture
University of Nottingham School of Agriculture
Sutton Bonington, Loughborough, Leic. U.K.

Table 1 shows the area, average yield and production of faba beans (Vicia faba) in the major producing countries of Europe and compares these with European and world figures.

These data indicate that Europe grows about 5.4% of the world area and produces about 8% of world production. The major producers in order of importance are Italy, Spain, United Kingdom, Czechoslovakia, Portugal, France, Greece, and East and West Germany. Other European countries such as Sweden (5000 ha), Denmark (110 ha), The Netherlands (18000 ha), Belgium/Luxemburg (1000 ha), Switzerland (155 ha), also grow limited areas.

According to the literature, important research and breeding work is being carried out in Italy, Spain, United Kingdom, France, East and West Germany, Sweden, Switzerland, Denmark, Austria, Netherlands, Greece and Ireland.

TABLE 1   Area grown, average yield and production of dried broad beans (Vicia faba L.) grown in countries in Europe, 1980.

	Area grown 1000 ha	Average yield t/ha	Production 1000 t
Czechoslovakia	41	1.8	72
France	20	3.2	64
German Democratic Republic	6	2.2	13
German Federal Republic	4	3.2	14
Greece	11	1.4	15
Italy	172	1.3	228
Portugal	35	0.5	16
Spain	84	1.2	104
United Kingdom	46	3.1	149
Europe total	374	1.4	529
World total	6980	0.9	6709

Source: FAO, 1980

P.D. Hebblethwaite, T.C.K. Dawkins, M.C. Heath and G. Lockwood (eds.)
Vicia faba: Agronomy, Physiology and Breeding. ISBN 90-247-2964-5.
© 1984, Martinus Nijhoff/Dr W. Junk Publishers. Printed in The Netherlands.

Consequently, in 1982 I was commissioned by the Commission of European Communities to review the present situation of Vicia faba in Europe and the research being carried out, and to make recommendations to the Commission on the future of the crop and its research. The full report has been published elsewhere (Vicia faba research in Europe, Report EUR 8649 EN)[1]. This chapter repeats much of the discussion, conclusions and recommendations section of this report.

There is no doubt that the faba bean crop has tremendous potential. As a legume it can produce its own nitrogen supply and also some for the following crop in the rotation. It can play a major role as a break crop in intensive cereal systems and in so doing reduce nitrogen (energy) costs, decrease disease, improve soil physical conditions and decrease certain weed populations. The energetic efficiency of faba bean production is clearly shown in Table 2.

TABLE 2   Inputs of fertiliser N and support energy to some U.K. crops

	kg crude protein/ kg fertiliser N	kg crude protein/ 10^3 MJ support energy
Wheat	4.8	30
Barley	4.6	26
Grass (UK average)	4.9	40
Rapeseed	2.2	19
Field beans	30.0	74

Data from J.C.O. (1976)

As a medium protein and energy crop it has the potential for reducing the need for imported protein and energy feedstuffs. The E.E.C., in fact, utilise 18150,000 t of oilseed cake and meals and 2094,000 t of animal meals into animal compound feedstuffs annually and approximately 65% of this is imported from non E.E.C. countries. Both soya and fishmeal prices can fluctuate widely, for example, in 1972 the price of fishmeal doubled from £123 to £246 per t and soya trebled from £80 to £250 per t. This situation resulted in trade embargoes and

[1] Available from: Office for Official Publications of the E.E.C., Batiment Jean Monnet, Luxembourg.

eventually a licensing system. Such a situation could occur again and consequently the E.E.C. must be interested in any crop that has a potential for protein supply.

The E.E.C. Commission was thus prompted to examine the problems of protein supply and their report (Improvement of the Common Agricultural Policy Com (73) 1850 Final) stated that:

(1) A dependence on large amounts of imported plant proteins for animal foodstuff was unavoidable,

(2) sources of imported protein should be diversified and supply guarantees obtained, and

(3) the capacity for plant protein production in the E.E.C. should be increased to a level sufficient to provide a buffer in the event of another world shortage.

Since the report there has been an increase in research in the E.E.C. at aiming to increase the productivity of protein crops and this report examines some of this research on Vicia faba. There has also been more tangible support to encourage growers in the form of subsidies.

In spite of the apparent potential of the faba bean, data from all E.E.C. and other European countries, except during the last 2 or 3 years, has indicated a steady decline in the area of Vicia faba over the last century. Figure 1 illustrates this decline for major grower countries and European countries as a whole from 1961 to 1981 and Figure 2 shows the decline in area for the United Kingdom from 1866. The rapid decline in area in Italy is highly significant and is dealt with in the Italian section.

A large number of reasons have been given for this decline in Europe but the most important seem to be:

1. A gradual change in human consumption away from legume protein sources.

2. Low yields in comparison with cereals (including maize).

3. Low profitability compared with cereals.

4. Unstable yields compared with most other crops.

5. In cool temperate countries difficulties in harvesting due to uneven maturity.

6. In some years and areas catastrophic disease problems can occur, e.g. Botrytis and Orobanche.

300

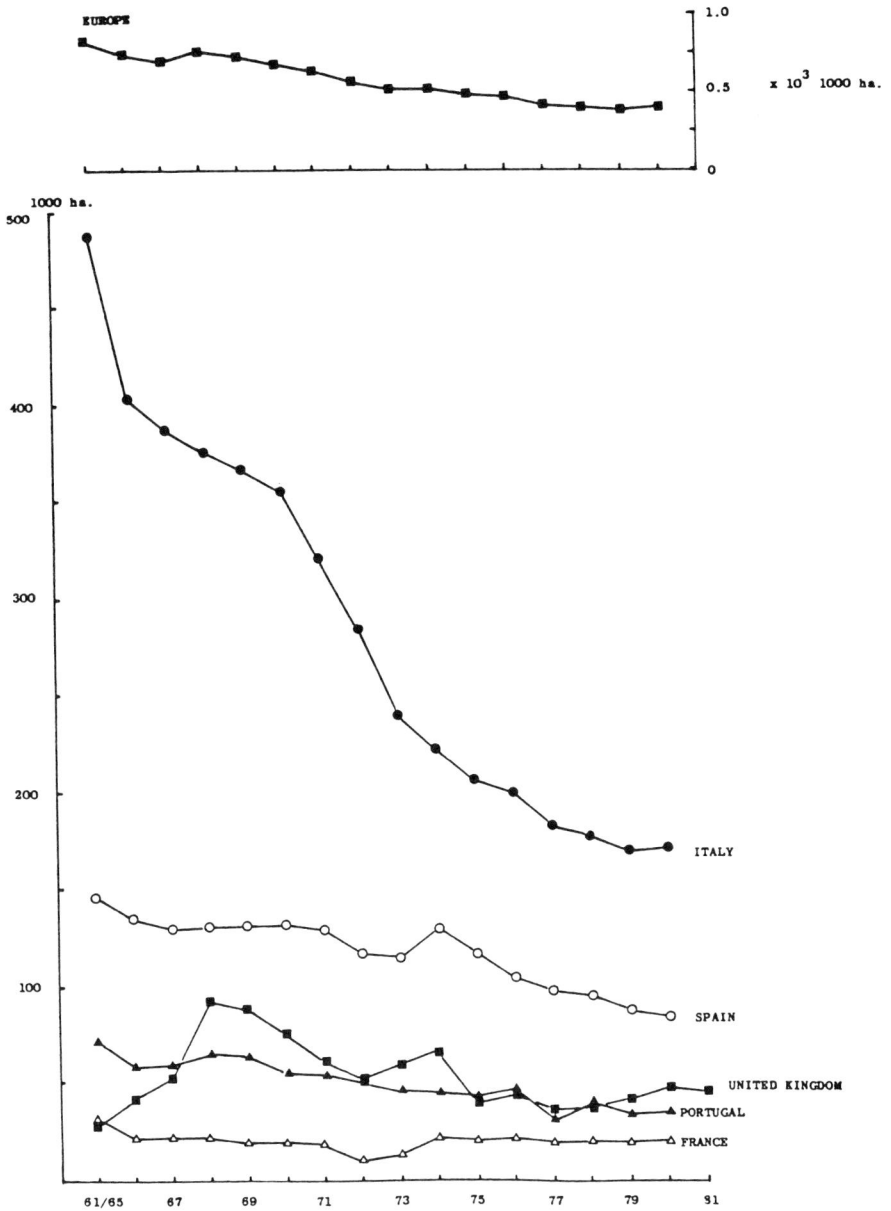

Fig. 1   Area of faba beans in Europe and selected European countries.
Source: FAO Production Yearbook

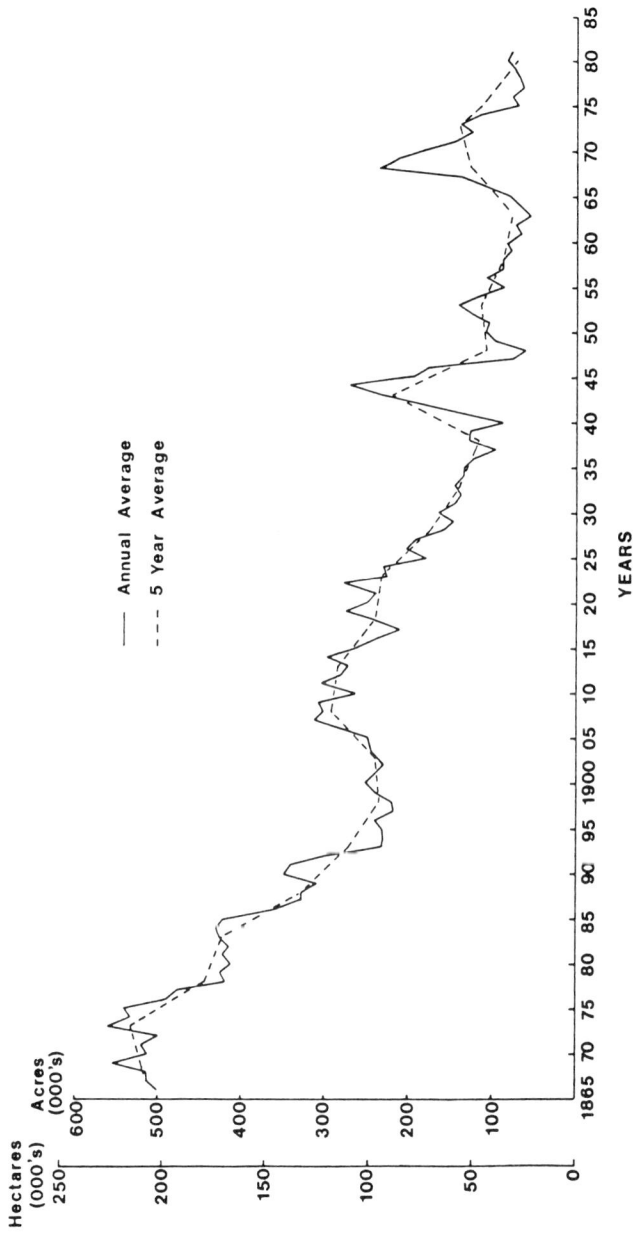

Fig. 2   Area of faba beans in the United Kingdom, 1865 to 1981.
Source: Agricultural Statistics

7. The crop has problems in relation to utilization and marketing.

8. Rapid expansion of other more profitable break crops for cereals
    such as oilseed rape at the expense of faba beans.

A reduction in the human consumption of grain legumes as a protein
source is common to most developing countries where alternatives are
available. Historically, grain legumes are considered as the 'poor
man's' food. Consequently any medium term expansion in the crop is
likely to .be in the medium protein animal-feed sector. It is recognised
that the crop could also play a minor role in relation to texturised and
functional protein products for human consumption.

Figure 3 compares average European faba bean yields from 1961 to
1981 for faba beans with that of wheat. Not only are yields much lower
for faba beans but also increases in yields have been virtually
non-existent while increases in wheat and other cereal yields have been
substantial. For instance, close examination of Figure 3 shows that
average yields for Europe in 1967 were 1.3 t/ha for faba beans and 2.4
t/ha for wheat. However, by 1981 faba bean yields were still 1.3 t/ha
but wheat had increased by 50% to 3.6 t/ha. Figure 4 gives a further
example of different changes in yield as between these crops but looks at
data from 1885 to 1980 in the U.K. In 1885 the yields of both crops were
just over 2 t/ha but by 1981 yields of faba beans had increased by 35% to
2.7 t/ha and wheat by 145% to 5.8 t/ha. Data from most other major
growing E.E.C. countries show similar trends.

It can therefore be clearly seen that the differential in yield of
cereals and faba beans is widening and on a profit basis the faba bean
crop must be increasingly less attractive to grow unless subsidised. It
is important to note that increases in cereal yields have come about only
as a result of much greater inputs than with faba beans. However, there
is evidence that for the U.K. about 50% of the cereal yield increases is
due to improved varieties and the rest to improved husbandry. Thus the
faba bean has become less profitable, and data on gross margins from
economic surveys support this.

Figures 3 and 4 also indicate that yields for faba beans are more
unstable than for wheat. The greater residual standard deviation of the
U.K. data indicates that faba beans are 60% more unstable than wheat.
Stability indices data in Table 3 also show that the national average
U.K. yield of faba beans is more unstable than all other major arable
crops.

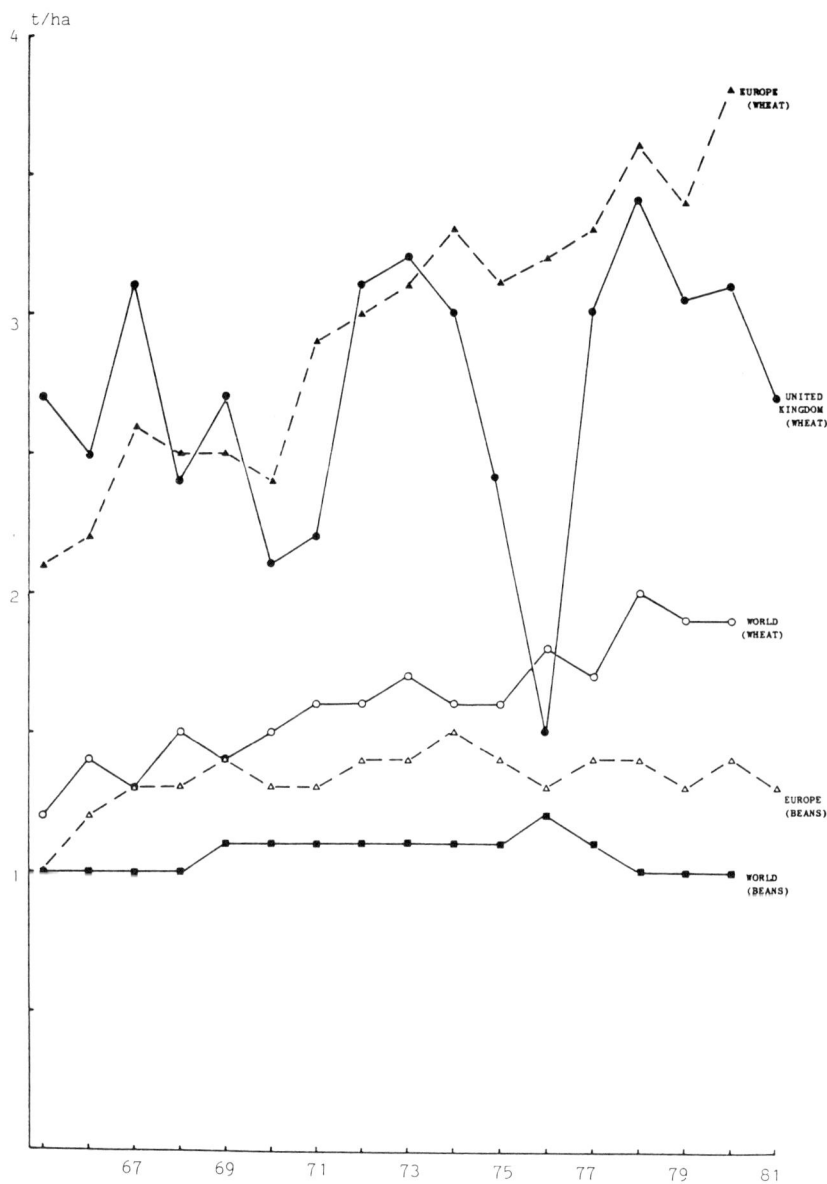

Fig. 3   Average yields of faba beans and wheat in Europe and the world
          1966 to 1981.
          Source : FAO Production Yearbook

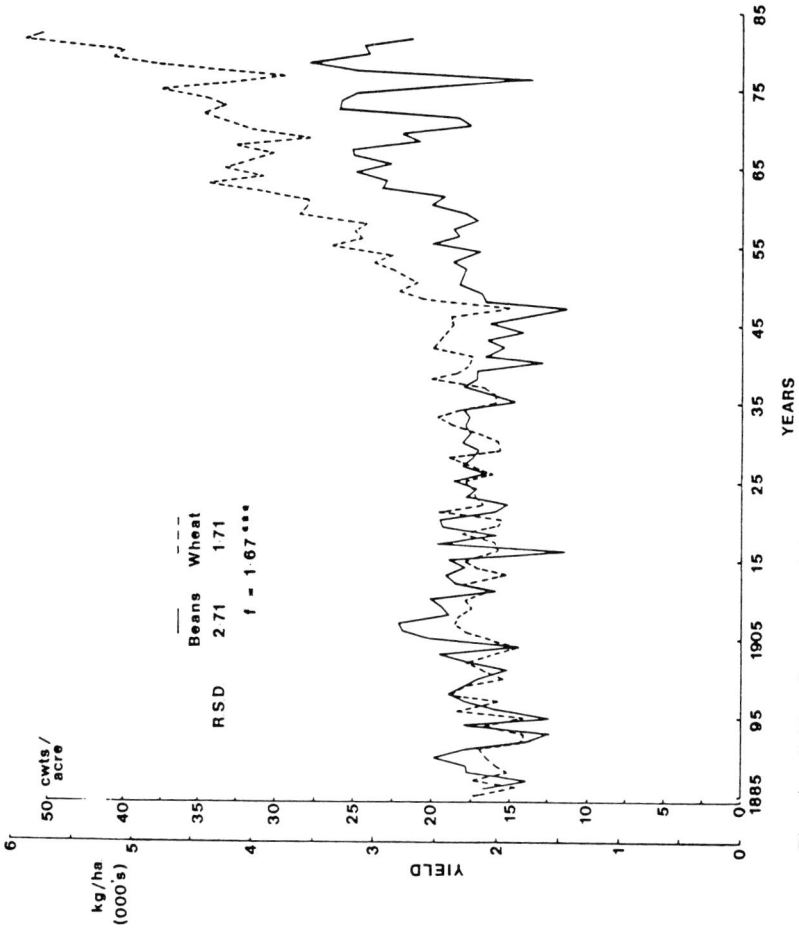

Fig. 4 National average yields of wheat and faba beans in the United Kingdom 1885 to 1981
RSD = residual standard deviation of this data.   Source : Agricultural Statistics.

TABLE 3 Yield stability index (%) of various crops in
the U.K. 1866 to 1979

Crop	Stability Index [1] 1866 to 1979
Faba beans	137
Wheat	53
Barley	35
Oats	33
Potatoes	71

[1] Calculated by expressing the annual deviation
from a five-year moving average as a % of this
moving average. These percentages are then
squared to remove negative signs.

Examination of national average yield data from other European
countries would indicate that this was the case in most of these
countries. It is important to note that some authors claim that in
trials faba bean yields are no more unstable than those of cereals, it
being only in agriculture where levels of husbandry given to the two
crops differ, that the differences in stability appear. If this is true
the instability is not inherent in the bean plant but only in the culture
of it. However, most scientists in Europe feel the instability is in the
plant itself and this is made worse by poor husbandry.

Low and unstable yields in comparison to other crops can lead to
lower profit margins and growers, if given the option, grow more
profitable cereals or other break crops such as rapeseed. Coupled with
this are the major problems of possible catastrophic disease problems
like Botrytis and Orobanche which can result in no yield at all.
Furthermore, a large number of present day western and northern European
varieties are indeterminate and tall growing and this leads to uneven
maturity which can result in major problems at harvest.

The crop also has minor utilization problems in relation to
difficulties in handling, grinding, storing and nutritional factors
particularly in relation to monogastric animals. However, these problems
could fairly easily be overcome if adequate supplies of faba beans at the
right price were consistently available. A declining area of this crop
and low and unstable yields do not lead to this being the case.

For the crop to fulfil its potential and expand substantially in area all or most of these problems must be overcome. It is therefore important to examine in more detail some of the factors that are responsible for low and unstable yields and attempt to indicate possible approaches to improving and stabilizing yields in Europe.

1. Breeding

The majority of breeding programmes in Europe were small and many had only been on a serious basis for 5 to 10 years. Most lacked adequate financial backing. It is no wonder therefore that cereal breeders with their large numbers and relatively large programme had made so much more progress. Most bean breeders agreed that much genetic variation existed but that adequate resources were just not available to make maximum use of this variation. Practical breeding of Vicia faba is also more difficult because it is a partly cross-pollinating species and for this reason alone progress is likely to be slow. The number of options available to the breeder are greater and more complex than in inbreeding species. Thus, it is difficult to chose the strategy which will make the most rapid progress. Background information on the efficiency of different breeding methods is just not available and obviously work is needed in this area.

In recent years progress has and is still being made particularly in relation to genetic yield stability, short-stature and determinate types, improved harvest index, autofertiliy, resistance to disease, resistance to pests and improvement in earliness. But progress is limited by the size of each individual breeding programme.

Undoubtedly breeders hold the key to major improvement in the future. But no breeder was optimistic about a major breakthrough. Most felt that yield increase would be provided by gradual progress proportional to the resources available. Throughout Europe a number of breeding methods were being used and opinions obviously differed on which may be the most efficient. Assuming that each programme is likely to remain small, it is necessary to ask how breeders can rapidly improve efficiency and output. A number of options seem to be available:

1. Breeders (possibly State-owned in the main) could co-operate and combine their efforts. Some co-operation is already taking place and has come about mainly through the efforts of the E.E.C. protein

group. However, perhaps there is a place for formal co-operation
even to the extent of financing a Euro-breeder who would be
responsible for the co-ordination of programmes. This sort of
co-operation would of course result in many problems particularly in
relation to plant breeders' rights etc. and the relationship with
private breeders. However, one cannot help but feel that with such
small scales of operation, progress will be very slow and
consequently this is a case where co-operation on a formal basis is
essential if more rapid progress is desirable. This type of approach
is made by ICARDA therefore the Headquarters make the crosses and
selection to F_3 then the unfinished gene pools are distributed to
different locations for final selection according to the importance
of different characters in different environments.

2. This co-operation should extend to physiologists, agronomists,
pathologists and other specialised workers. Many breeding programmes
rely on their own breeders' knowledge of physiology etc. It is of
course possible that the sort of co-operation that has been suggested
under No.1 above may also lead to co-operation in these attendant
disciplines. Whole-crop physiological work assessing the efficiency
of possible changes in canopy structure produced by new varieties is
one example of the sort of co-operation that is needed.

3. Substantially more financial resources must be made available to
breeders particularly if formal co-operation takes place. This
cannot be over-emphasised as a reliable high yielding variety is the
major priority quoted by all scientists and growers engaged in this
crop. Good breeders are working on the crop and much genetic
variation is available but finance is just not adequate if more rapid
progress towards this aim is to be made.

Such resources should not only be available for applied breeding
but also for fundamental work such as studies on breeding methods,
obligate autogamy, independent vascular supplies, control of
cytoplasmic male-sterility, the production of computer based genetic
maps, cytogenetics and possibly genetic engineering/nucleic acid
manipulation. Co-operation between Institutes and countries in these
more fundamental aspects because of their specialisation and possible
high cost would be even more essential.

4. Priorities in applied breeding would seem to be to select in

Northern Europe for improved pod set and harvest index so as to obtain earlier maturity without loss of yield, greater tolerance to excess water and resistance to chocolate spot and in drier areas for drought tolerance and resistance to Orobanche.

5. A number of breeders stated that greater co-operation with Southern hemisphere countries may be worth considering. This would enable two generations to be produced annually.

6. A number of State owned Breeding Institutes do not produce varieties but carry out breeding research and process genetic variation and hand this to private breeders. Consequently this policy in relation to Vicia faba should be reviewed and evaluated in relation to possible Euro-breeding co-operation and to private companies. In this context it is important to start discussions on whether private breeders should be encouraged to participate more in E.E.C. collaborative projects.

7. This work should also be in close mutual co-operation with ICARDA and other interested organisations or countries.

2. Weather conditions and its effects on water availability and disease

2.1. Water availability

Undoubtedly present cultivars respond greatly to water. Western and Northern European types become very tall and indeterminate when excess water is available. The crop is then prone to poor pod set (as vegetative growth becomes dominant), lodging, to uneven maturity and in winter beans to chocolate spot. Most Western and Northern European countries regularly have such conditions. Future selections and breeding for these countries should therefore concentrate on the production of semi-dwarf determinate types with low vegetative response to excess water. Research should also be conducted on the physiology and mechanism of this response in order to aid breeders.

Drought is possibly the most important factor which limits and causes variation in yield in Vicia faba within Europe. Faba beans are inherently shallow rooting (80 to 90 cm) compared with cereals (120 to 200 cm). Consequently the crop has less water available to it. Future variety selection should therefore try to identify deeper rooting types with greater water use efficiency. Breeding work at ICARDA takes these

factors into account and this work does suggest that there is some
genetic variation in relation to root size, depth and volume. Much more
physiological and agronomic work is needed in this area particularly in
relation to the response of new genotypes to water stress and irrigation.

The ideal future variety should therefore be determinate with
tolerance to water stress and excess water. It should be able to set
pods and retain them at a range of available water capacities.

It is strongly recommended that a new E.E.C. joint field bean test
should be started and this should include 5 or 6 countries with
contrasting climatic conditions particularly in relation to water
availability. The test should include the most promising genotypes
(ranging in plant model type) and they should be grown under differing
water regimes. This programme should possibly be co-ordinated from one
centre with the co-ordinator making reasonably regular visits to other
co-operating countries. This would enable the data to be collated and
published or circulated as quickly as possible so that breeders could
obtain first hand information with the minimum of delay.

2.2. Disease

As previously stated disease has a major influence on yield
variation in Vicia faba. Work is therefore needed in winter beans on the
major diseases such as Chocolate Spot, Ascochyta and rust, and in spring
beans, on viruses, root rot complexes and on Alternaria in both types.
Improved chemical methods for control of chocolate spot and an ever
continuing search for resistance are essential. Work on the
standardisation of screening techniques would be a great help to future
breeding work in all the above diseases.

Work is also needed on the relationship between differing husbandry
techniques and disease as well as on fundamental aspects of diseases.

The parasite Orobanche is the factor which most limits the growing
of faba beans in Mediterranean countries. Further work is therefore
needed on the parasite itself, agronomic techniques which avoid its
development, control measures and the breeding for resistance.

Where practicable, future research programmes on disease should be
linked to the new joint field bean test and to the faba bean programme at
ICARDA.

3. Area stability of the crop, the need for more research and dissemination of research findings

In many countries unstable yields result in unstable areas grown and this in turn results in inadequate grower experience in relation to the husbandry of the crop. The amount of research information available on the crop in many countries is very limited and the dissemination of what is available is, in many cases, inadequate. This leads to crops being grown with little or no knowledge of optimum techniques.

The possible production of new improved varieties with higher and more stable yields is likely to make the crop more competitive with cereals and as a result area expansion and improved stability will be likely. This will require further applied research on these new types. For example semi-dwarf determinate self-pollinating varieties may have differing responses to agronomic factors (such as plant population, sowing date, water etc.) than conventional types. Research programmes will therefore be needed to evaluate the agronomy of these new types. This work should be in advance of the production and release of new cultivars so that dissemination of research findings can take place at the same time as the variety is released.

4. Reliance on insect pollination

Present varieties are partly cross pollinated and bees assist in self pollination, thus maximum yields and vigour of bean stocks may depend on adequate populations. In many areas where the crop is grown bee populations are not always adequate. Future breeding work should therefore include selection for auto-fertility. The improvement of closed-flower types should be included in such work since this could lead to more effective breeding methods. Fundamental work is also needed in conjunction with these programmes.

5. Low priority crop

In a number of countries the crop is grown on marginal land where cereals will not normally be considered. This includes, especially in the Mediterranean area, growing the crop under olive trees, vines and under conditions of low available water. Furthermore, the crop is often considered as low priority in comparison to cereals and as a result optimum agronomic techniques are not carried out e.g. late sowing, inadequate weed and pest control etc. Marginal soils and marginal husbandry is therefore a major factor causing low yields in many areas of Europe. The production and availability of new varieties may not improve this level of husbandry. However, breeding should possibly include the production of types better suited to marginal land in countries where the crop needs to be grown in such areas e.g. under dryland conditions.

Improved varieties with higher yields are likely to improve the crop's image and consequently growers are more likely to consider the crop as a priority in its own right.

6. Harvesting conditions

During wet summers which are common in Northern and Western Europe the crop can cause major harvesting problems due to excess growth. Consequently crop losses can be much higher than in cereals. In dry conditions shattering can also be a problem. Future breeding programmes should therefore aim to produce varieties with less response to excess water in vegetative growth, shorter stems and reduced shattering. New semi-dwarf types, with much improved harvest index, pod set, pod retention and improved podding position, if produced, in the future would do much to improve harvesting. Size of seed is also important in relation to mechanical harvesting. It may therefore be of help to breeders if an optimum size and shape of bean could be defined in relation to possible utilization.

7. Control of grass and other weeds

Faba beans can suffer from infestations of Agropyron repens and broad leaved weeds that present chemicals are unable to control satisfactorily. Furthermore, in a number of European countries no chemical weed control is used and since the faba bean crop offers little suppression of weeds in the early and late part of the growing season, yields can be substantially reduced.

Further work is therefore needed not only on chemical weed control but on weed competition studies in order to define possible critical competition stages and the most competitive weed species. Such information would enable scientists to optimise chemical type, timing of application etc.

8. Virus diseases

Unlike cereals and many other major arable crops, faba beans suffer from a large number of virus diseases. The effects of some of these are well known but many have only just been identified. The problem is further complicated as plants can be attacked by a number of viruses at the same time. To growers, symptoms are also often very similar to other causal factors such as drought, other environmental stresses, herbicide effects, etc. Further work is therefore needed in this area particularly at a fundamental level and in relation to the effects of viruses on yield, their distribution and importance, their control by the use of chemicals on vectors and seed hygiene. The production of new varieties and the possible expansion of the crop may highlight the importance of viruses in this crop. Breeders should if possible select for resistance.

9. Pests

The faba bean is attacked by many pests but little is known about the effects on growth development and yield of minor pests e.g. nematodes and weevils. It is possible that an expansion of the crop may result in some of these minor pests becoming major problems. The effects of possible crop expansion on major pests such as aphids should also be recognised. Our present level of research particularly in relation to pest forecasting may not be inadequate for the future.

10. The utilization of the crop has little relationship with yield and yield stability but it does affect continuity of supply. Furthermore, new varieties or types will need to be quantified in relation to their nutrient content and potential for livestock feed. Breeders should plan to put the selection for feed quality on their priority list. Percentage crude protein and the quality of protein of future varieties could play an important part in the ability of the crop to compete with imported protein and home grown cereals. It may be possible that breeders should aim at increasing crude protein levels to 35 or even 40% and in addition they should try to improve quality. Consequently research work on the screening for quality by analytical techniques and via the animal is important. The breeder must be able to obtain feedback information in this area as rapidly as possible. The selection for higher yielding, low tannin, white flowered types, lower vicine and convicine contents, low

anti-trypsic factors etc. are important in this context. Much financial saving could be made if co-operation between countries in the above research areas are made. Analytical systems and animal feeding trials are expensive yet they are not influenced by the environment. Consequently, information from one country is very appropriate to most other countries. The E.E.C. Commission could play a major role in encouraging such co-operation.

For the future it may also be worth considering the production of varieties with lower levels of protein and greater energy production. Selection for protein levels of, say 20%, with higher energy yields could result in higher grain yields and yet this level of protein is high enough for direct feeding with little or no supplementation. The photosynthate needed to produce one unit of starch energy is about half that needed for protein. In this way breeders may make more rapid progress.

Conclusions on present research

This report shows that research is taking place on the crop in many countries. At first sight the amount of work seems very large. However, it is important to point out that most of the programmes are very small and only play a very minor role in the overall research of each Institute. The amount of research relative to that carried out on crops such as cereals, sugar beet, potatoes and grassland is small. However, it is noted that in relation to the size of the crop it is a fair distribution of research effort. Nevertheless, if the E.E.C. wishes to substantially increase home grown protein supply then faba beans seem to have the potential for such expansion. This expansion is likely to be successful only when varieties with higher and more stable yields are found. It may therefore be reasonable to recommend substantial financial backing for breeding and research on the crop in the first instance, so that such varieties can be produced as rapidly as possible, and the knowledge on their production obtained, before substantial subsidy is given to encourage the growing of these varieties. It may also be profitably suggested that the subsidy finance which is given to growers and users of the crop at the present time may be more usefully utilized in the search for 'new bean' and the definition of its optimum husbandry.

The discussion of this report has indicated a large number of possible areas for research. However, it is recognised that some order of priority should be stated.

Priority research areas for the next 5 years

1. The setting up of co-operative breeding programmes between E.E.C. countries. This should include germplasm collection and maintenance, studies on breeding methods and on cytogenetics with a view to obtaining interspecific hybrids. Substantial financial support is recommended for such work.

2. This programme should be linked to whole crop and plant physiological work carried out on new types and varieties. This should include work on the nitrogen fixation ability of new genotypes.

3. A further link on the agronomy of these varieties should be made. This programme would include irrigation, water use, and water stress studies as a priority. Such work would include the importance of water stress in changing vegetative to reproductive growth. Nitrogen fixation ability, plant spacing, sowing date and other agronomic factors may also be important in relation to new models.

4. Integration with disease and pest programmes would be essential particularly in relation to genetic resistance, pest control and forecasting.

5. Assessment of quality in relation to utilization within these programmes would be important particularly in relation to vicine, covicine, anti-trypsic factors, lectins, tannins, protein level itself.

6. Fundamental work on the physiology and biochemistry of pod set, pod retention and assimilate transfer should be continued and enlarged. This should include work on synchronous flower development, independent vascular supply systems and hormone balance in the plant. This programme should be co-ordinated with the breeding and other programmes above.

7. It is strongly recommended that a faba bean scientist (possibly on a part-time secondment basis) should be appointed to co-ordinate these linked programmes within Europe.

The 'setting up' of the above programmes should seek to encourage individual countries to continue to enlarge their present on-going programmes.

The past and present role of the E.E.C. Commission via the protein group

Undoubtedly the E.E.C. Commission has played a major role in setting the scene for the above co-operative programme. It has enabled the relatively small number of faba bean scientists to meet, and therefore to interchange knowledge, ideas and enthusiasm without barriers. This has been achieved through conferences, seminars, workshops and the exchange of scientists. It has allowed a tremendous amount of knowledge to accumulate and be interchanged during the last 6 years.

It is recommended that funds should be available for the continuation of such meetings etc.

CYTOGENETICS

Vicia faba has been since the beginning of cytology a subject of classical interest and its literature is compendious. Recently, the breeding of Vicia faba has assumed increasing importance. The purpose of this meeting was to explore how far it was possible to pool these hitherto divergent interests, together with the more recently developed techniques of molecular biology in a new approach to Vicia faba cytogenetics.

The principal conclusions of the meeting were as follows.

1. The publication 'Genetic Variation within Vicia faba' would be revised and expanded using Database technology under the guidance of Dr. F.A. Bisby of the University of Southampton to include a more complete account of chromosome aberration stocks and (as it becomes available) data on DNA sequencing, for example.

2. We would aim to establish a model of chromosome structure that finally resolves the problem of multistrandedness and reconciles data from optical and electron microscopy and molecular biology.

3. Readily available stocks, both of tester translocations and a complete set of trisomics would be assembled so that in several institutes a coordinated organisation of linkage data could take place. Professor Cubero undertook to organise this although the meeting recognised that emphasis in his Department was on chromosome aberration rather than on maintaining a large collection of mutants for which other institutions would be responsible.

4. A reasonable target for linkage data was over five years, upwards of 30 gene sites with some emphasis on characters for stature, growth habit and disease resistance. It was decided that while recognising the pioneering work of Dr. J. Sjodin of Svalof in allocating genes to chromosomes in this species we would adopt the chromosome notation of Michaelis and Rieger (1959, 1968) for further linkage studies.

5. In the longer term there should be a coordinated approach to the DNA sequencing of preferential breakage sites and an attempt to relate the composition of genetically active DNA to known gene sites and to chromosome banding patterns.

6. Interesting mutant forms would be made available to breeders, physiologists, agronomists and pathologists so as to widen our knowledge of plant efficiency.

7. A small organising committee of five people, was to be established to coordinate these various objectives. Dr. G.P. Chapman at Wye was

P.D. Hebblethwaite, T.C.K. Dawkins, M.C. Heath and G. Lockwood (eds.)
Vicia faba: Agronomy, Physiology and Breeding. ISBN 90-247-2964-5.
© 1984, Martinus Nijhoff/Dr W. Junk Publishers. Printed in The Netherlands.

elected Convenor. Professor Cubero was also appointed together with Professor Rieger. Two other invitations are pending at present.

The meeting was one that brought together scientists of diverse backgrounds and was essentially exploratory. It was, however, soon apparent that Vicia faba cytogenetics represents an area of research highly relevant to the future development of the crop and it is planned to hold a larger, more widely publicised meeting at Wye in (probably) the Spring of 1984. Enquiries and offers of papers and posters should be sent to the writer.

The organisers of this first meeting express their thanks to all the participants, not least to the Zeiss Company of West Germany for its demonstration at the meeting of the EM. 109 electron microscope and a range of optical microscopes.

References

Michaelis, A. and Rieger, R. (1959). Structurheterozygotie bei Vicia faba Zuchter 29 354-361.

Michaelis, A. and Rieger, R. (1968). On the distribution between chromosomes of chemically induced chromatid abberrations: studies with a new karyotype of V. faba. Mutation Res. 6 81-92.

G.P. Chapman
Wye College (University of London)
Ashford, Kent. TN25 5AH.

GENERAL DISCUSSION AND RECOMMENDATIONS

Mrs. Wauters of the E.E.C. Commission opened the discussion with an explanation of the possibilities for financing of projects in the framework of the new programme on Plant Productivity which is scheduled to run for a five year period (1984-1988). The current programme on the improvement of the production of plant proteins shall come to an end in December 1983 after five years of operation. Although the new programme has a much wider scope than the latter, it is likely that a substantial part of the available finances will still be destined to "Protein".

In the context of the 1984-1988 programme two types of activities will continue to be sponsored by the Community: co-ordination activities (e.g. workshop and seminars, exchanges of scientists, publications) and Common Research Programme Contracts. Both types of activity shall be supervised by the Standing Committee for Agricultural Research (SCAR). At a more technical level, there is a further Advisory Committee called the "Programme Committee for Plant Productivity" which comprises two scientists per country. One of the tasks of the Committee as well as of the SCAR is to distribute information relevant to the Research Programme at national level. The lists of the membership of both Committees is freely available to all those who may be interested in receiving them.

The meeting agreed that the structure of North-West European agriculture requires a legummous protein source. There was a lively debate about the future of the faba bean crop and the relative priority of breeding and other fields of research.

The majority felt that faba beans have a good long term future and are a rival to peas. The key is finding the right plant model in relation to agro-climatic area. Higher and more predictable yields are the priority objectives for breeders, physiologists and agronomists to tackle in order to secure this future. Two major views were put forward as to how to achieve these objectives. The first favoured a detailed physiological and biochemical elucidation of yield-limiting factors in order that the breeder could then select for specific characters using rapid screening methods and produce more stable and reliable varieties. A specific example of this approach was research at Durham which had shown that independent vascular supply to each flower within a raceme could improve pod retention on individual plants and this character could be readily identified in a breeding programme. Before a major programme

P.D. Hebblethwaite, T.C.K. Dawkins, M.C. Heath and G. Lockwood (eds.)
Vicia faba: Agronomy, Physiology and Breeding. ISBN 90-247-2964-5.
© 1984, Martinus Nijhoff/Dr W. Junk Publishers. Printed in The Netherlands.

involving such traits could be justifiably financed, it is, however, essential to show that such characters promote improved yield and reliability under field crop conditions.

The alternative strategy favoured the continuation of short term approaches with closely defined objectives as priorities. This approach was argued on the basis that unless further specific agronomy and pest and disease control work were done to improve yields within the next few years then the crop in Europe may be overtaken by other legumes including imported seeds and meals. This was particularly urgent in relation to some of the new varieties which are and will be available in Europe. This argument pointed out that physiological findings so far tended to be of a general nature. If new specific information that could be used by the breeder became available then it should be made available to scientists so that its value could be judged in relation to priority for finance.

In spite of different assessments of priorities, the meeting agreed that there is an urgent need for increasing research through both fundamental and short term approaches. The meeting was concerned that the gap between cereal and faba bean yields was widening each year and it was agreed that the gap could be closed only through the co-operation of physiologists, breeders and agronomists from all countries. E.E.C. Meetings and Seminars already catalyse substantial co-operation between breeders and physiologists in Europe and it was thought to be essential that such meetings should continue.

Most scientists agreed that breeding must take top priority but breeders needed physiological information to help in their selection. Provenance testing of new types should be as rapid as possible and agronomical information in relation to new genotypes must also be provided rapidly. It was felt that the Joint Field Bean Test could play an important role in this context and it should therefore be extended and adequately financed.

Discussion took place about optimising use of the limited resources available for breeding. It was agreed that breeders in different countries should collaborate closely, and that development of regional gene pools should be considered. Predictive modelling might help in defining ideotypes for different areas and then use of physiological screens might improve the efficiency of selection. However, because agro-climatic conditions and the role of breeding institutions differ greatly between countries, it will continue to be necessary for most countries to produce their own finished varieties. In the short term, improvements in germplasm handling, gene pool organisation and the use of physiological screens will play the most important part in developing better varieties; but further discussion would be necessa y before a more unified breeding strategy could be adopted.

All scientists agreed that collaboration with ICARDA had been of great benefit and it was urged that collaboration should continue both in relation to exchange of scientists and gene pools.

A brief discussion took place in relation to a meeting that had been held the evening before on distinctness, uniformity and stability (D.U.S.). This meeting recommended that "The E.E.C. Faba Bean Seminar notes that the natural breeding system of the faba bean necessitates use of breeding methods which rely on heterogeneity to maximise yield. Therefore, there should be no constraint on the breeder imposed by standards of uniformity for qualitative characters of no known economic or agricultural value." This recommendation was agreed by the meeting and it was suggested that it should be forwarded to the Committee on Seed Production and Subventions who would in turn forward it to the National Seed Testing Authorities.

There was insufficient time to discuss the role of biotechnology in faba bean improvement although this topic was explored during informal meetings outside the seminar. Application of biotechnology will depend on the availability of techniques for recovering intact plants from protoplast, cell and explant cultures. Since such technology does not exist for faba beans, its development merits high priority for research funding.

The meeting ended with the proposal and agreement that the recommended priority research list given by Dr. Hebblethwaite in his report on Vicia faba research in Europe should be adopted with minor modifications in relation to priority and with the exclusion of recommendation 7. The modified list of priority research for the next five years was as follows:-

1. The setting up of co-operative breeding programmes between E.E.C. countries. This should include germplasm collection and maintenance, studies on breeding methods and on cytogenetics with a view to obtaining interspecific hybrids.

2. This programme should be linked to whole crop and plant physiological work carried out on new types and varieties. This should also include fundamental work on the physiology and biochemistry of pod set, pod retention and assimilate transfer. This would include work on synchronous flower development, independent vascular supply systems and hormone balance in the plant.

3. A further link on the agronomy of these varieties should be made. This programme would include irrigation, water use, and water stress studies as a priority. Such work would include the importance of water stress in changing vegetative to reproductive growth. Nitrogen fixation ability, plant spacing, sowing date and other agronomic factors may also be important in relation to new models.

4. Integration with disease and pest control programmes would be essential particularly in relation to genetic resistance, pest control and forecasting.

5. Assessment of quality in relation to utilization within these programmes would be important particularly in relation to vicine, convicine, anti-trypsic factors, lectins, tannins, and protein level itself.

P. Hebblethwaite
G. Lockwood
M. Heath
T. Dawkins

DR. IR. G. DANTUMA 1924 - 1983

It was with great sorrow that we learned of the death of Gerrit Dantuma on 21 September only 3 days after the E.E.C. faba bean seminar.

Gerrit had a distinguished career in agricultural research and plant breeding. He was born on 15 February 1924 at Leeuwarden, The Netherlands. After finishing High School in 1942 he attended an Agricultural School in Groningen. After the war he studied at the Agricultural University in Wageningen where he graduated in 1950 specialising in plant breeding and pathology. He then joined the Foundation for Agricultural Plant Breeding (S.V.P.) as a cereal breeder. As head of the cereal section he started an extensive wheat and barley breeding programme. In 1958 he obtained his Doctorate from the Agricultural University on a thesis entitled "Breeding of wheat and barley for winter hardiness". From 1958-59 he spent one year at Lethbridge, Alberta, Canada working on cereal breeding in the Prairie Regions. From 1962-63 he organised a project in Chile for speeding up cereal breeding programmes by growing two generations per year. By circumstances he was obliged to terminate his cereal breeding work and in 1964 he joined the Institute for Biological and Chemical Research on Field Crops and Herbage in Wageningen (now centre for Agrobiological Research, CABO). By 1971 his cereal work was diminished and he concentrated on the physiology and breeding of legumes particularly Vicia faba. He was a member of several working groups and from 1974-82 was Chairman of the physiological section of Eucarpia.

It was in 1975 that his ideas on increasing faba bean production became known to other faba bean scientists in Europe, and in 1976 he instigated the first series of E.E.C. Joint Faba Bean Tests which ran from 1977 to 1979. The results were reported in 1982 (Dantuma et al, (1983) Z. Pflanzenzuchtung, 90, 85-105). He continued as co-ordinator of the second series, 1980-82, (results of which are reported elsewhere in these Proceedings) and was enthusiastically planning the next series for 1984-87.

Gerrit contributed regularly to E.E.C. faba bean meetings, in organisation, formal presentations, as Chairman of sessions and informal discussions. All will remember his ability to fluently converse in most European languages.

Physiologists and agronomists will remember him for his planning and

P.D. Hebblethwaite, T.C.K. Dawkins, M.C. Heath and G. Lockwood (eds.)
Vicia faba: Agronomy, Physiology and Breeding. ISBN 90-247-2964-5.
© 1984, Martinus Nijhoff/Dr W. Junk Publishers. Printed in The Netherlands.

contribution towards joint agronomic/physiology trials, an example being the identical trials on irrigation and water stress carried out at CABO and the University of Nottingham in 1983. This project also involved exchange of students from the U.K. and Netherlands and was planned to be continued until 1986. A large number of students benefited greatly from his experiments and his wealth of experience which was shared generously with all who sought his advice.

Dr. Dantuma's contributions on 'The Whole-crop Physiology and Yield Components' and 'Grain and Whole-crop Harvesting, Drying and Storage' to the Faba Bean - a basis for improvement, published by Butterworths, London are a living memorial to his enthusiasm and achievement.

Gerrit was good company, he had a delightful sense of humour, an honest and courageous character. He was a kind and helpful friend to us all and he will be greatly missed, but he laid the foundation of knowledge and enthusiastic enquiry upon which we can all build.

We all pass on our deepest sympathy to his wife and three sons.

LIST OF PARTICIPANTS

A. ALMOND

University of Nottingham,
School of Agriculture,
Sutton Bonington,
Loughborough,
Leics. UK.

M. ANDREWS

Department of Biological Sciences,
University of Dundee,
Dundee, Scotland.

E. AUGUSTINUSSEN

Statens Førsogsstation,
Ledreborg Allé 100,
DK-4000 Roskilde,
Denmark.

K. BAEUMER

Institut fur Pflanzenbau und
Pflanzenzuchtung,
Universitat Gottingen,
Von-Siebold-Strasse 8,
D-340 Gottingen,
Federal Republic of Germany.

W.H. BAIER

Pflanzenzucht Oberlimpurg,
Dr. Franck,
D-7170 Schwabisch Hall,
Federal Republic of Germany.

M.P. BAILEY

Department of Biological Sciences,
Wye College, (Univ. of London),
Wye, Nr. Ashford,
Kent. TN25 5AH. UK.

D.A. BAKER

Department of Biological Sciences,
Wye College, (Univ. of London),
Wye, Nr. Ashford,
Kent. TN25 5AH. UK.

P. BERTHELEM

Station d'Amelioration des Plantes,
INRA, BP29,
35650 Le Rheu,
France.

R. BISTON

Station de Haute Belgique (C.R.A.
Gembloux),
48, rue de Serpont - 6600,
Libramont-Chevigny,
Belgium.

D.A. BOND

Plant Breeding Institute,
Maris Lane,
Trumpington,
Cambridge, CB2 2LQ. UK.

D. BOULTER

University of Durham,
Department of Botany,
Science Laboratories,
South Road,
Durham. DH1 3LE. UK.

D. BOURDON

INRA - Station de Recherches sur
  L'Elevage des Porcs,
Saint-Gilles,
35590 L'Hermitage,
France.

J.C. BRERETON

University of Nottingham,
School of Agriculture,
Sutton Bonington,
Loughborough,
Leics. UK.

M.E. CAMMELL

Dept. Pure and Applied Biology,
Imperial College,
Silwood Park,
Ascot, Berkshire. UK.

G.P. CHAPMAN

Wye College, (Univ. of London),
Wye,
Nr. Ashford, Kent. TN25 5AH. UK.

Ms. C. CHATHAM

University of Nottingham,
School of Agriculture,
Sutton Bonington,
Loughborough, Leics.
UK.

Miss M. CHERRY

"Byeways",
Hook Norton,
Banbury, OX15 5LG. UK.

A.J. COCKBAIN

Rothamsted Experimental Station,
Harpenden, Herts. UK.

Mrs. C. CONICELLA

Cattedra di Miglioramento Genetico
    delle Piante Coltivate,
Universita di Napoli,
Portici,
Italy.

G. CROFTON

National Institute of Agricultural
Botany,
Huntingdon Road,
Cambridge. UK.

J.I. CUBERO

Departmento de Genetica,
Escuela Tecnica Superior de Ingenieros
Agronomos,
Apartado 3048,
Cordoba,
Spain.

G. DANTUMA

Centre for Agrobiological Research,
P.O. Box 14,
6700 AA Wageningen,
The Netherlands.

T.C.K. DAWKINS

University of Nottingham,
School of Agriculture,
Sutton Bonington,
Loughborough, Leics.
LE12 5RD. UK.

Miss S. DEELEY

Butterworth-Scientific Ltd.,
Borough Green,
Sevenoaks, Kent. UK.

H. DEKHUIJZEN

Centre Agrobiological Research,
Postbox 14,
6700 AA Wageningen,
The Netherlands.

G. DUC

INRA Station d'Amelioration des
Plantes,
BV 1540 21034 Dijon Cedex,
France.

E. EBMEYER

Institut fur Pflanzenbau und
Pflanzenzuchtung,
v. Siebold Str. 8,
D-3400 Gottingen,
Federal Republic of Germany.

J. ELSTON

The University of Leeds,
Department of Plant Sciences,
Agricultural Sciences Building,
Leeds. LS2 9JT. UK.

A. FALISSE

Faculte des Sciences Agronomiques
de L'Etat a Gembloux,
5800 Gembloux,
Belgium.

J.S. FAULKNER

Department of Agriculture, N.I.,
Plant Breeding Station,
Manor House,
Loughgall, Armagh,
Northern Ireland, BT61 8JB.

E. FILIPPONE

Cattedra di Miglioramento Genetico
delle Piante Coltivate,
Universita di Napoli,
Portici,
Italy.

L. FRUSCIANTE

Institute of Agronomy and Plant Breeding,
University of Naples,
Portici,
Italy.

P.J. GATES

University of Durham,
Botany Department,
Durham. DH1 3LE. UK.

C. GREEN

University of Nottingham,
School of Agriculture,
Sutton Bonington,
Loughborough,
Leics. UK.

D.W. GRIFFITHS

Welsh Plant Breeding Station,
Plas Gogerddan,
Aberystwyth,
Dyfed. UK.

M.C. HEATH

University of Nottingham,
School of Agriculture,
Sutton Bonington,
Loughborough,
Leics. UK.

P.D. HEBBLETHWAITE

School of Agriculture,
University of Nottingham,
Sutton Bonington,
Loughborough, Leics. LE12 5RD. UK.

R.J. HERINGA

Foundation of Agricultural Plant Breeding,
Laboratory de Haaff,
P.O. Box 117,
6700 AC, Wageningen,
The Netherlands.

J. HIGGINS

National Institute of Agricultural Botany,
Huntingdon Road,
Cambridge. CB3 0LE. UK.

D.G. HILL-COTTINGHAM

Long Ashton Research Station,
Long Ashton, Bristol, BS18 9AF. UK.

J.D. IVINS

University of Nottingham,
School of Agriculture,
Sutton Bonington,
Loughborough,
Leics. UK.

A. KARAMANOS

Laboratory of Crop Production,
75 Iera Odar,
Athens 301,
Greece.

E.R. KELLER

Institut fur Pflanzenbau,
ETH - Zentrum,
CH-8092, Zurich,
Switzerland.

M. KELLERHALS

Institut fur Pflanzenbau,
ETH - Zentrum,
8092 Zurich,
Switzerland.

H. KERR

University of Nottingham,
School of Agriculture,
Sutton Bonington,
Loughborough,
Leics. UK.

S. KERR

University of Nottingham,
School of Agriculture,
Sutton Bonington,
Loughborough,
Leics. UK.

E. von KITTLITZ

Universitat Hohenheim,
Landessaatzuchtanstalt,
Postfach 700562,
D-7000 Stuttgart 70,
Federal Republic of Germany.

D. H. LAPWOOD

Plant Pathology Department,
Rothamsted Experimental Station,
Harpenden, Herts.
AL5 2JQ. UK.

J.F. LE GUEN

INRA - Station d'Amelioration des
Plantes,
Le Rheu,
France.

C.P. LLOYD-JONES

Long Ashton Research Station,
Long Ashton,
Bristol. BS18 9AF. UK.

G. LOCKWOOD

Plant Breeding Institute,
Maris Lane,
Trumpington,
Cambridge. CB2 2LQ. UK.

A. MARTIN

Dpt. de Genetica, Escuela Tecnica
Superior de Ingenieros Agronomos,
Apartado 3048, Cordoba,
Spain.

L.M. MONTI & Mrs. L.M.	Istituto di Agronomia Generale e Coltivazioni Erbacee, Cattedra di Miglioramento Genetico Delle Piante Coltivate, Portici, Naples, Italy.
Mrs. M.T. MORENO	Unidad de Leguminosas, Centro Regional de Andalucia, Inia, Apartado 240, Cordoba, Spain.
M.V. MURINDA	International Center for Agricultural Research in the Dry Areas (ICARDA), P.O. Box 5466, Aleppo, Syria.
K. NAGL	Bundesanstalt fur Pflanzenbau in Wien, Postfach 64, 1201 Wien, Austria.
J. PICARD	Institut National de la Recherche Agronomique, Station D'Amelioration des Plantes, I.N.R.A. B.V. 1540, 21034 Dijon, France.
B. PICKERSGILL	Department of Agricultural Botany, University of Reading, Whiteknights, Reading. RG6 2AS. UK.
Ph. PLANCQUAERT	Institut Technique des Cereales et des Fourrages, 8 Avenue du President Wilson, 75116 Paris, France.
D. POULAIN	INRA - Laboratoire d'Agronomie, 65 Rue de Saint Brieuc, 35042 Rennes Cedex, France.
A G. PRENDERGAST	An Foras Taluntais, 19 Sandymount Avenue, Dublin 4. Republic of Ireland.

G. RAMSAY

Dept. of Agricultural Botany,
Plant Sciences Laboratories,
University of Reading,
Whiteknights,
Reading, RG6 2AS. UK.

E.H. ROBERTS

Dept. of Agriculture and Horticulture,
University of Reading,
Plant Environment Laboratory,
Shinfield Grange,
Cutbush Lane,
Shinfield,
Reading. RG2 9AD. UK.

L.D. ROBERTSON

ICARDA,
P.O. Box 5466,
Aleppo,
Syria.

G.G. ROWLAND

Crop Development Centre,
University of Saskatchewan,
Saskatoon,
Saskatchewan,
S7N 0W0 Canada.

F. SACCARDO

ENEA - FARE - BIOAG.,
Casaccia,
S. Maria di Galeria,
Rome,
Italy.

J. SJODIN

Swedish Seed Association,
Svalof AB,
S-26800, Svalov,
Sweden.

M. L. SMITH

Edinburgh School of Agriculture,
West Mains Road,
Edinburgh. EH9 3JG.   UK.

B. SNOAD

John Innes Institute,
Colney Lane,
Norwich. NR4 7UH. UK.

H. SORENSEN

Chemistry Department,
Royal Veterinary and Agricultural
University,
40 Thorvaldsensvej,
DK-1871 Copenhagen V,
Denmark.

M.J. STANDISH

Little Butlers,
26 Church Street,
Wye,
Kent.   UK.

Miss M.H. STEWART	Department of Agricultural Botany, Plant Science Laboratories, University of Reading, Whiteknights, Reading. RG6 2AS. UK.
F.L. STODDARD	Plant Breeding Institute, Maris Lane, Trumpington, Cambridge. UK.
R. STULPNAGEL & Mrs. R.	Gesamthochschule Kassel, Fachbereiches Landwirtschaft, Norbahnhofstabe 1a, 3430 Witzenhausen 1, Federal Republic of Germany.
R.J. SUMMERFIELD	University of Reading, Dept. of Agriculture & Horticulture, Plant Environment Laboratory, Shinfield Grange, Cutbush Lane, Shinfield, Reading. RG2 9AD. UK.
T.M. THOMAS	Agriculture Institute, Oak Park, Carlow, Republic of Ireland.
R. THOMPSON	Scottish Crop Research Institute, Invergowrie, Dundee, DD2 5DA. UK.
P. TURCOTTE	Universite Laval, Faculte des Sciences de l'Agriculture et de l'Alimentation, Cite Universitaire, Quebec, Quebec, Canada. G1K 7P4.
Mrs. M. WAUTERS	Commission of the European Communities, D6 VI - F4, 86 Rue de la Loi, Bureau 3133, B-1040 Brussels, Belgium.
Miss G. WHITE	University of Durham, Botany Department, Durham. DH1 3LE. UK.

W.J. WHITTINGTON

University of Nottingham,
School of Agriculture,
Sutton Bonington,
Loughborough,
Leics. UK.

W. WILLIAMS

Agricultural Botany,
University of Reading,
Whiteknights,
Reading. UK.

G. ZERBI

C.N.R. Irrigation Institute,
Via Argine 1085,
80147 Naples,
Italy.